GEOLOGY AND HISTORY OF SOUTHEASTERN SAN DIEGO COUNTY, CALIFORNIA

Oblique aerial view of 2006 field trip area, with major areas of interest indicated. Photo by Phil Farquharson, 5 May 2003. Pilot, Frank Morris.

GEOLOGIC TIME SCALE			
ERA	**PERIOD**	**EPOCH**	**Millions of Years Ago**
CENOZOIC	QUATERNARY	Holocene	0.011
		Pleistocene	1.8 (2.67)
	TERTIARY NEOGENE	Pliocene	5.3
		Miocene	23.8
	PALEOGENE	Oligocene	33.7
		Eocene	55
		Paleocene	65
MESOZOIC	CRETACEOUS	Late	99
		Early	142
	JURASSIC	Late Middle Early	205.7
	TRIASSIC	Late Middle Early	250
PALEOZOIC	PERMIAN	Late Early	290
	CARBONIFEROUS PENNSYLVANIAN	Late	323
	MISSISSIPPIAN	Early	354
	DEVONIAN	Late Middle Early	417
	SILURIAN	Late Early	443
	ORDOVICIAN	Late Middle Early	495
	CAMBRIAN	Late Middle Early	545
	PRECAMBRIAN		
ORIGIN OF EARTH			4,550

From Abbott, Patrick L., 1999. *The Rise and Fall of San Diego: 150 Million Years of History Recorded in Sedimentary Rocks.* San Diego: Sunbelt Publications

Geology and History of Southeastern San Diego County, California

San Diego Association of Geologists
Field Trip Guidebook for 2005 and 2006

Includes: 2005 Road Log, Geology and Lore Along Old Highway 80 in the San Diego Back Country
and
2006 Road Log, Where the Pavement Ends: A Geologic Excursion in Canyon Sin Nombre and the Volcanic Hills

Philip T. Farquharson – 2005 Field Guide Editor
David M. Bloom – 2006 Field Guide Editor
Carole L. Ziegler – Managing Editor

SAN DIEGO ASSOCIATION OF GEOLOGISTS

Distributed by Sunbelt Publications
www.sunbeltbooks.com

Geology and History of Southeastern San Diego County, California
San Diego Association of Geologists' Geologic Field Trip Guidebook for 2005 and 2006

 First edition 2006, fourth printing 2018.

2005 Field Trip Leader – Philip T. Farquharson
2006 Field Trip Leader – David M. Bloom
Managing Editor – Carole L. Ziegler
Book design and composition by Philip T. Farquharson
Cover design and layout by Philip T. Farquharson
Printed in the United States of America

Please direct comments and inquiries to:
San Diego Association of Geologists
P.O. Box 191126
San Diego, California 92159-1126
www.sandiegogeologists.org

or the distributor:
Sunbelt Publications
1250 Fayette St.
El Cajon, California 92020
(619) 258-4911, fax: (619) 258-4916
www.sunbeltpublications.com
info@sunbeltpub.com
20 19 18 6 5 4

Library of Congress Cataloging-in-Publication Data

Bloom, David M., 1963-
Geology and history of southeastern San Diego County, California : San Diego association of geologists 2005 and 2006 field guides / David M. Bloom, Philip T. Farquharson, Carole L. Ziegler.
p. cm.
ISBN-13: 978-0-916251-78-9
ISBN-10: 0-916251-78-0
1. Geology—California—San Diego County—Guidebooks. 2. History—California—San Diego County. 3. San Diego County (Calif.)—Guidebooks. I. Farquharson, Philip T. II. Ziegler, Carole L. III. Title.

QE90.S116B66 2006
557.94'98—dc22

2006030685

Cover photograph courtesy of:
Philip T. Farquharson, oblique aerial view of 2006 field trip area. Frank Morris, pilot.
Front inside — Schematic geologic map of Peninsular Ranges Batholith and adjacent forearc sedimentary strata, David Kimbrough (2002). Modified by Philip T. Farquharson.
Back inside — Geologic Map of Arroyo Tapiado: Carrizo Mountain NE, and parts of Harper Canyon, Borrego Mountain SE and Sweeney Pass, 7.5 Quadrangles, Winker (1987). Redrafted by David M. Bloom.
Back — David M. Bloom, panorama view north from head of southeast fork of Canyon Sin Nombre across Borrego Badlands to Whale Peak; Diana Lindsay, pictographs in Blue Sun Cave; Philip T. Farquharson, large yoni, a Kumeyaay fertility symbol.

San Diego Association of Geologists

	2005 Officers	2006 Officers
President	Margaret Eggers meggers@eggersenv.com	Philip T. Farquharson geoguy@geology-guy.com
Vice President	Philip T. Farquharson geoguy@geology-guy.com	David M. Bloom dbloom@geology.sdsu.edu
Secretary	David M. Bloom dbloom@geology.sdsu.edu	Scott Snyder Sdag2006@prodigy.net
Treasurer	Sarah Gray sgray@sandiego.edu	Sarah Gray sgray@sandiego.edu

Publications Manager	Lowell Lindsay llindsay@sunbeltpub.com
Guidebook Managing Editor	Carole L. Ziegler cziegler@swccd.edu
Webmaster	Carolyn Glockhoff carolyn@caro-lion.com

SDAG Corporate Sponsors
2005 and 2006

Geotechnics, Inc. – Super Sponsor

Ninyo & Moore – Support Services Sponsor

George Copenhaver – Field Trip Sponsor

Dr. Richard Berry
Sherry Bloom, Coldwell Banker Real Estate
Patrick Brooks and Julie Crosby
Construction Testing & Engineering, Inc.
Joe Corones, City of San Diego
Robert Crisman and John Franklin, GeoSoils, Inc.
Dr. Margaret R. Eggers, RG, CHG,
Eggers Environmental, Inc.
William J. Elliott, Engineering Geologist
Philip T. Farquharson, CG-Squared Productions
First National Oil, Inc.
Katherine Freese
Geotechnical Exploration, Inc.
Carolyn Glockhoff, Caro-Lion Enterprises
Jonathan Goodmacher, Ninyo & Moore
Dr. Sarah Gray, University of San Diego
Robert Walters Equipment Rentals
Helenschmidt Geotechnical, Inc.
John Hoobs, Geocon, Inc.
James Knowlton, Geopacifica, Inc.
Leighton & Associates, Inc.
Lowell Lindsay, Sunbelt Publications
Laurie Maghsoudlou, EnviroMINE, Inc.
Dr. Monte Marshall
Diane and Monte Murbach
Mark Palmer, Hargis + Associates, Inc.
Doug Peeler, Bechtel Environmental, Inc.
John Peterson, Peterson Environmental Services
Tom Rockwell, Earth Consultants International
Southern California Soil & Testing, Inc.
Southwest Geophysics, Inc.
Donald Sorben
David and Jan Steller
Dr. Anne Sturz, University of San Diego
Sue Tanges, Southland Geotechnical Consultants
Lee Vanderhurst, Geotechnics, Inc.
Vinje & Middleton Engineering, Inc.
Christian Wheeler Engineering
Jerry Wilson, Fugro West, Inc./Fugro Pelagos, Inc.
Carole L. Ziegler, Southwestern College

The generosity of these corporate sponsors allows our organization to continue to keep membership dues low, sponsor reduced rates for students, produce the monthly newsletter, purchase supplies, host the annual field trip, produce our geologic guidebooks, and provide scholarships to local geology students. To contact any of the above sponsors, please visit our website at www.sandiegogeologists.org.

Table of Contents

List of Current Publications — 2006

San Diego Association of Geologists www.sandiegogeologists.org

2006 — *Mines and Geology of the Randsburg Area: An Historical Gem of the Mojave Desert*

2004 — *Mining History and Geology of Joshua Tree National Park*

2003 — *Geology of the Elsinore Fault Zone, San Diego Region*
"Hot Springs and Tourmalines of Eastern San Diego County"

2001 — *Coastal Processes and Engineering Geology of San Diego, California*

2000 — *Geology and Enology of the Temecula Valley, Riverside County*

1999 — *Water for Southern California: Water Resources Development*

1998 — *Geology and Geothermal Resources of the Imperial and Mexicali Valleys*

1997 — *Santa Cruz Island Geology*

1995 — *Paleontology and Geology of Anza-Borrego Desert State Park*

1994 — *Geology and Natural History, USMC Camp Pendleton*

1992 — *Natural History of the Coronado Islands, Baja California, Revisited*

1991 — *Environmental Perils, San Diego Region*

1990 — *Geotechnical Engineering Case Histories in San Diego County*

1989 — *Seismic Risk in the San Diego Region*

1988 — *Landslides in Crystalline Basement Terrain — San Diego and Riverside Counties*

Related Publications www.sunbeltbooks.com

Patrick L. Abbott (1999) *The Rise and Fall of San Diego: 150 Million Years of History Recorded in Sedimentary Rocks*

Michael L. Cassiliano (2002) *The Stratigraphic Nomenclature of the Plio-Pleistocene Palm Spring Group, Anza-Borrego Desert*

George T. Jefferson and Lowell Lindsay, eds. (2006) *Fossils Treasures of the Anza-Borrego Desert: The Last Seven Million Years*

Scott E. Johnson and others, eds. (2003) *Tectonic Evolution of Northwestern México and the Southwestern USA*

Diana Lindsay (2001) *Anza-Borrego A to Z: People, Places and Things*

Lowell and Diana Lindsay (2006) *The Anza-Borrego Desert Region: A Guide to the State Park and Adjacent Areas of the Western Colorado Desert*

Victoria R. Todd (2004) *Preliminary Geologic Map of the El Cajon 30′ x 60′ Quadrangle, Southern California, Version 1.0*

Michael J. Walawender (2000) *The Peninsular Ranges: A Geological Guide to San Diego's Back Country*

Michael J. Walawender and Barry B. Hanan, eds. (1991) *Geological Excursions in Southern California and Mexico*

Christopher Wray (2004) *The Historic Backcountry: A Geographic Guide to the Historic Places of the San Diego County Mountains and the Colorado Desert*

Christopher Wray (2005) *Highways to History: A Driving Guide to the Historic Places of the San Diego County*

Preface

This guidebook accompanies the San Diego Association of Geologists' 2005 and 2006 field trips. In 2005, SDAG vice president Phil Farquharson led us on an exploration of the Peninsular Ranges Batholith in southern San Diego County along historic Old Highway 80 and the San Diego & Arizona Eastern Railway. In 2006 we continue east over the Laguna Mountain Escarpment into the breakaway zone of the Salton Trough. Most of the 2006 stops take us off the pavement along rough jeep trails. At several points we leave our vehicles behind and hike into the geology.

This volume is organized into two parts, road logs and papers. Both sections are ordered from west to east. Thus, the 2006 road log follows the 2005 road log. Papers also are roughly organized from west to east. If you were to drive both trips from the beginning at Laguna Summit to the end at Agua Caliente Regional Park, you would find the papers in the order of your drive.

Most of the papers are new, based on original research. We chose to reprint several published papers because they are relevant and important. We have also included papers on the rich history of this area. They include railroad, highway, and mining history. Many of our member geologists work for these industries. We felt these papers are of sufficient interest to be included in this guidebook.

David M. Bloom, CEG
Certified Hydrogeologist
San Diego, California

Philip T. Farquharson, M.S.
CG-Squared Productions
San Diego, California

Carole L. Ziegler, M.S.
Southwestern College
Chula Vista, California

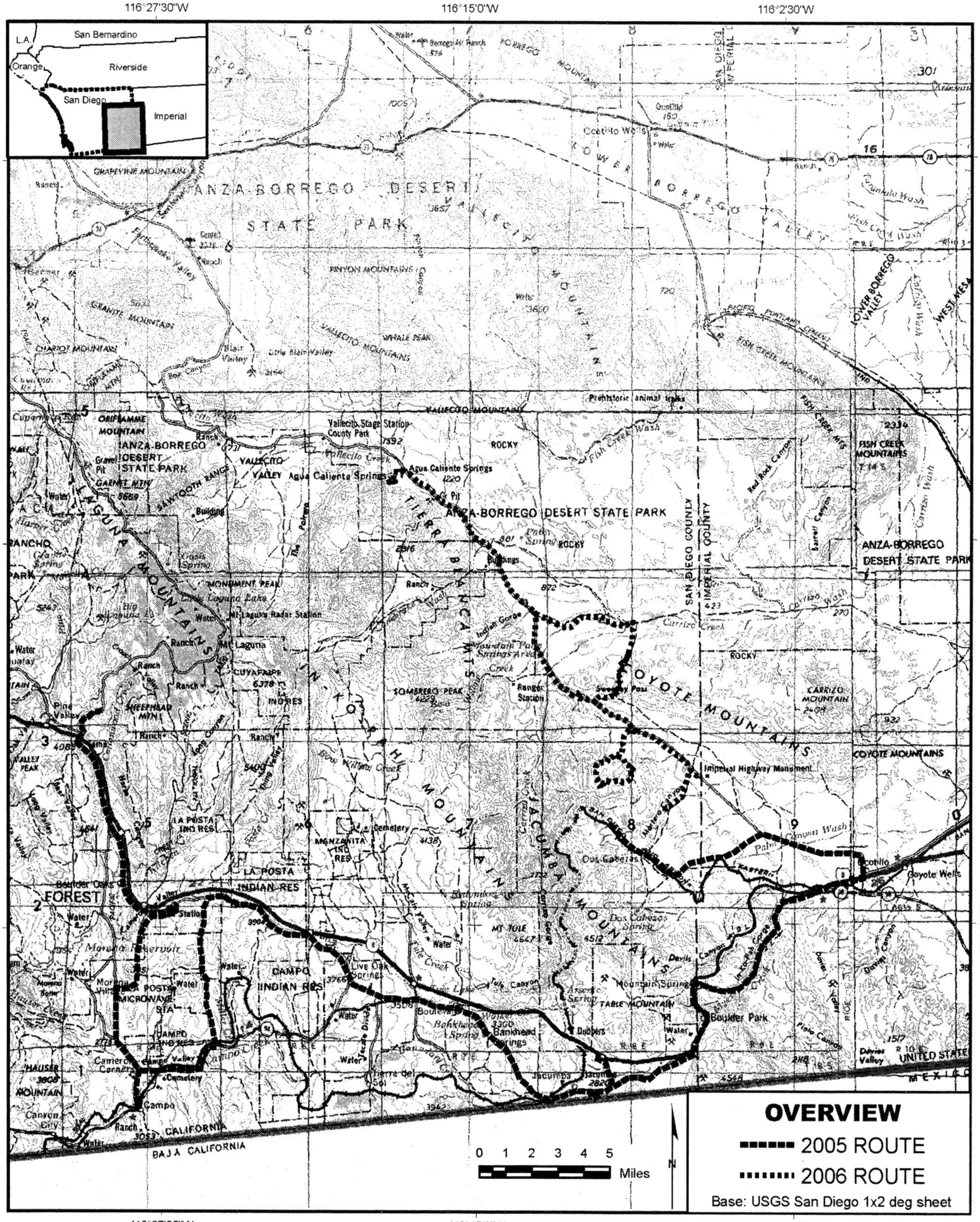
OVERVIEW
2005 ROUTE
2006 ROUTE
Base: USGS San Diego 1x2 deg sheet
0 1 2 3 4 5
Miles
N
116°27'30"W
116°15'0"W
116°2'30"W
33°7'30"N
32°55'0"N
32°42'30"N
San Bernardino
L.A.
Orange
Riverside
San Diego
Imperial
ANZA-BORREGO DESERT STATE PARK
ANZA-BORREGO DESERT STATE PARK
SAN DIEGO COUNTY
IMPERIAL COUNTY
COYOTE MOUNTAINS
LAGUNA MOUNTAINS
JACUMBA MOUNTAINS
CAMPO INDIAN RES
LA POSTA INDIAN RES
Agua Caliente Springs
Ocotillo
Coyote Wells
Jacumba
Campo
Boulder Park
Imperial Highway Monument
BAJA CALIFORNIA
CALIFORNIA
MEXICO
UNITED STATES

Part I. Road Logs

The San Diego Association of Geologists' 2005 Field Trip Road Log

Geology and Lore Along Old Highway 80 in the San Diego Back Country

and

The San Diego Association of Geologists' 2006 Field Trip Road Log

Where the Pavement Ends: A Geologic Excursion in Canyon Sin Nombre and the Volcanic Hills

WARNING
This book describes driving and hiking situations and other potentially hazardous activities that are the sole responsibility of the reader. Proper planning and knowledge of safe behavior in the mountain and desert areas described is essential for preserving health and enjoying the places described. Private property, including railroad right-of-way, must be respected. The hiking routes described herein, while relatively short, may be unmarked, expose the unprepared reader to extreme environmental conditions, and put inexperienced and experienced explorers alike in potentially harmful situations. Safe exploring of mountains and deserts can be found in many good references, including *The Anza-Borrego Desert Region*.

The San Diego Association of Geologists' 2005 Field Trip Road Log

Geology and Lore Along Old Highway 80 in the San Diego Back Country

"Trains, Veins, and Oikocrysts"

Philip T. Farquharson — Field Trip Leader

Introduction and Acknowledgments for the 2005 Field Trip

Welcome to the San Diego Association of Geologists' 2005 version of the annual field trip. We are about to take a trip back in time — not just the usual trip back into geological time, but a trip touching on the humans who ventured into the back country of what is now San Diego County. Native Americans called this area home as long as 9,000 years ago, while Europeans arrived on the scene much later. We'll look at some of the transportation corridors, especially the route of the San Diego and Arizona Railroad and U.S. Highway 80 and its predecessors, as well as its successor. Along the way, we'll look at the geology and mineral resources of the area.

The subject of this trip occurred to me around the campfire at Joshua Tree National Park on the 2004 field trip. Some folks had mentioned to me that they wished they knew more about the geology in their own back yard, and it occurred to me that the Interstate 8 transect field trip had not been done since the 1991 GSA meeting, when Gary Girty and Mike Walawender led such a field trip. Many of our newer SDAG members (including me) had not yet arrived in San Diego at that time, so this subject is a fresh one. Also, in our hurried, harried lives, we have a tendency to get on the freeway with blinders on, trying to make it to our work or play destination, without being able to stop and examine road cuts, which we geologists are wont to do. So I have tried to keep our path off the "superslab" as much as possible on this trip.

I hope this will be a tantalizing overview, encouraging all to learn more about the geology and history of southeastern California. Among the great publications describing the geology and history is Mike Walawender's book, *The Peninsular Ranges: A Geological Guide to San Diego's Back Country*, which contains field excursions and geological history of 30 sites. We are fortunate to have Dr. Walawender along to personally narrate several of these histories, and a new article is included in this publication. More history of the San Diego and Arizona Railroad is included in *A Brief Guide to San Diego's Railroad History*, by Jim Lundquist, which is available at the Pacific Southwest Railway Museum in Campo. Chris Wray has written and published several books on the history of the back country, and after you hear him speak at the Desert View Tower, you may want to procure one or more of these. We are lucky to have Diana Lindsay teaching us about the Native American presence in Anza-Borrego Desert State Park, since she has "written the book," literally. When I went on my first SDAG field trip, in 1998, to Guadalupe Canyon, her talk on the Kumeyaay was the highlight of the trip for me.

I definitely did not put this thing together by myself. Dave Bloom, that amazing bundle of energy and ideas, prodded me incessantly, beginning as early as a year ago, and proved just how great a secretary he is by capturing all my off-hand comments in the road log which follows. Monte and Diane Murbach's dedication to all things pertaining to San Diego geology helped me overcome various sticking points in the quest to see this work to fruition. George Copenhaver's enthusiasm for the SD&AE railroad helped me keep the train in the mix when we were obviously NOT going to achieve our original goal of riding the rails through Carrizo Gorge. "Carrizo" George has provided an excellent geological tome for the guidebook, which contains the sort of classical geology unseen since Powell's journals of the 1869 Grand Canyon expedition. Chuck Houser inadvertently introduced me to Jacumba by hiring me to do graphics for him on his Jacumba groundwater projects. Bob Mitchell and Todd DeJarlais, our hosts at Jacumba Hot Springs Spa and Cabana Club, have been extremely accommodating and generous, and all of our planning sessions in the restaurant have been quite pleasant. Ben Schultz, owner of the Desert View Tower, is living a dream I've had for quite some time, at his roadside attraction. Thanks, Ben, for your hospitality. Thanks to Carl Calvert of the Motor Transport Museum, and Chris Wray, for taking the time to speak with us. Thanks to Lowell Lindsay for placing such confidence in me, even suggesting that I write a certain book in the near future (after this guidebook and the 2002 guidebook are put to bed, of course).

And last, but not least: a big thanks to the membership and past and present officers of San Diego Association of Geologists, who have helped me immeasurably to get my geological career back on track after an extended life hiatus. It is important to remember that I was directed toward SDAG by the faculty and staff of Geological Sciences at San Diego State University, back in 1998. And Marie Grace was responsible for me finding out about an open teaching slot at Southwestern College in January of this year, exactly 40 years after I started my geological odyssey at the University of Wisconsin.

Welcome to another unforgettable San Diego Association of Geologists Field Trip!

Phil Farquharson
October 22, 2005

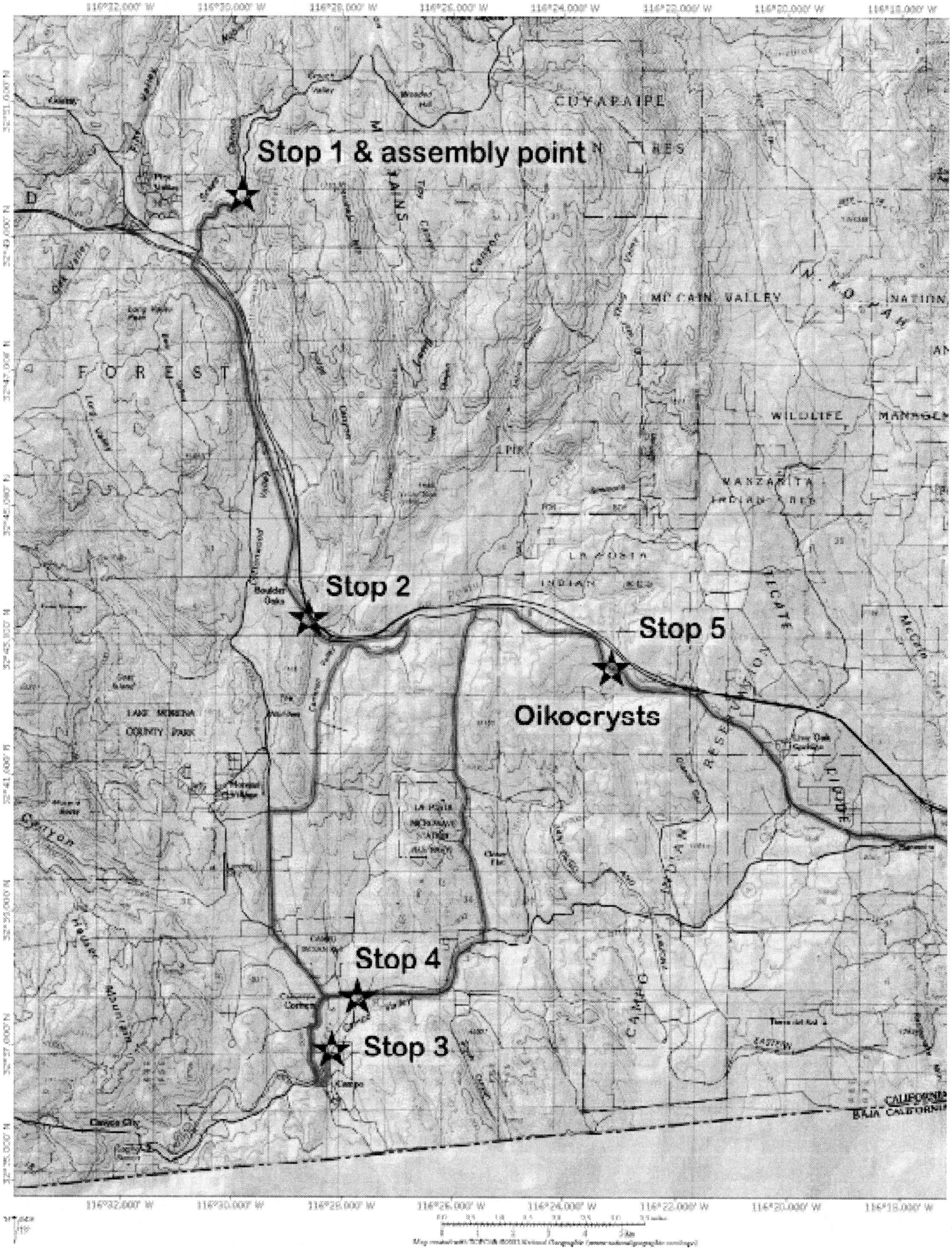
Stop 1 & assembly point
Stop 2
Stop 5
Oikocrysts
Stop 4
Stop 3
CUYAPAIPE
FOREST
MC CAIN VALLEY
WILDLIFE
LAKE MORENA COUNTY PARK
CALIFORNIA
BAJA CALIFORNIA

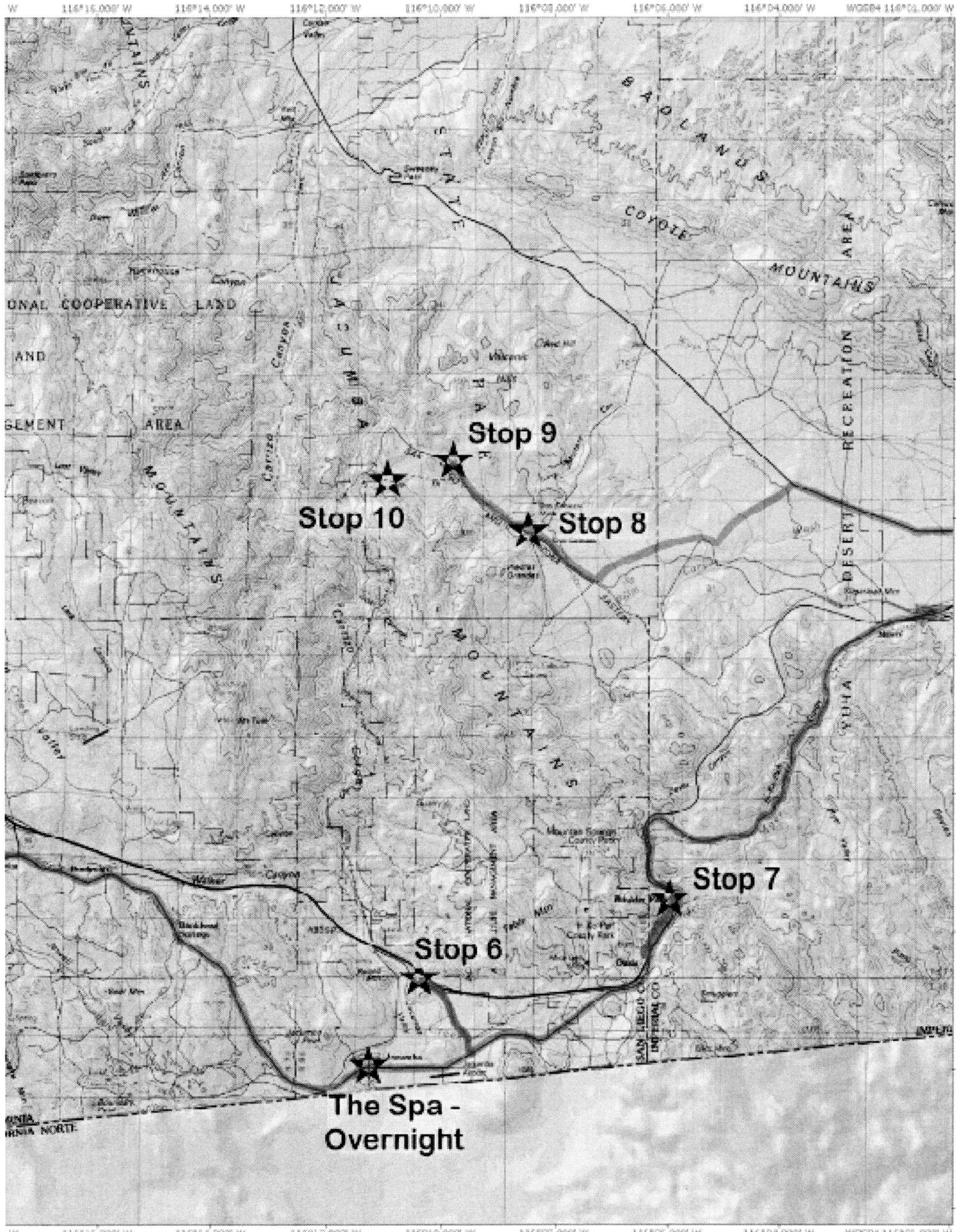
Stop 9
Stop 10
Stop 8
Stop 7
Stop 6
The Spa - Overnight
COYOTE MOUNTAINS
DESERT RECREATION AREA

2005 San Diego Association of Geologists' Field Trip Road Log

Geology and Lore Along Old Highway 80 in the San Diego Back Country

"Trains, Veins and Oikocrysts"

Philip T. Farquharson — Field Trip Leader

DAY NO. 1

Stop 1. (Assembly Point) Sunrise Highway (County Highway S1) turnout, 1.8 miles north of Highway I-8 Exit 47. Stop 1 is just about a quarter of a mile north of "Mile post 15." A lone picnic table is present at the turnout. Meet here for continental breakfast and presentation at 7:30 a.m. Set odometer to 0.0.

Stop 1 Speakers:
Mike Walawender and Chuck Houser

Mike will discuss the geologic setting of the Peninsular Ranges Batholith (PRB) and contrast the western and eastern zones of the Peninsular Ranges Batholith. A major discontinuity separates the two. We are standing in the western batholith, at its eastern boundary. Mike will discuss how the two zones differ geologically, petrologically, and geochemically, and how they are related tectonically. Excellent details are contained in Mike's publication, *The Peninsular Ranges: A Geological Guide to San Diego's Back Country*, published in 2000 by Kendall/Hunt Publishing Company.

Chuck will discuss the groundwater setting in Pine Valley, which is the valley below us to the west. Leaking underground fuel storage tanks have affected groundwater in the area. Chuck will give us his insight into the extent of the problem and offer some solutions.

Odometer (Trip) Mileage

0.0 Turn south on Sunrise Highway toward I-8. Stay off the Interstate! We will be on old highways and back roads the rest of today.

1.8 Turn left after crossing over I-8 onto Old Highway 80. Interstate 8 replaced this highway in 1974. This section of the old highway was constructed in the late 1920s to early 1930s.

2.8 Glencliff Fire Station (USFS).

3.5 Large landslides on the right.

3.6 Bridge over creek. This is one of Chuck Houser's favorite fishin' holes.

4.1 Border Patrol checkpoint. The authorities are checking vehicles heading west. We should not need to stop.

5.7 Four-way stop at Buckman Springs Road. Continue straight.

7.9 Boulder Oaks Campground.

8.5 Stop 2. Pull as far to the right as you can along the road cut.

Stop 2 Speaker:
Mike Walawender

Transition zone from western to eastern PRB. The recent rains this year have cleaned decades of dust and colluvium from this circa 1930 road cut, revealing part of the transition zone. Toward the west end of the road cut, there is a uniform granitic body of grossly "western" character. About one-third of the way along the cut to its eastern end, vertical deformation banding and widely varied rock types mark the transition.

In the road cut across Interstate 8, the transition zone is visible. In good light an outcrop of white and gray banded "zebra striped" rocks can be seen. At the eastern end of the I-8 road cut is the La Posta Pluton, age-dated at about 94 Ma. Western PRB plutons typically yield dates of 110-120 Ma.

10.2 La Posta Diner.

10.5 Turn right onto Cameron Valley Road. This USFS

road is easily drivable by sedans, provided care is taken where it is uneven. Please drive slowly to avoid damage. Look back toward Kitchen Creek Road and the Cuyamaca-Laguna Mountain Shear Zone. For the next 5 miles, we will be driving along Cameron Valley, one of the many "onion skin" valleys of the La Posta Pluton formed by zoning and jointing in this massive pluton. The tonalite of this pluton is zoned and has been described as having four facies from outer to inner: hornblende-biotite, large biotite, small biotite, and muscovite-biotite. We are driving through the hornblende-biotite zone. There are many side dirt roads. Keep to the main dirt road.

15.6 Turn left onto Buckman Springs Road (paved).

19.0 SR 94. Turn right. The Campo Valley has a long history, occupied by the native tribes, settled by the Spanish then Mexico, and continuing today.

20.5 On your right, pass the Gaskill Stone Store, now operated as a museum. Cross over the railroad tracks.

20.6 Turn left onto Forrest Gate Road. Turn left again on a dirt road leading to the railroad station. Follow the signs for the "Pacific Southwest Railway Museum."

20.8 Keep right at the sand loading operation.

21.0 Stop 3. Campo Station: 90-minute train ride, leaves at 11 a.m. We will have lunch fixin's and beverages on the train.

Stop 3 Speaker:
George Copenhaver

Bedrock along the railroad during this ride is granitic rock of the La Posta Pluton. George will discuss various aspects of the geology along this and other parts of the railroad, including engineering and economic geology of a section through Carrizo Gorge. An article by "Carrizo" George is enclosed in this volume, including map plate.

21.0 Drive back to Forrest Gate Road.

21.3 Turn right on Forrest Gate Road. Turn right immediately onto SR 94.

23.0 Continue straight, past Buckman Springs Road. Campo Valley.

23.6 Stop 4. Turn into the driveway at the Feldspar Mill, now operated as the Motor Transport Museum. No, it's not a junkyard!

Stop 4 Speaker:
Carl Calvert

Carl will present a historical overview of the Feldspar Mill built here in the 1920s to grind feldspar powder for ceramics. Feldspar was mined from an underground mine on Barrett Butte overlooking Hauser Canyon.

See "History of Campo Mill," by Carl Calvert, reproduced in this publication.

24.5 Turn left on La Posta Road.

25.1 Exfoliation weathering of La Posta Pluton.

25.2 Pass the Shockey Truck Trail. Find Aussies and Border Collies here.

26.3 Turn left. Road still named La Posta Road.

26.4 Cross railroad tracks.

26.8 The antenna you see at the La Posta Naval Microwave Station was built by the Naval Electronics Laboratory in 1964. Since 1986 the Navy has used this 1,079-acre facility for training its SEALS in mountain warfare.

28.9 Rancho de los Sobra. Good view of microwave tower to left. The 60-foot-diameter dish is stored "vertically" when not in use. In its heyday, the "La Posta Astro-Geophysical Observatory" played a major role in solar radio mapping, studies of environmental disturbances, and development of a solar optical videometer for microwave research. The observatory has been inactive for many years.

30.4 Lazy Bear Ranch.

31.2 View of Interstate 8 viaduct.

32.2 Right turn on Old Highway 80. For his 1991 field trip, Mike Walawender stopped here to observe the small biotite facies of the La Posta Pluton. Note the degree of weathering in this cut.

33.0 We're driving out of another "onion skin" valley. Consider the La Posta Pluton as an onion, cut at the surface. Each layer of onion weathers in this pattern. Laguna Mountains are visible on the left (north) in the distance.

35.0 Stop 5. Oikocrysts!

Stop 5 Speakers: Mike Walawender and Phil Farquharson

Pull all the way off the pavement. Observe the large exfoliation of the relatively fresh tonalite in the cut on the opposite site of the highway. Be extremely cautious crossing the road! Note the many core holes. They represent oriented samples collected by researchers for paleomagnetic studies. Some samples may have been subjects of radiometric dating. If sunny, look for the albite oikocrysts. These optically continuous feldspars surround biotite and quartz and can be quite large, sometimes 10 cm or more across.

This is also one of the locations sampled for SDSU's petrologic suite of the Peninsular Ranges, which was the subject of a poster session at the Denver GSA meeting in 2004, entitled "Peninsular Ranges Batholith Teaching Suite: A Petrology Laboratory Case Study of Continental Margin Magmatism."

"Thank you for visiting Stop 5. Come for the exfoliation, stay for the oikocrysts."

35.0 Continue east on Old Highway 80. The La Posta Pluton continues south into Mexico. The large steel electrical transmission pylons are just on the U.S. side of the border.

35.1 Miller Valley Road.

36.7 Turn right at Crestwood Road. Continue on Old Highway 80. Notice the low relief of the plateau in this area. This is one possible airport site under consideration by the San Diego Airport Authority.

37.1 Golden Acorn Casino.

37.3 At 9 o'clock, you can see a set of new windmills for a 50-megawatt electric generation project on the Campo Indian Reservation. Last month (September 2005) there were four windmills and three more under construction. Two weeks ago there were 15 windmills with 5 more under construction. How many are there today?

According to reporter Chet Barfield in the *San Diego Union-Tribune*, March 20, 2005, the Campo Band of Kumeyaay Indians contracted Superior Renewable Energy to build and operate 25 turbines along the Tecate Divide ridgeline. "Each turbine will have three 140-foot blades mounted atop a tubular pole 218 feet tall," he reported.

38.6 Live Oak Springs.

40.4 Tierra del Sol Road at the Tecate Divide (3,890 feet). This divide separates streams that flow west to the Pacific Ocean and east into the Colorado Desert—specifically, west-flowing Campo Creek to Tijuana River, and east-flowing Walker Creek to Carrizo Creek and Salton Sink. We now have a view of rocks other than tonalite. Mt. Tule ahead at 11 o'clock is shot through with pegmatites. Table Mountain is composed of generally andesitic Miocene rocks of the Jacumba volcanics, dated at 19 to 18 Ma. There is a sharp point of volcanic cone on Table Mountain called Squaw Tit. The bouldery hills just ahead of us are composed of a true granite or monzogranite. The Jacumba Mountains on the skyline behind Mt. Tule are partially composed of tonalite of the La Posta Pluton. Carrizo Gorge follows a metamorphic pendant, located between Mt. Tule and the Jacumba Mountains.

41.2 Junction with Highway 94. Continue straight on Old Highway 80.

41.5 Boulevard. Mapquest.com finds only one city or town named "Boulevard" in the U.S.

41.9 Junction with Highway 94. Continue straight on Old Highway 80.

42.6 Calexico Lodge.

42.8 Chef's Hat Grill.

43.1 Another view of Tule Mountain and the Jacumba Mountains.

43.7 McCain Valley Road. Post "Mile 28.5."

44.0–44.4 This section of highway was built around 1930 to straighten an older segment which was built in the 1910s. Various other sections of Old Highway 80 were also straightened in the 1930s. The 1910 version of the old highway was also narrower than the "newer" Old Highway.

44.5 Bankhead Springs, "an unincorporated community of the County of San Diego." Here is a good outcrop and hill of true granite, similar in composition and age to the Indian Hill Pluton we'll see tomorrow.

46.4 View into the Jacumba Valley. This valley is a graben characterized by the most abundant package of Miocene volcanic rocks in San Diego County.

48.7 The white building that has a red tiled roof and is enclosed by a masonry wall is in Mexico. This military garrison was constructed when the U.S. Border Patrol stepped up activities along this border. The posts and rails of the international border fence are made of steel railroad rails.

The border patrol has placed numerous motion sensors and other remote sensing devices along the border in this area. Historically, until just a couple decades ago, the native peoples moved back and forth across the border that divides the valley into Jacumba on the U.S. side and Jacume on the Mexico side.

The name Jacumba comes from the Kumeyaay language, meaning "Look at the Water."

49.0 Highway bridge over the railroad.

49.6 On the left, observe wells that provide the town of Jacumba's water supply. The ruins of an old swimming pool from an earlier version of the Jacumba Hot Springs resort can also be seen through the willow trees. Water supply wells cannot exceed 40 feet, as deeper wells yield water unsuitable because of the presence of high levels of sulfur and the high temperature.

49.8 Jacumba Hot Springs and Spa Cabana and Resort. Turn left into the parking lot. This is our overnight stop.

Evening speakers:
Bob Mitchell and Chuck Houser

Bob will present the history of the hot springs and resorts of the Jacumba Valley. Bob is our host tonight as the owner of Jacumba Hot Springs Resort.

Chuck will present his findings from hydrogeologic studies in the Jacumba Valley. His work has included an evaluation of a fuel leak from a gasoline station adjacent to I-8 at the Jacumba exit. Chuck's PowerPoint will contain numerous aerial photographs taken from a Piper Archer piloted by Chuck, with Phil "Professor Wrinklescreen" Farquharson pushing the button on his Nikon D-100.

DAY NO. 2

Reset odometer to 0.0. After breakfast, we'll collect our bag lunch and leave the Jacumba Hot Springs Resort.

0.0 Turn left out of the parking lot onto Old Highway 80.

0.3 The surrounding volcanic rocks provide the Jacumba Valley with rich soil. The valley has been farmed for many years. About every six weeks, the farm on the left harvests "baby greens" that are sold to local high-end restaurants. Round Mountain, the large dome, is comprised of stratified volcanic rock and is at 9 o'clock. John Minch and Pat Abbott described this peak in 1973 as a volcanic neck or plug. Others have suggested that the hill is a down-dropped block separated by a fault from the bluff to the west. Granitic rock has been noted near the summit. Joan Kimbrough says the exposed granitic rocks in the notch at the summit contain brick-red biotite indicating hot volcanic flows across existing granitic rock that had been exposed at the surface. Table Mountain tops out at over 4,000 feet in elevation and is at 10 o'clock. The Jacumba Mountains in the distance are from 11 o'clock to 2 o'clock.

1.3 The Jacumba "International" Airport is literally located on the international border with Mexico. This airport is San Diego County's least-used airport. Currently the border fence ends at the base of the Jacumba Mountains. The "Valley of The Moon" is located just east of the crest of the ridge. Valley of the Moon is a high desert mountain valley, notable for Smuggler's Cave and other large granitic boulders reminiscent of Joshua Tree National Park. Bandits are rumored to have hidden in the area, including those involved in a series of Campo raids and shootouts of the 1870s. Valley of the Moon is mostly in the central muscovite-biotite facies of the La Posta Pluton. There are thin metamorphic fragments of a roof pendant which contain mineralized zones, some of which have been prospected, including the Elliot Mine.

1.8 Turn left onto Carrizo Gorge Road.

2.3 Mt. Tule, shot through with light-colored pegmatites, is straight ahead.

3.0 Junction to I-8. Keep going straight past the Shell Station on the well-graded dirt road, marked "Not a Through Road." Frequently used by the visitors to De Anza Resort, this road can develop a pronounced washboard.

3.4 Stop 6. Turn right into a wide turnout that should accommodate all our vehicles. Circle in as far as possible.

Stop 6 Speakers:
Mike Walawender and Phil Farquharson

Stop 6 is the low cliff-forming pink outcrop of volcanic breccia about 100 feet north of the turnout. Look at the rock and see how many different rock types you can find.

Phil will also discuss the Round Mountain controversy. This road ends before Carrizo Creek begins its drop into Carrizo Gorge. The railroad can be seen at the base of Round Mountain. The Carrizo Gorge Railway uses the Titus siding to store double-decker commuter railcars formerly of Chicago Transit. The railroad, completed in 1919, descends east along Carrizo Creek into Carrizo Gorge. The *carrizo* (Spanish for "reed") grows abundantly in this perennially watered valley.

Mike will discuss the nature and evidence for a large fault in Carrizo Gorge. Although the fault has not been directly observed, geobarometric and other lines of evidence support its presence. His paper describing the evidence will be in the final guidebook, and is summarized in this document.

We'll return back to Old Highway 80 via Carrizo Gorge Road.

5.1 Turn left onto Old Highway 80.

5.9 Good view of Table Mountain at 10 o'clock. Behind it at 9 o'clock is a sandstone unit under-/overlain by the Boulevard monzogranite. In 1913 the first motor highway (unpaved) to pass through this area descended into Mountain Springs after curving around the north side of Table Mountain.

6.1 Bridge over Carrizo Creek. On the right is a shooting range into the hillside. Don't worry, the shooters fire to the south toward Mexico.

6.5 To the left are remnants of a roof pendant of gneiss, amphibolite, and schist. We'll see more of these rocks types later today.

7.9 Border Patrol checkpoint. We're "coming in," so be prepared to stop. You may be asked your citizenship.

8.5 The dirt road to the right takes you to Valley of the Moon.

8.7 Turn left onto In-Ko-Pah Park Road. Do NOT enter the freeway (yet)! Go under the freeway.

8.8 Bear right after the freeway, onto In-Ko-Pah Park Road, a section of Old Highway 80. Some may remember the In-Ko-Pah County Park located near here. San Diego County shut the park down decades ago, reasoning it was too far away to maintain park facilities here. In-Ko-Pah is Kumeyaay for "the meeting place" or "place of the mountain tribes."

9.1 Imperial County line.

10.0 Gate entrance to Desert View Tower.

10.1 Stop 7. Desert View Tower. Elevation 3,000 feet.

Stop 7 Speakers:
Chris Wray, Ben Schultz, and George Morgan

Historian Chris Wray will discuss the transportation history from the wagon trails of the 19th century to the modern freeway. His books are on sale in the Tower gift shop: *The Historic Backcountry: A Geographic Guide to the Historic Places of the San Diego County Mountains and the Colorado Desert* — and *Highways to History: A Driving Guide to the Places of the San Diego County Mountains.*

Ben will talk about the history of the Desert View Tower from the 1920s to the present, and tell us more about the desert. Ben is the owner of the Desert View Tower. An article on this subject by Mike Retz of the Mountain Empire Historical Society is reprinted in this volume.

George Morgan discussed his research on the Painted Gorge fault, on the eastern margin of the Coyote Mountains. This article is contained in this guidebook. *(Mention of his talk was not included in the original road log.)*

This is a good place to view the Salton Trough. Were it not for the massive load of sediment in the Colorado River, this valley would be the northern extension of the Gulf of California. The northern extent of the East Pacific Rise is creating new oceanic crust. High heat flow in the center of the valley is a source of geothermal energy, hot springs, and San Diego's nearest recently active volcanoes, the obsidian hills of the Salton Buttes volcano. The escarpment we are standing on top of was formed about 3 to 6 million years ago when this western side of the Salton Trough separated from the North American continental mainland.

We are still in the La Posta Pluton, which continues east, disappearing into the alluvium of the Colorado Desert at the base of the escarpment. View to the north and east. Whale Peak, Vallecitos Mountain, and Coyote Mountains are visible in the background. The modern freeway cuts through Old Highway 80 in many places as it descends to the desert floor. The 1930s version of the old highway is seen next to the earlier narrower 1910s version, the first paved highway. To the north a trail is seen rising from the canyon above the freeway before it curves down into Mountain Springs. That is the old toll road built in the 1850s. Ruins of the stone corrals are visible just to the left of the freeway as the trail makes the steep ascent. The stone corrals held oxen, which travelers could hire to assist in ascending the route. Those who wished to forgo paying the assist fee were still assessed a toll to use the road.

The road ends here, so return back to Interstate 8. Go under the freeway.

11.3 Turn left onto Interstate 8 eastbound and down the hill.

12.6 Boulder Creek. We're in the muscovite-biotite zone of the La Posta Pluton. This zone is the middle of four zones described by Mike Walawender.

13.4 San Diego County line.

13.7 Imperial County line.

13.9 On the left are the ruins of the stone corrals from the 1850s. Also note the old wagon roads rising steeply up the hill. You'll have a better view of them on the way back up. The stone corrals are located at emergency water tank 15.

14.5 Continue down the freeway past Mountain Springs Road exit. We pass over a section of Old Highway 80 here.

15.4 Runaway truck ramp.

16.6 Curving section of Old Highway 80 on left above the modern freeway.

16.9 Section of Old Highway 80 on left below the modern freeway. The old highway was a steep and curving road. This part of the Interstate was constructed in the late 1960s.

18.8 Myer Creek.

19.5 Elevation 1,000 feet.

19.7 Sugarloaf, a hill of alluvial gravel and sand, is located next to the railroad. Historically, the railroad quarried material from this hill.

20.7 We're entering the Yuha desert. Plaster City is visible in the distance at 11 o'clock. The Coyote Mountains are to the left at 9 o'clock. We pass a road cut of alluvium and slopewash on our right.

21.7 Railroad bridge.

22.8 Imperial Valley College Desert Museum ("Opening Soon").

23.0 Exit here: "Ocotillo / Imperial Highway."

23.2 Turn left toward Ocotillo on Imperial Highway (Highway S2).

23.6 Lazy Lizard Saloon. View of Coyote Mountains straight ahead.

24.2 Bear left.

24.6 Shell Canyon Road. This road takes you to the shutter ridges of the Elsinore fault and to Fossil Canyon. A subject for another field trip, see the SDAG 2003 field trip volume.

25.0 View up In-Ko-Pah gorge and Devils Canyon at 10 o'clock.

25.6 View of gravel quarry and Elsinore fault at 3 o'clock.

26.5 Cross under the high-tension power lines. Look at the topmost rise on the road we are traveling. That is the point where we'll turn left onto a dirt road.

28.0 A yellow diamond highway sign indicates the road will curve to the right.

28.2 Turn left at the top of the hill. Be cautious of oncoming traffic! High-clearance vehicles are recommended from this point.

29.1 Dos Cabezas water tower is visible.

29.5 View up Devils Canyon at 9 o'clock, where Interstate 8 westbound rises from the desert floor.

30.0 Turn right at the BLM sign "EC 158" and parallel the railroad.

30.3 Keep straight on EC 158.

31.2 Entering Anza-Borrego Desert State Park. Collecting is prohibited in the Park.

32.3 Stay right and rejoin the railroad track-side road in 100 feet.

33.5 Two marble quarries at Dos Cabezas in middle distance at 1 o'clock.

33.7 **Stop 8.** Water Tower at Dos Cabezas. This is our lunch stop.

Stop 8 Speakers:
Richard Borstadt and Phil Farquharson

Rich will tell us about the railroad history and the importance of Dos Cabezas in the construction and operation of the railroad. Rich is director of the Mountain Empire Historical Society.

Phil will tell of hiking trails of interest in the immediate vicinity, especially the trail up Mortero Wash and over into Carrizo Gorge, where the world-famous trestle spans Goat Canyon.

The hill of marble float is where marble was loaded onto the train.

34.0 Continue around the marble hill along the road beside the railroad track.

34.0 Trestle.

34.2 Mile post marker "109" in white on the railroad.

34.7 Note railroad cuts through metamorphic rocks to left.

35.2 Cross a dry wash. Note mile post "108" along the railroad.

35.7 View to left of Indian Hill. Note that this is actually "false" Indian Hill. The real Indian Hill is located behind it one-quarter mile farther to the south.

35.9 Stop 9. Amphibolite, gneiss, and quartzite.

Stop 9 Speaker:
Brent Foster

Brent will tell us about the metamorphic rocks here and their relationship to the Indian Hill Pluton, and how these rocks are related to the intrusion of the La Posta Pluton and the larger magmatic and tectonic context of the eastern peninsular range batholith.

Stop 10. Hike to Indian Hill.

Stop 10 Speakers:
Diana Lindsay and Eleanora I. Robbins

Walk with Diana Lindsay along the tracks east to the end of the cut, then leave the tracks to the right, heading south toward Indian Hill. It is about a mile. Diana will talk about the 5,000-year history of Native American occupation in this secluded valley. There is a rich assemblage of evidence of prehistoric human presence, including over 300 individual ceremonial fertility sites. See Diana's article included in this publication.

Norrie Robbins will tell us about the Kumeyaay language, puberty rites, and ethnobotany. She is a geologist, palynologist, and microbiologist, retired in 2001 from the USGS. She is learning the Kumeyaay language at Sycuan,

A Visit to Indian Hill's Rock Shelter, Pictographs, and Yoni Sites

Trip to Indian Hill from Dos Cabezas Siding

Entry from *The Anza-Borrego Desert Region*, 5th edition, 2006

Diana Lindsay — Field Trip Leader

Trip To Indian Hill (dirt road)

0.0 Dos Cabezas siding near the loading ramp.

Drive northwest (right) up the dirt road paralleling the railroad, past abandoned Dos Cabezas Mine on the right. Limestone used for roofing materials was quarried here. Continue northwest along the railroad to…

1.3 Indian Hill trailhead.

Indian Hill is about one mile west (left) across the railroad tracks. Hike up the obvious wash. The hill, one of several large rock outcrops in this upland valley, is in the northeast corner of section 29, UTM 5-77.6E 36-24.1N. It is misidentified on the topographic map.

The huge granite boulders of Indian Hill have tumbled together to form usable caves, which made the area attractive to Kumeyaay and earlier people. Numerous pictographs, smoke-blackened caves, pottery fragments, and unique ceremonial yonis attest to concentrated Native American use of this area. It has the distinction of being one of the oldest occupied areas in Anza-Borrego Desert State Park. This is a sensitive area because of the cultural resource. Be respectful. Do not move or remove any Indian artifacts as its original location is important for future study. Do not touch any rock-art sites because contact will destroy the pigment.

Indians first sheltered here about 5,000 years ago. These early Indians predated the Kumeyaay by several thousand years and differed from the later Indians in that they did not use pottery or bows and arrows—they used darts instead.

Over the low rise on the northwest edge of the upland valley are the circa 1910 remains of the main construction camp for crews working the lower portion of the Carrizo Gorge route. An old jeep trail, just north of the Indian Hill marked on the topographic map, ran directly east to west and connected the RR camp with the road from Dos Cabezas.

Farther northwest, below the railroad bed and about 1.5 miles from Indian Hill, is an east fork of Carrizo Canyon with several palm tree groves. Jerry Schad describes a hike through this tributary and Choral Pepper places the Reedy Lost Gold nearby.

Road ends at a wash at mile 2.2. A hike southwest from this point is an alternate approach to Indian Hill.

GPS Waypoint Locations

Name	Description	Position - WGS 84	Position - WGS 84	Altitude
		Lat/Lon hddd°mm.mmm'	UTM	
Stop One	Assembly Point - Sunrise Hwy	N32 49.609 W116 29.766	11 S 547164 3632201	4436 ft
Stop Two	Walawender - PRB Boundary	N32 43.230 W116 28.545	11 S 549127 3620424	3313 ft
Stop Three	Copenhaver - Campo Station	N32 36.786 W116 28.350	11 S 549491 3608519	2574 ft
Stop Four	Calvert - Feldspar	N32 37.811 W116 27.585	11 S 550678 3610419	2654 ft
Stop Five	Oikocrysts	N32 42.652 W116 23.227	11 S 557439 3619400	3902 ft
Camp	Jacumba Spa	N32 37.099 W116 11.342	11 S 576083 3609265	2827 ft
Stop Six	Jacumba Volcanics	N32 38.387 W116 10.410	11 S 577522 3611656	2770 ft
Stop Seven	Desert View Tower	N32 39.541 W116 05.975	11 S 584437 3613845	2919 ft
Stop Eight	Dos Cabezas Siding	N32 44.781 W116 08.380	11 S 580599 3623495	1712 ft
Stop Nine	Brent Foster - Amphibolite	N32 45.886 W116 09.998	11 S 578057 3625516	1969 ft
Stop Ten	D. Lindsay - Indian Hill	N32 45.427 W116 10.446	11 S 577363 3624664	2100 ft

Indians of the Anza-Borrego Desert

Selected and edited entries from *Anza-Borrego A to Z*, 2001

Diana Lindsay, *Sunbelt Publications, 1256 Fayette Street, El Cajon, California 92020*
DLindsay@SunbeltPub.com

10,000 B.C. ± TO THE PRESENT

Archaeologists have identified four distinct cultural sequences in the Anza-Borrego area. The earliest are paleo-Indians, represented by the San Dieguito who lived in the area more than 9,000 years ago when the climate was milder and lakes and streams were plentiful. They were primarily fisherman who lived on high ground above lakes and streams. They were also hunters who used projectile points and atlatls. San Dieguito sites have been found along the Kane Spring Road west of Ocotillo Wells and near the Truckhaven Trail in Imperial County as well as on the mesas near Sweeney Pass in the southern portion of Anza-Borrego. This cultural sequence has not been found at Indian Hill.

The next cultural sequence is known as desert archaic or milling archaic and is represented by the Pinto Basin Indians who existed in the area from 7,000 to 3,000 years ago. They were gatherers who used milling stones to process food. Basketry was used but not pottery. Projectile points were attached to darts and spears for hunting. Bows and arrows had not been introduced yet. Possibly the first petroglyphs or etchings on stone appeared during this time. Evidence of their presence has been found at Indian Hill, which is one of the oldest Indian occupied areas of ABDSP.

The third cultural sequence is known as prehistoric or pottery archaic and is represented by Yuman and Shoshonean speaking Indians (the Kumeyaay and the Cahuilla) who were attracted to the Anza-Borrego area by fresh-water Lake Cahuilla roughly 2,000 to 1,200 years ago. These were true desert Indians who gathered and processed local seeds, fruits, flowers, stalks, roots, and fibers. They used basketry and pottery and hunted with bows and arrows. They also used rabbit sticks, netting, and digging sticks. Rock art is found near or at their seasonal village sites. Evidence of this cultural sequence is found at Indian Hill.

The last cultural sequence is known as the historic period and is represented by the Kumeyaay and Cahuilla and their neighbors, the Cupeño and Yuma. This period begins with first European contact and includes the replacement of native tools and foods by European cultural elements. It also includes the rapid deterioration of the native culture due to diseases, imposed foreign culture, encroachment on native lands, and disruption of subsistence patterns. Only vestiges of native culture remain. In historic times, Kumeyaay Indians occupied the Indian Hill area.

PINTO BASIN INDIANS

The oldest identifiable true desert culture dates from 7,000 years ago when the desert archaic or milling archaic tradition began in the Great Basin and spread to California. A dryer climate and disappearance of larger game brought more dependence upon a gathering economy that used mortars and pestles to process foods. Malcolm Rogers, from the San Diego Museum of Man, called this cultural sequence Amargosan, after a river in the Mojave Desert. His Amargosa II included the Pinto Basin Indians.

Pinto Basin Indians followed the desert tradition in the Anza-Borrego area some 5,000 years ago. Archaeologists from the University of Southern California excavated a Pinto Basin Indian site at Indian Hill in 1958. The report, written by the team director William J. Wallace, concluded that Pinto Basin aborigines used the Indian Hill rock shelter as a temporary residence. The cave shelter was further excavated in the late 1980s by archaeologists from the University of California at Riverside. This research uncovered pre-pottery storage vessels made of rock-lined caches.

The excavations conducted in the area showed evidence of continued seasonal occupation. Milling stones, including mortars and pestles, and some Olivella shell beads and bone awls were found. Judging from these implements, Indians gathered wild seeds and vegetable foods in the area. Archaeologists believe that the overall population of the Pinto Basin culture was greater than that of the earlier San Dieguito cultural period, due largely to fluctuating climate conditions, which brought increased rainfall and lower temperatures.

As in the San Dieguito culture, projectile points were used to tip darts rather than arrows, and no pottery was used. The name Pinto comes from the distinctive cut marks on the projectile points initially associated with this group. Since the original research, the points have been reclassified as Elko-eared. The spear-like darts were used in hunting deer and bighorn sheep and were thrust with the help of an atlatl or spear-thrower. The Pinto Basin culture declined some 2,000 or 3,000 years ago.

MORTERO

Morteros are deep grinding holes in granitic rocks and boulders made by Indians while pounding and grinding seeds and desert plants. An elongated rock or pestle was used to pound the vegetative matter in the bedrock mortar or mortero. Often in association with morteros are metates or slicks, which were used with a small smooth rock or mano to grind seeds with a roller-pin action into a flour. Metates are smooth or slick areas found on rock surfaces. Morteros and metates are commonly found throughout the park where Indian habitation occurred. Mortero Wash/Canyon and South Mortero Wash, off Highway S2 near the San Diego and Imperial County line, takes its name from the close approximation to many Indian sites found at Indian Hill and Piedras Grandes. Mortero Palms, found south of Highway S2 and Piedras Grandes, has over 100 palms and an intermittent small waterfall at the end of the grove. The bedrock mortar holes, which are the namesake of the canyon, are found about 300 feet down slope from the fronds. The McCain cattlemen drew water from Mortero Palms for their cattle which grazed in the Dos Cabezas area.

KUMEYAAY INDIANS

The Yuman-speaking Indians of San Diego County have been formally called Kumeyaay since the early 1970s. Previous to that time, they were referred to as (San) Diegueño, the name given to them by the Spaniards for the Indians who belonged to the parish of Mission San Diego de Alcalá, which was established in their territory in 1769. Eventually, the name referred also to culturally and linguistically related bands south and east of the mission.

Three distinct dialects have been identified and are known now as Ipai, Tipai, and Kumeyaay (formerly known as Northern Diegueño, Southern Diegueño, and Mountain or Eastern Diegueño). Indians of Anza-Borrego spoke the Kumeyaay dialect basically south of what is known now as Highway 78.

The Kumeyaay were the first of the historic Indian groups to enter the Anza-Borrego area 1,500–2,000 years ago from the Colorado River. They may have been attracted to the area by freshwater Lake Cahuilla. They brought with them the knowledge of pottery making. When the lake evaporated, around 1500 a.d., many bands dispersed west into San Diego County.

Bands were autonomous and self-governing with clearly defined territories of about 10–30 square miles along tributaries and streams that included collective and individual gathering areas. Villages were inhabited seasonally to accommodate the annual gathering cycle. Key foods included acorns, mesquite beans, pinyon nuts, agave hearts, and chia seeds. The land was managed. Valleys and groves were periodically burned to keep them healthy. Women were principally involved in the food gathering and production process. Men hunted rabbits and other small mammals, quail and other birds, deer, antelope, and bighorn sheep.

The Kwaaypaay, a hereditary position, held clan leadership. The tribal chief or Kwataay was elected. A shaman or Kuseyaay, whose position was usually hereditary, led religious ritual and practiced medicine. Puberty ceremonies were very important. Young boys were given *Datura* (jimsonweed) to produce visions, and young girls were ceremonially heated in a pit, after which they were tattooed on their chin. Yoni sites found at Indian Hill may have been ceremonial sites involved with fertility and/or power centers for the shaman.

Estimates of the Kumeyaay population of San Diego County vary from 9,000 to 20,000 when the Spanish arrived in 1769. By 1860, however, the population had dropped to 2,800 from forced labor, exposure to disease, and dislocation from their environmental and economic resources. The destruction of their spiritual foundation through conversion to Catholicism was compounded with the death of the elders who had kept the history, culture, and traditions preserved in songs which were hundreds or even thousands of years old. Today, grinding holes, pottery fragments, stone tool chips, and rock art silently attest to the once thriving presence of the Indians of the Anza-Borrego Desert region.

PIEDRAS GRANDES

The jumble of massive granitic boulders, south of the old Dos Cabezas station and north of Dos Cabezas Spring is called Piedras Grandes or "large rocks." Among the rocks are old Indian shelters and good wind-sheltered sites for today's campers. This area contains the famous Piedras Grandes pictograph. The pictograph features the only representation of a man on a horse known in Kumeyaay rock art.

DOS CABEZAS

Dos Cabezas is Spanish for "two heads," named for two towering bald-headed boulders that jut out from the rock outcropping against the skyline of the Jacumba Mountains. Dos Cabezas spring lies below the two heads, south of the railroad tracks and Mortero Palms. This area was known by this name at least as far back as 1891 when "Dos Cabasos" was mentioned in the *San Diego Union*.

The railroad station near Mortero Wash was also called

Dos Cabezas because of the proximity to the large boulders. However, railroad personnel later designated other outcrops as Dos Cabezas that could be seen from the train. Representatives from the San Diego Railroad Museum estimate that there may have been as many as six sites that were designated as the namesake Dos Cabezas.

Dos Cabezas station was the last watering stop before westbound trains entered the Carrizo Gorge (see SD&AE). It was also used as a fueling stop. The loading ramp was used to load cars with dolomite from the Dolomite Mine up North Mortero Wash on the slope of the Coyote Mountains.

DOLOMITE MINE

The largest and most accessible deposits of dolomite limestone in San Diego County are found in a low group of hills near the Dos Cabezas siding and in the southwest flank of the Coyote Mountains. Also called dolomite marble or magnesium calcium carbonate, it was used "mainly as sources of crushed and broken stone for use as roofing granules, poultry grit, and decorative stone." The Coyote Mountains deposit is the one referred to as the Dolomite Mine. It was located the furthest from the rail siding at Dos Cabezas and had to be trucked eight rugged miles to the siding.

Fred M. Elliot and James Elliot of San Diego owned the dolomite deposit, which was first claimed by the family in 1924. They leased the claim in 1951 to the Milroy Roofing Company of South Gate, who worked the deposit for about six months. The Elliots also worked the claim briefly during 1952-53. By 1957, only a small shack and a loading bin remained on the property. According to Horace Parker, "The big expense in dolomite mining is sorting and when Mexican labor is cheap the dolomite mines flourish(ed)."

After the Dolomite Mine was abandoned, a group of hippies lived near the site in the 1960s and early 1970s. They painted a corrugated iron paneled building at the mine site psychedelic colors. The building was still standing in the early 1980s.

The two major Dos Cabezas deposits of dolomite were named the Heathman and the Golden State. The Heathman Quarry was located only a quarter-mile from the Dos Cabezas siding. It was named for R.W. Heathman in 1924. The claim remained idle until the 1950s when it was leased to the Campo Milling Corporation, which processed the dolomite limestone for roofing granules, chicken grit, and agricultural use. The property then remained idle from 1957 to 1960 when it was leased to Don Weaver of Jacumba who mined it for several years for decorative stone. It was worked from two locations, about 500 feet apart. Most of the dolomite was crushed, screened, and bagged adjacent to the railroad siding.

The Golden State deposit was located about a mile northeast of the Dos Cabezas siding. B.A. Sweet of the Pine Tree Portland Cement Company owned it. It was first prospected in the 1920s and worked in the 1950s.

A NOTE ABOUT YONIS

According to avocational archaeologist Robert Begole, "The whole subject of yonis was 'brought out of the closet' by Charlotte McGowan" in 1979 when she raised "the Victorian curtain" to discuss the female aspect of rock art—the vulva or yoni symbol (Begole 1984). Yoni is a Sanskrit word for the female organ of generation given to those rock formations that resemble female genitalia. The Kumeyaay enhanced these natural formations found on granitic boulders in the foothills of the Peninsular Range (extending south into Baja California) and used them in ceremonies to remove sterility or produce fertility. Ethnographic information indicates that shamans viewed these sites as sources of power. McGowan states that it was the woman, and symbolically the vulva, that was seen as the source of generative power—a common theme in many early cultures found throughout the world. She states "that yonis in Kumeyaay territory represent vaginal openings into the uterus of Mother Earth and that spirits associated with them were called upon to increase the fertility of Kumeyaay women." (McGowan 1982) The yonis also may have had a role in the girls' puberty ceremony.

REFERENCES

Begole, Robert S. "Fertility Symbols in the Anza-Borrego Desert." *Pacific Coast Archaeological Society Quarterly*, Vol. 20, No. 4, pp. 13-28.

California State Parks. "Anza-Borrego Desert State Park Volunteer Handbook." (Borrego Springs: Anza-Borrego Desert State Park, 1997), copy in author's files.

Gough, Galal. "The Indian Hill Ceremonial Fertility Site Complex." In Ken Hedges, ed. Rock Art Papers No. 5, San Diego Museum Paper 23 (San Diego: Museum of Man, 1987), pp. 55-60.

Gudde, Erwin G. *California Place Names: The Origin and Etymology of Current Geographical Names*. 3rd ed. (Berkeley and Los Angeles: University of California Press, 1969.

Hedges, Ken. "The Indians of the Anza-Borrego Desert." Unpublished manuscript prepared for Sunbelt Publications. Copy in author's files.

Jorgensen, Mark. Letter and resource data sent to the author, 2 February 2000.

Knaak, Manfred. *The Forgotten Artist: Indians of Anza-Borrego and Their Rock Art* (Borrego Springs: Anza-Borrego Desert Natural History Association, 1988).

Landis, Virginia Bridge, ed. Handbook on the Indians of Southern California with Emphasis on the Kumeyaay. (San Diego: Museum of Man, 1992).

Lindsay, Lowell and Diana Lindsay. *The Anza-Borrego Desert Region*, 4th ed. (Berkeley: Wilderness Press, 1998).

Lindsay, Diana. *Anza-Borrego A to Z: People, Places, and Things* (San Diego: Sunbelt Publications, 2001).

_____. *Our Historic Desert: The Story of the Anza-Borrego Desert* (San Diego: Copley Books, 1973).

McGowan, Charlotte. "Using Ethnology to Identify and Interpret a Ceremonial Fertility Site in San Diego County." Paper presented at the meeting of the Society for American Archaeology, Santa Barbara, California (April 1979).

_____. "Ceremonial Fertility Sites in Southern California." San Diego Museum Papers, No. 14. (San Diego: Museum of Man, 1982).

Shipek, Florence Connolly. *Pushed Into the Rocks: Southern California Indian Land Tenure*, 1769-1986. (Alpine, CA: Viejas Band of Kumeyaay Indians, n.d. Reprint. Lincoln and London: University of Nebraska Press, 1987).

Wallace, William J. and Edith S. Taylor. "The Indian Hill Rockshelter, Preliminary Excavations." Southwest Museum *Masterkey*, Vol. 34, No. 2 (April-June 1960), pp. 66-82.

_____. "The Surface Archaeology of Indian Hill." Southwest Museum *Masterkey*, Vol. 34, No. 1 (January-March 1960), pp. 4-18.

Weber Jr., F. Harold. "Mines and Mineral Resources of San Diego County, California." County Report 3. (San Francisco: California Division of Mines and Geology, 1963).

White, Phillip M. and Stephen D. Fitt. Bibliography of the Indians of San Diego County: The Kumeyaay, Diegueño, Luiseño, and Cupeño. Native American Bibliography Series, No. 21 (Lanham, MD: The Scarecrow Press, Inc., 1998).

WEBSITES

http://www.kumeyaay.info/museums/about/indianhill.html.

http://www.petroglyphs.us/photographs_ pictographs _anza_borrego_IH.htm.

The big shelter from a distance – basically the boulder pile you see as you approach Indian Hill.

Indian Hill rock shelter.

Indian Hill yoni series – a series of small yonis on a boulder next to the rock shelter.

Dos Cabezas yoni – one of the most visited yonis on a boulder in Dos Cabezas near the famous pictograph of a man on a horse.

Pictographs found in the Blue Sun Cave – the shaman's cave.

Facing page: Mile 1.5 (page 30) Field trip speaker Charlie Winker discusses his findings in the well-exposed Plio-Pleistocene sedimentary units in Canyon Sin Nombre. Charlie has been mapping the geology of the Coyote Mountains and Carrizo Basin to the north. JR Morgan and David Bloom flank Charlie. Photo taken October 28, 2006 by Philip T. Farquharson.

The San Diego Association of Geologists' 2006 Field Trip Road Log

Where the Pavement Ends: A Geologic Excursion in Canyon Sin Nombre and the Volcanic Hills

"I had a trip, and No Name was its name."

David M. Bloom — Field Trip Leader

Introduction and Acknowledgments for the 2006 Field Trip

It is with great pleasure that we present the 2006 San Diego Association of Geologists' field trip. The trip focuses on the Coyote Mountains, which remain a little studied and incompletely understood complex. I was formally introduced to the Coyote Mountains of southwestern Imperial County on the 2003 SDAG field trip. My curiosity about this complex recurved ridge was piqued. I returned the next year with the San Diego State University Geology alumni to learn that the obvious "lava flow" at its southeastern end was actually an extensive hash of fossil oyster shells. Lesson learned: Don't identify rocks while 3 miles away on a freeway.

Dr. Monte Marshall had presented paleomagnetic data that he and his students gathered in the Coyote Mountains. The rotations suggested that the Coyote Mountains hold answers to questions about the tectonic history of the Pacific-Farallon-North American Plate convergence, western North American extension, Gulf of California rifting, and San Andreas transtension. Dr. Marshall and his students have provided an original paper on their paleomagnetic findings with "big picture" tectonic interpretations. These data, first collected, in 1981, are compiled for the first time in this guidebook.

The Coyote Mountains are here in part because transpression on a bend in the Elsinore Fault forced up the Coyote block. The mountain core now exposes crystalline basement, which we will see contacting the Elsinore Fault at the head of Canyon Sin Nombre. Flanking the mountains are arkosic red beds of the Red Rock Formation and basalt flows of the Alverson Volcanics. Well known in Split Mountain Gorge and Red Rock Canyon to the north, and Painted Gorge on the southern end of the Coyote Mountains, the only outcrops of these units found on the north flank are inaccessible by car. The 1.5-mile hike up a little-known side canyon will take us down a nearly complete late Miocene to early Pleistocene stratigraphic section. Charlie Winker mapped this canyon as part of his 1987 dissertation. He will speak about some of his findings and additional as-yet unpublished mapping.

George Morgan and JR Morgan have been mapping in the Coyote Mountains for several years. They have been meeting the challenge posed by this complex structure and abruptly changing stratigraphy. The Morgan brothers provide a glimpse of this complexity in their new paper based on original mapping of the Painted Gorge Wash Fault, at the extreme eastern end of the Coyote Mountains.

The Carrizo Badlands Overlook provides a panorama of the field trip from a single point. To the west, the long, forested ridge of the Laguna Mountains rises abruptly over 4000 feet above the alluvial fans of the desert floor. Flanked on the south by the Coyote Mountains and to the north by the Vallecitos Mountains and Fish Creek Mountains is the Carrizo Basin, a 5000-meter-thick package of fossiliferous strata. The arid climate causes these soft sedimentary units to erode to badlands. Much of the sediment derived from local streams and the ancestral Colorado River in the late Miocene, Pliocene and early Pleistocene epochs. George Jefferson and Lowell Lindsay's new paper discusses how these rocks are being used to teach natural science.

Extensive, nearly pure anhydrite and gypsum deposits found in the Fish Creek Mountains formed shortly after rifting 6 million years ago began opening the Gulf of California. David Peterson and George Jefferson contributed a new paper to the guidebook providing additional evidence for a marine hydrothermal vent source for the gypsum.

The Coyote Mountains form a long ridge truncated by the Elsinore Fault. Highway S2 parallels the fault in an alluviated basin, which itself is bordered on the southwest by the black lavas of the Volcanic Hills. We will attempt to tease some sense out of alluvial terraces, lava flows, and crystalline basement. Charlie Lough will share his thoughts on the significance of the older terraces of the Mesa Conglomerate. We will see these terraces capping older formations nearly everywhere we go.

West of the Volcanic Hills is the jagged ridgeline of the Jacumba Mountains. Lavas similar to those we will see are found in the Jacumba area. We have reprinted John Minch and Pat Abbott's important 1973 paper and map on the post-batholithic geology of this area. We are especially pleased to include Pat's report on his subsequent findings that show that the Table Mountain Gravels are Miocene in age. Pat added, "My major contribution to this paper was three semesters of running SDSU advanced field-geology classes there, and doing mapping myself." Current workers in the area well know that careful recording of observations and good geologic mapping such as this are invaluable. The similarities between the volcanic rocks in the two areas suggest that the lavas beds may have formed together at roughly the same elevation. The later development of the escarpment vertically separated the volcanic deposits by about 500 meters.

Len Sinfield's new paper based on his original mapping shows zoning here in the eastern part of the La Posta Pluton is similar to that seen in the western portion.

At the northern end of the La Posta Pluton, hot springs at Agua Caliente occur at the base of the Tierra Blanca Mountains. Carole Ziegler discusses the geochemistry of the drinking water at Agua Caliente Regional Park in her new paper. We may choose to drink an alternate beverage after reading about the fluoride levels.

The Elsinore Fault crosses almost through our campsite at Agua Caliente Regional Park. This fault is one of a family of extensive, active right-lateral faults that have obviously shaped the landscape in the last million years. Equally important are more enigmatic detachment faults that have been found within these mountains and basins. These older structures indicate that Miocene rifting and extension fractured the crust, rotating the fragments into mountains and valleys. The valleys became basins, which filled with deposits from the erosion of local mountains and the prograding Colorado River Delta. The Carrizo Basin is the most prominent of these basins. It is especially important for its rich, nearly continuous late Miocene to early Pleistocene fossil assemblages.

Diana Lindsay provides a fascinating look into the history of the often-curious place names encountered on this trip. Diana literally wrote the book on Anza-Borrego. She introduced me to the wonders of Canyon Sin Nombre and the Carrizo Badlands a year ago. She planted the idea that this area was worthy of visiting and revisiting.

Diane and Monte Murbach have stepped in again to share their wealth of experience gained from over 15 years of organizing geology field trips. Their expertise has provided much of the wisdom that should be apparent in this volume.

Lowell Lindsay's persistent optimism was enough to complete this guidebook before the field trip. We all must thank him for getting this volume into your hands on October 28.

Not enough thanks go to Carole Ziegler who stepped in when we needed her most to attend to guidebook managing editor duties. Her dogged persistence and determination to extract text from your editors, read and edit the text, collect the graphics, and generally poke and prod us into submission is the kind of tough love we truly appreciate. We do. We really do!

Special thanks to my co-editor Phil Farquharson for his passionate love of geology, enthusiasm for exploring, and vast knowledge of all kinds. Phil bubbles with ideas, many of which are found in the guidebook. He provided the high-quality graphics that grace this volume. I am especially grateful for the superb cover graphics. We hope that you find the combined 2005 and 2006 field trips in one volume as satisfying as we have had in producing it.

Charlie Lough is my true inspiration and guiding force behind this field trip. His knowledge, curiosity, patience, thoughtfulness, and integrity inform all aspects of this trip. It is from his suggestions that we selected the field trip stops. It is my hope that his 20 years of mapping in this area will be the subject of a field trip in the near future.

My greatest hope is that you will gain something truly meaningful and lasting from exploring the natural wonders of the places described.

David M. Bloom, CEG
Vice President
San Diego Association of Geologists
San Diego, California

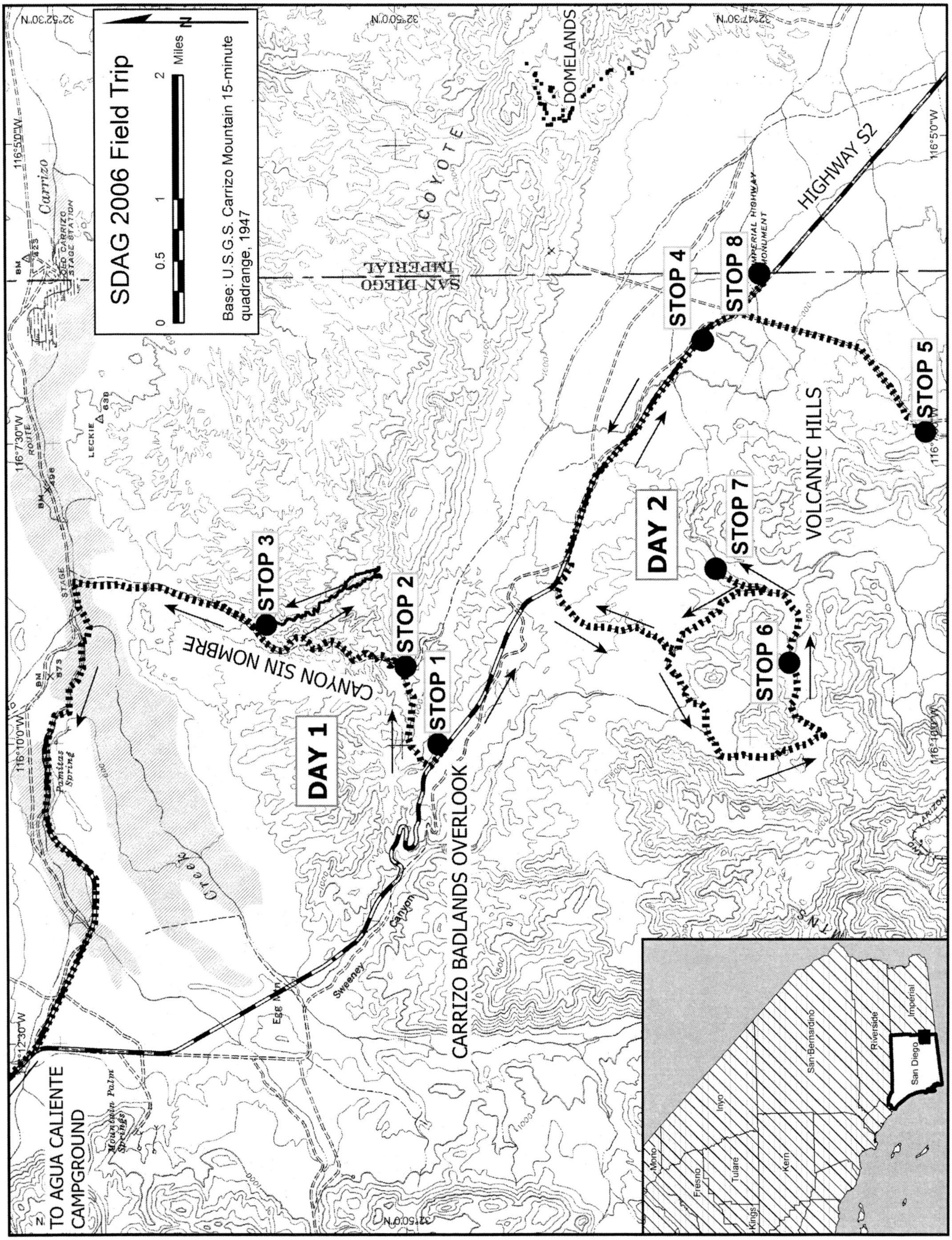
SDAG 2006 Field Trip
0 0.5 1 2 Miles
N
Base: U.S.G.S. Carrizo Mountain 15-minute quadrange, 1947
DAY 1
DAY 2
STOP 1
STOP 2
STOP 3
STOP 4
STOP 5
STOP 6
STOP 7
STOP 8
CANYON SIN NOMBRE
CARRIZO BADLANDS OVERLOOK
VOLCANIC HILLS
DOMELANDS
HIGHWAY S2
IMPERIAL HIGHWAY MONUMENT
SAN DIEGO
IMPERIAL
COYOTE
TO AGUA CALIENTE CAMPGROUND
Carrizo
OLD CARRIZO STAGE STATION
Pamitas Spring
Mountain Palm Springs
Sweeney Canyon
Egg Mtn
Creek
LECKIE
STAGE ROUTE
32°52'30"N
32°50'0"N
32°47'30"N
116°5'0"W
116°7'30"W
116°10'0"W
116°12'30"W
Mono
Fresno
Kings
Tulare
Inyo
Kern
San Bernardino
Riverside
San Diego
Imperial

2006 San Diego Association of Geologists' Field Trip Road Log

Where the Pavement Ends: A Geologic Excursion in Canyon Sin Nombre and the Volcanic Hills

"I had a trip, and No Name was its name."

David M. Bloom — Field Trip Leader

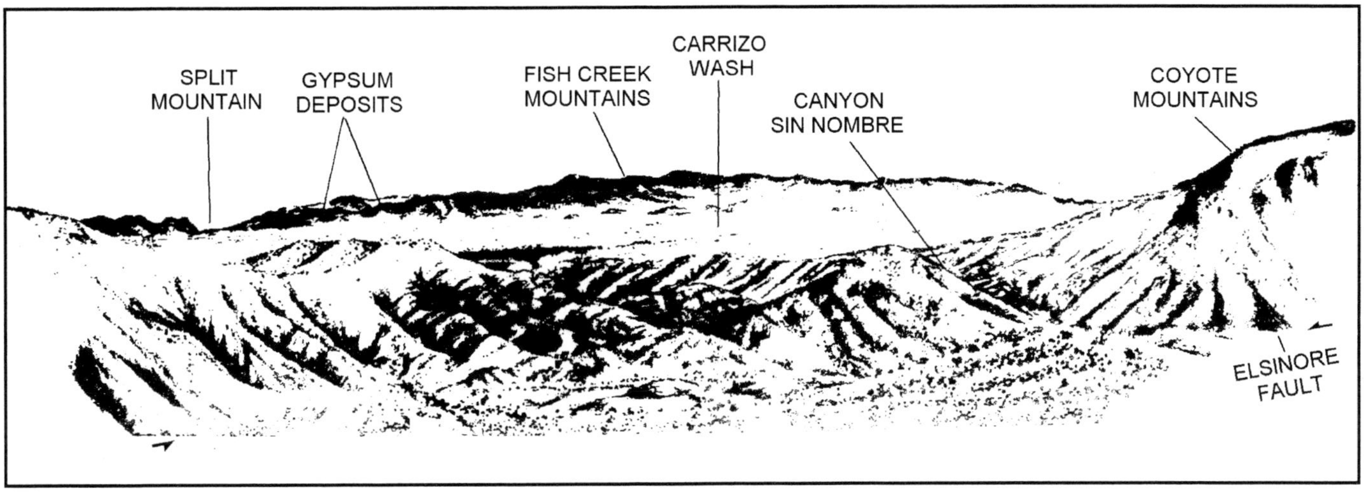

THE CARRIZO BADLANDS...a walk through time

Marine reefs, lake deposits, and silt from the Colorado River delta have been cast into layers over millions of years.

These deposits, once horizontal, have been tilted up and exposed like pages of a book. The elements of erosion have sculpted the desolate badlands in front of you.

Paleontologists studying the fossils of the Carrizo Badlands have discovered this is one of the richest deposits of its kind on Earth. The time period of one-half million to five million years ago is represented by the hardened bones of mastodons, llamas, horses and tapirs. Grasslands and scattered forests supported large animal herds.

In the last 20,000 years this landscape has become a desert. The large herds are gone, as are the streams and lakes, but what is left will fascinate park visitors and scientists for years to come.

—Interpretive panel, Carrizo Badlands Overlook

DAY NO. 1

Stop 1. (Assembly Point) Carrizo Badlands Overlook. County Highway S2 (Sweeney Pass Road) dirt turnout loop, 13.2 miles southeast of Agua Caliente Campground. For those driving from I-8, drive 12.6 miles northwest on Highway S2 from the Ocotillo exit. At the overlook, reset odometer to 0.0. After this point we will be on rough dirt roads. High clearance vehicles are required and four-wheel drive is strongly recommended.

Stop 1 Speakers:
Charlie Lough and Monte Marshall

From this viewpoint on the western edge of the Salton Trough, we can see, in a general way, all the geology we will explore in more detail the rest of the trip. Let us begin looking in a northeasterly direction and turn slowly 360 degrees to the right. Directly below us is Canyon Sin Nombre. The Elsinore Fault strikes northwest across the head of the narrows. The close mountains to the right of the narrows are the Coyote Mountains. They form a long ridge bounded by the Elsinore Fault. A small alluviated basin to the west is bordered on its southwest side by the Volcanic Hills whose black lavas stand out in contrast to the light gray alluvium. At the bases of the mountain blocks are ubiquitous older alluvial terraces sometimes called the Mesa Conglomerate.

Volcanic Hills. In-Ko-Pah Mountains in distance.

The irregular skyline in the distance is the boulder strewn In-Ko-Pah Mountains. The eastern portion of the zoned La Posta Pluton makes up most of these mountains. The La Posta Pluton is some 30 miles wide. We are looking at just the eastern 5 miles. The La Posta Pluton is even long-er than it is wide. It extends into the Tierra Blanca Mountains to the northwest. The northern contact of the pluton lies near Agua Caliente Regional Park to the northwest.

The La Posta Pluton is the best studied of a family of similar granitic bodies that formed about 94 million years ago in a rough line from southern California to central Baja California (see figure inside cover). The compositional and textural zoning from edge to center of this single massive pluton has been recognized in other plutons which are affectionately called La Posta-type. These magmas were calcium-rich tonalites that became granodiorites when the mafic mineral composition changed as the magma cooled from the outside in.

The Laguna Mountains to the west rise 4,000 feet along the Peninsular Ranges Escarpment. There is no equivalent topographic break on the east side of the Salton Trough. The geometry of imbricated slabs rotated by detachment faulting provides a useful model to explain this asymmetry.

The Vallecitos Mountains to the north and Fish Creek Mountains to the northeast form the northern edge of the Vallecitos and Fish Creek Badlands, which merge south into the Carrizo Badlands. The southern edge of the badlands comes almost to our feet, and defines the Carrizo Basin, a greater than 15,000 feet thick package of marine and nonmarine strata. Much of the sediment is derived from local streams and from the deltaic succession of the late Mio-Pliocene Colorado River.

Whale Peak is known for the easternmost Eocene fluvial deposits containing Poway-type cobbles and boulders containing a rhyolite of unique composition and texture. A large river is thought to have carried this material from a source now located 250 miles southeast in northern Sonora, Mexico.

The Fish Creek Mountains yield extensive, nearly pure anhydrite and gypsum deposits that formed shortly after rifting 6 million years ago. United States Gypsum Company extracts a large portion of our nation's production of gypsum from their quarry.

BASIN

Rift crust, create sea
Fill basin with river sands
Creation is destruction

—Rocky Sands, Jr.

Victoria Todd (2004) concisely describes the tectonic setting in the pamphlet accompanying her excellent compilation *Preliminary Geology Map of the El Cajon 30' x 60' Quadrangle, Southern California.* Extracting liberally (pp. 4-5):

TECTONIC HISTORY OF
WESTERN SALTON TROUGH

In early to middle Miocene, continental rifting and extensional faulting produced half-graben basins and basement horsts in the western Salton Trough (Winker, 1987; Kerr and Kidwell, 1991). The onset of strike-slip tectonism, which produced major northwest-southeast dextral strike-slip faults and antithetic northeast-southwest sinistral strike-slip faults (Winker, 1987), has not been dated precisely, but is estimated to have begun between the late Miocene (about 10 Ma) and early Pliocene (about 5 Ma) (Kerr and Kidwell, 1991). Late Pliocene or Pleistocene uplift, tilting, and about 35° clockwise rotation of the basin fill in the western Salton Trough (Johnson and others, 1983) was accompanied by folding on east-west axes and listric-normal and detachment faulting (Winker, 1987).

Evidence that the Fish Creek-Vallecito Basin was initiated in early to middle Miocene includes sedimentary and volcanic rocks of this age that were deposited upon, and subsequently deformed along the flanks of basement horsts such as the Vallecito and Fish Creek Mountains. West of the Salton Trough, the Peninsular Ranges block apparently began to rise at about the same time, elevating Miocene volcanic rocks about 500 m above the desert floor in the Jacumba area. Smaller half-graben basins in the northwestern part of the Elsinore Fault Zone (Mason and Vallecito Valleys and the northern part of Carrizo Valley) are probably younger than the Fish Creek Basin because the oldest strata in these basins, assigned to the Pliocene and Pleistocene Palm Spring Formation, are confined to basin floors and faulted along their margins.

...[L]ate Pliocene and Pleistocene uplift of the Fish Creek-Vallecito Basin is recorded by large folds on east-west axes in basin fill, and by basement-cored folds on northwest axes in the western Coyote Mountains. Both sets of folds involve rocks as young as the Palm Spring Formation. In the western Coyote Mountains, the prominent fault that bounds the southwest flank of the range is a reverse fault that dips from 47° to 55° NE and carried crystalline rocks southwestward over a late Pleistocene to Holocene alluvial fan (Hoggatt, 1979).

Look for this structural relationship at Stop 2.

Kerr and Kidwell (1991) cited the persistence of coarse basin-marginal deposits throughout the Neogene history of the region as suggesting continuous rejuvenation of local source areas. This is indicated in the quadrangle by the occurrence of Miocene to early Pliocene megabreccia, or landslide deposits in the Vallecito Mountains and by similar deposits that involve the Palm Spring Formation in the northeast wall of Mason Valley and in Sandstone Canyon. Quaternary landslides are mapped along the northeastern margin of Mason Valley and the northern margin of Vallecito Valley (Hart, 1991).

Extensional faulting continued into Holocene time in the southern part of the Elsinore Fault Zone. Predominantly high-angle normal faults cut late Pleistocene to Holocene alluvial fans and terraces and locally cut older and younger alluvium (e.g., Mason and Vallecito Valleys; frontal fault zone of the Tierra Blanca Mountains; northern Jacumba Mountains). Along the Laguna Salada Fault, which is the extension of the Elsinore Fault Zone into Baja California, Mexico, Isaac and others (1986) described Plio-Pleistocene detachment faulting that is associated with low- and high-angle extensional faults. The authors proposed that these extensional structures were exposed by left-stepping of recent dextral faults.

The Carrizo Badlands from about where we stand northeast to the Vallecito Mountains and Fish Creek Mountains mark the Plio-Pleistocene Carrizo Basin, also known by other variants such as the Fish Creek-Vallecito Basin and Vallecito-Fish Creek Basin.

We will explore this depositional center at length at Stop 3.

Continuing to quote from Todd's excellent and useful work:

FISH CREEK-VALLECITO BASIN

The Neogene sedimentary and volcanic rocks that are exposed along the western margin of the Salton Trough comprise the most complete stratigraphic record of the northern Gulf of California in southern California (Winker, 1987; Kerr and Kidwell, 1991). As a result of the Neogene transition from a subduction plate boundary to the Gulf of California-San Andreas rift-transform system (Kerr and Kidwell, 1991; Atwater, 1992), the Salton Trough had become the site of rifting, extensional faulting, basaltic volcanism, and marine transgression by Miocene time.

During Miocene through Pleistocene time, a thick (5-6 km) section of continental and marine clastic rocks characterized by complex facies changes and local unconformities accumulated in subsiding basins such as the Fish Creek-Vallecito Basin.... The oldest strata in the Fish Creek-Vallecito Basin are early and middle Miocene fluvial, alluvial-fan, and rock-avalanche deposits mainly of local derivation that were deposited nonconformably upon crystalline basement of the Peninsular Ranges Batholith (Kerr, 1984). Volcanic rocks that range in age from 22 to 14 Ma (Hawkins, 1970; Minch and Abbott, 1973; Hoggatt, 1979; Ruisaard, 1979) are interstratified with these continental strata at several horizons and also occur as clasts within them.

In late Miocene and early Pliocene, the Fish Creek-Vallecito Basin underwent rapid subsidence and was invaded by a northward transgression of the Gulf of California (Johnson and others, 1983). Basin-marginal alluvial fan deposits grade laterally to shallow-marine evaporites (Dean, 1988), sandstone, and conglomerate, and, as subsidence continued, to deeper-marine turbidites (Winker, 1987; Kerr and Kidwell, 1991). Early Pliocene marine strata record gradual shoaling as the basin was rapidly filled by the influx of large volumes of fine-grained sand and silt supplied by the Colorado River. The initiation of the Colorado River as a major sediment source for the Salton Trough was soon followed by widespread deposition of thick (about 3.5 km) Pliocene progradational deltaic deposits (Kerr and Kidwell, 1991).

Pliocene through Quaternary strata record a transition from deltaic deposition to largely nonmarine fluvial and basin-marginal alluvial fan deposits whose clasts were derived from local basement uplifts. The Late Pleistocene and Holocene were marked by widespread erosion of folded Pliocene to middle Pleistocene strata and the formation of extensive alluvial fans and terraces (Woodard, 1963).

Odometer (Trip) Mileage

0.0 Head north on Highway S2.

0.3 Turn right at the sign for "Canyon Sin Nombre."

1.2 Stop 2. Pull off to the shoulder, either the right or left, so that traffic can get by. This is State parkland so please do not drive off the road.

Stop 2 Speaker:
George Morgan

The Elsinore Fault is an active strike-slip fault. There is a significant component of dip-slip as seen by crystalline basement thrust over the Pliocene to Recent sedimentary units. The crystalline rocks are highly deformed and recrystallized gneisses, with cross-cutting pegmatite dikes. The protoliths of the metamorphic rocks are probably Paleozoic sediments.

1.5 Prominent folds in the Olla Formation. The Olla is mostly fluvial deposits from local streams and the Colorado River. The folding probably occurred before the Olla sediments lithified. These beds slumped and were squeezed by the force of gravity.

3.0 Stop 3. Turn sharply to the right and pull forward to the "Vehicles Prohibited" sign. We will hike up this side canyon of Canyon Sin Nombre for a total of 3 miles (1.5 miles each way) with an elevation gain of 530 feet, from 730 to 1,260 feet (mean sea level). We will be away from the cars for up to 4 hours. Carry water and (optional) your

Basement complex thrust over Tertiary sedimentary units at the head of Canyon Sin Nombre. White band of hydrothermally altered rock dominates the fault zone. Photo: Phil Farquharson.

lunch. Use sunscreen and wear a hat. Shade is limited.

LESSON

This is the desert
Hot dry, wear hat, drink water
Live another day

—Rocky Sands, Jr.

Folded Olla Formation. Photo: Phil Farquharson.

Stop 3 Speaker:
Charlie Winker

At Stop 3, we are surrounded by nonmarine clastic sedimentary strata of the Plio-Pleistocene Palm Spring Group, dipping about 30 degrees north into the badlands. Capping these outcrops is an extensive system of terrace deposits (Qt). We are parked on Quaternary alluvium (Qal). The canyon sides are Hueso Formation (PQh), a locally derived suite of nonmarine sandstone with conglomerate and siltstone. Charlie Winker mapped the geology in this area. See a redraft of his geologic mapping on the inside back cover.

0.0 mile. Do *not* reset odometers. This is hiking mileage! As we begin hiking up the stream channel, the walls of the canyon narrow. Where the walls become conveniently close to our path about 500 feet from the cars, Charlie Winker's map shows that we have crossed a contact into the Olla Formation (Po), a fluvial suite of interbedded local and Colorado River derived sandstone, siltstone, claystone, and conglomerate. The Z-jagged contact lines represent gradational contacts and are diagrammatic. Look for signs of local sources such as abundant biotite, and coarser clasts of granodiorite. Thin muddy beds without biotite indicate a Colorado River source.

0.1 mile. Turn left to stay in the main channel.

We will walk one-half mile through the Olla before encountering the contact with the next lower unit. In the meantime, observe the rock types of the cobbles and boulders in the bottom of the channel. These rocks are apparently derived from the basement of the Coyote Mountains we saw at the previous stop. Can you find granites? pegmatite? gneiss? amphibolite? Look for basalt, marble, and pink conglomerate. We may find the sources of these rocks toward the end of the hike.

Look at the sedimentary structures in the channel. Notice a series of alluvial terraces that have been deposited, in some places just above the canyon bottom, in other places halfway up the steep eroded side, or at the top. These lower terraces appear to have been recently eroded by the present channel. Find paleochannels and other sedimentary structures such as clast imbrication, scour marks, graded beds, and cross bedding.

0.4 mile. To the left is a good view of an angular unconformity between the alluvial terrace deposit and the dipping strata of the Palm Spring Group.

After having fun identifying rocks, notice that the dip has steepened to about 45 degrees as we encounter the Arroyo Diablo Formation (Pd), a nonmarine suite of sandstone, siltstone, and claystone. The Arroyo Diablo is the upper part of the Colorado River deltaic succession. Notice the subtle changes such as the thin cross-bedded paleochannels and concretionary sandstone beds.

STRATA 2

Hueso tops Palm Spring
Then Olla and Diablo
Imperial below

—Rocky Sands, Jr.

0.75 mile. Keep to the right fork, the slightly larger channel. (Note: We will return by way of the left fork.)

About 200 feet from the fork, there is a good view of the basement rock. The white patches are the likely source of the marble boulders that are becom-

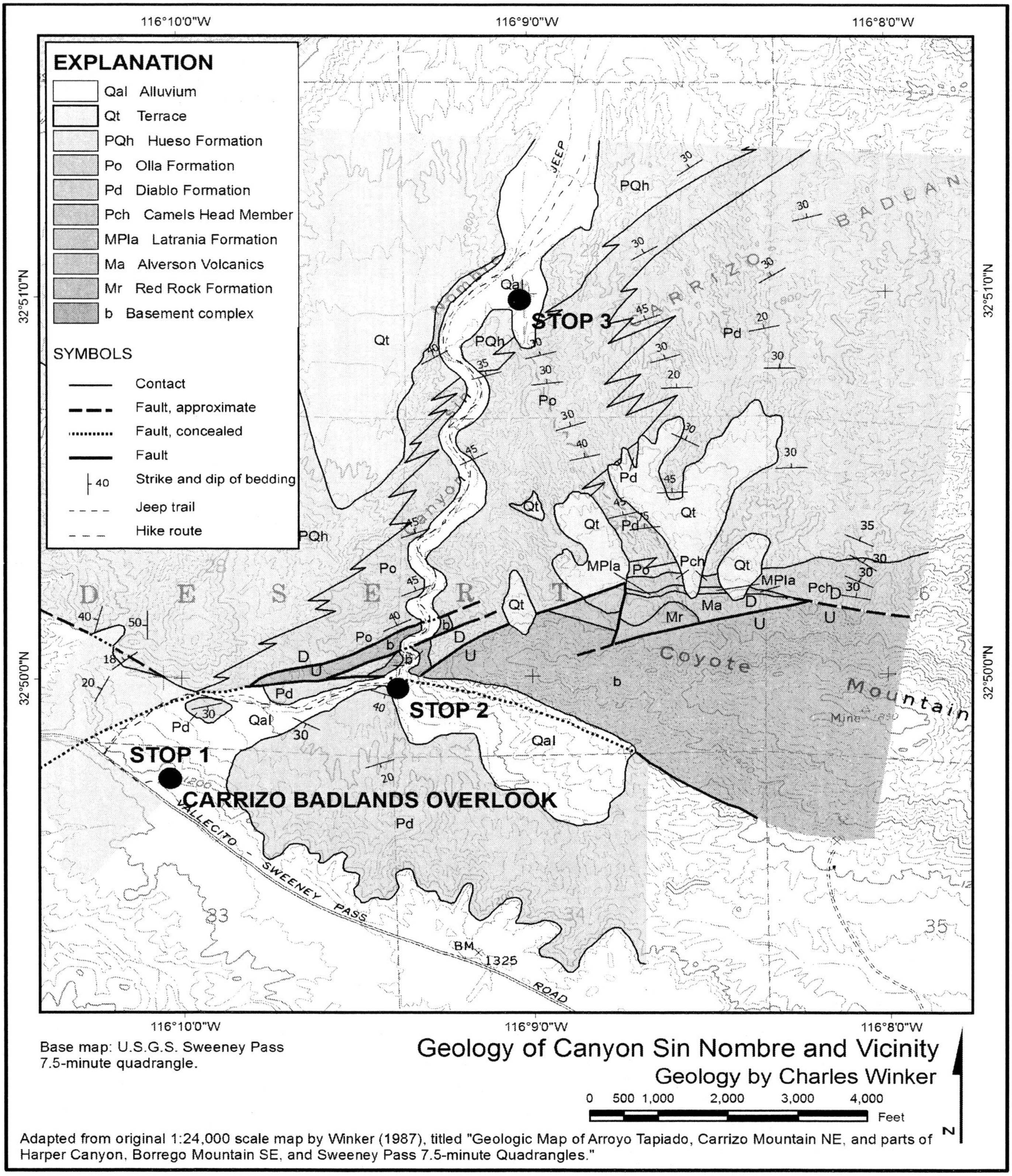

U.S.G.S. Topographic Map of Canyon Sin Nombre with Geologic Overlay.

1.1 miles. We are back into the Olla for about 200 feet.

1.2 miles. As we climb up the canyon, we drop downsection from nonmarine Palm Spring Group to the marine Imperial Group of fossiliferous clastics.

Terrace gravels in angular unconformity over Palm Spring Group. Photo: Phil Farquharson.

Don't miss the contacts. The units are thin at this point. Can you find the nonmarine-marine contact? The first Imperial unit we encounter is the Deguynos Formation, a suite of marine clastics of the Colorado River deltaic succession. Specifically, we are at the Camels Head Member (Pch), marginal marine sandstone, siltstone, and claystone interpreted to be tidal flats.

A hundred feet or so farther up we descend stratigraphically into the Latrania Formation (MPla), mainly composed of locally derived marine clastics. This unit is less than 100 feet thick whose contact should be obvious with underlying Alverson Canyon Formation (Ma), one of two formations of the Split Mountain Group we see in the canyon.

The Alverson Volcanics are mainly alkaline to tholeiitic basalt with volcaniclastics. Once called the Alverson Andesite, later analysis found little andesite.

The canyon becomes narrow and steeper. Take care to keep your balance and move slowly over the boulders.

Interbedded with and below the volcanics, are red beds of the Red Rock Formation (Mr), composed of locally derived alluvial sandstone and conglomerate. Below the lowest volcanic layer, there are some volcanic clasts in the upper portion of the red beds. Biotite-rich granitic pebbles are abundant.

Continue another 0.1 mile to the fault contact between the Split Mountain Group and the crystalline basement (b). There is one 6-foot-high dry fall but a large inclined boulder of Red Rock Formation allows for a three-point scramble to the top. We are now out of the narrow gorge. The fault appears to be a steeply dipping reverse fault. It is not clear from this view what the structural relations are. Let's get a better view from on top of the terrace to the east.

Turn left, and hike up the gradual slopewash hill. The alluvium is loose so check your footing. At the top of the terrace continue east to the next drainage. Observe the conglomerate of the Red Rock with metamorphic basement in fault contact to the right. The black and purple basalt flows and volcaniclastic beds of the Alverson lie above (see photo, back cover).

About 3.5 million years ago, this area was a river delta (Jefferson and Lindsay, this volume).

1.5 miles. This is the turnaround point. Drop into this canyon and hike down the canyon. We will see examples of the same formations as we walk back. We join our ascending route in about 0.25 mile, well within the Arroyo Diablo Formation.

Return to the cars. Continue the drive down canyon.

DIRECTIONS

Drive downstream fork left
Fork left, turn hard left, then straight
Vallecitos too far

—Rocky Sands, Jr.

4.1 Fork left.

4.4 Fork left.

4.7 Hard left. (Straight and slightly left is Vallecito Creek.)

5.0 Continue straight. This is Bow Willow Creek.

5.8 There is a shady tamarisk tree on the right. Continue straight. We are on the Great Southern Overland Stage Route of 1849.

9.1 Turn right onto the paved road. We are back on Highway S2, heading northwest.

9.4 Mile Post 46.

9.7 Well of Eight Echoes is on the right. Now a tourist attraction, these wells were wildcat oil wells drilled in 1914.

10.4 Mile Post 45. There is a good view straight ahead of Whale Peak (elevation 5,349 feet). It is from this peak that we get the name of the Eocene Ballena River. *Ballena* means "whale" in Spanish. Partway up the side of the mountain are Eocene deposits containing distinctive cobbles and boulders of rhyolite known as Poway-type clasts. The deposits located here are the farthest east known until their source area in the basement rocks of the weathered Jurassic foothills of Tajitos, Sonora, Mexico, over 250 miles southeast.

Whale Peak is one of the most popular peaks in Anza-Borrego Desert State Park for weekend hikers.

11.4 Canebrake Canyon to the left. Mile Post 44. To the right are the badlands of the Vallecitos-Fish Creek Basin. From this view we can see mostly Plio-Pleistocene Palm Spring Group non-marine sedimentary units capped by terrace deposits, which Hoggatt mapped as Mesa Conglomerate.

12.3 Canebrake. The homestead settlement of Canebrake is to our left.

12.4 The sign pointing right for "Palm Spring 1.6 mi" means the "real" Palm Spring, not the populated Palm Springs resort in Riverside County.

13.1 To the left is an unobtrusive power line. Currently this San Diego Gas & Electric Company easement consists of a single line of wooden poles with two nearly invisible wires. By 2010, these 40-foot sticks may grow to over 150 feet high. The Sunrise Powerlink Project, first proposed in 2005, would carry up to 1,000 megawatts on steel towers. Compare to the pylons of the 1980s Southwest Powerlink located between our last stop tomorrow and the town of Ocotillo.

16.4 Agua Caliente Airport. This airstrip serves park visitors and the residents of Canebrake.

16.8 Left turn toward the Agua Caliente Regional Park campground.

17.2 Agua Caliente Springs Store. Food, beverages, and curios may be purchased here. The owner of the Desert View Tower now operates the store.

17.6 Agua Caliente Regional Park campground entrance. Continue straight to the back of the campground and follow the road as it circles to the left.

The mountains rise steeply from the alluviated desert floor. These are the Tierra Blanca Mountains, the northern extent of the La Posta Pluton. Many of the outcrops of gra-

An eroding series of older alluvial terraces at the base of the Coyote Mountains, where the trace of the Elsinore Fault can be seen. Photo: David Bloom.

nodiorite are moderately to highly hydrothermally altered. The mafic minerals and plagioclase have been altered to clays.

Reeds, palms, and healthy green vegetation indicate the moist and wet areas of the springs. Restrooms and a large enclosed hot spring pool are on the left. The Moonlight Canyon Trail starts at Campsite 140, on your right. Hike along this trail to see a good exposure of a detachment fault dipping 30 degrees east with granodiorite on both the upper and lower plates with a well developed chlorite zone in the breccia of the upper plate.

17.9 Turn right into the "Caravan Area" group campsite.

18.0 Stop—Overnight Camp.

Enjoy the campsite located on a broad flat sandy wash with the altered granodiorite rising to the south. We have a full view to the north of Whale Peak with a broad alluvial fan at its base.

The Elsinore Fault strikes northwesterly through Agua Caliente Regional Park. The hot springs here are one of several hot springs locations found along this fault, notably Warner Hot Springs in northern San Diego County and Murrieta Hot Springs in Riverside County.

Evening Speaker:
Carole Ziegler — *"Sixty million years in sixty minutes"*

DAY NO. 2

Reset odometer before leaving the campsite. Circle back to the entrance and exit the campground. Turn right at the store, then right on Highway S2, heading southeast for 16.9 miles, or 0.5 mile beyond Mile Post 55.

Whale Peak is behind us and to the left. The Tierra Blancas are to our right. In the distance is Carrizo Peak (elevation 2,408 feet), the highest point in the Coyote Mountains.

1.4 Mile Post 40. Altered granodiorite of the La Posta Pluton in the Tierra Blanca Mountains to our right. On our left are gently dipping hogbacks of Palm Spring Group strata capped by Mesa Conglomerate.

3.2 Sweeney Pass is straight ahead, the low point on the horizon between the Coyote Mountains on the left and the Tierra Blanca and Jacumba Mountains on the right.

4.2 Canebrake homestead site turnoff is to the right.

4.4 "Palm Spring 1.6 mi" sign is to the left.

5.4 "Canebrake Canyon" to the right.

5.7 Carrizo Valley is verdant with vegetation to our left at 11 o'clock. The Carrizo Creek surfaces year-round at this location. In aerial images, where the rest of the area is light brown, this ribbon is consistently dark green.

7.7 The Great Southern Overland Stage Route of 1849 is to our left. Straight ahead is the mouth of Carrizo Gorge. A large unrecognized fault may exist in the gorge (See Walawender, this volume). The rocks of Carrizo Gorge consist of a pendent of older metamorphic rocks within the La Posta Pluton. The In-Ko-Pah Mountains are to the west and the Jacumba Mountains to the east.

8.6 Mountain Palm Springs is on our right.

9.3 Egg Mountain is the low hill on our right.

9.4 Mile Post 48.

9.8 Bow Willow campground is to our right.

11.4 Sweeney Pass. The road cuts expose a thickness of over 400 feet of gravel and cobbles of the Canebrake Conglomerate, part of the Palm Spring Group. The local mountains were the source for this unit, whose beds were deposited contemporaneously with the finer sediments of Hueso, Olla, and Arroyo Diablo formations. Sweeney Pass Road was the name of this road before it was paved. The first road through this pass was graded in 1933. It wasn't paved until the early 1960s when it was re-graded. The old road is still visible with even steeper curves than today's highway. Look across the ravine just as we begin our ascent. The route of the 1933 road is the steep path on the other side of the drainage. Before 1933, the main highway headed east along Carrizo Creek into Imperial County.

12.8 Canyon Sin Nombre is on the left.

12.9 Carrizo Badlands Overlook is on the left.

14.5 Decent from the terrace. The Volcanic Hills are straight ahead at the base of the Jacumba Mountains.

14.9 Jojoba Wash. We will return to the Jojoba Wash jeep trail for Stops 6 and 7 later today.

15.4 Mile Post 54.

15.6 Basalt flow of the Alverson Canyon Formation (or

EXPLANATION

Qs	Modern Alluvium, stream channel
Qual	Modern Alluvium, outwash fans
Qal	Older Alluvium
Qt	Terrace Deposits
Tcb	Canebrake Conglomerate
Tps	Palm Spring Formation
	Alverson Canyon Formation
Taf	Undifferentiated Flows
Tap	Plugs, dikes, and cinder deposits
Tafu	Upper Flows
Tal	Volcanic Mudflows
Tas	Interflow Sediments, undifferentiated
Tas	mudstone
Tas	sandstone
Tas	conglomerate
Tafl	Lower Flows
Tsm	Split Mountain Formation
Kgd	Basement complex

SYMBOLS

- Bedding
- Contact, approximate
- Contact
- Fault, U upthrown, D downdropped
- Fault, approximate
- Fault, concealed
- Field trip route
- Jeep trail
- Hiking route

VOLCANIC HILLS ROCK UNITS

Modern Alluvium (Qs, Qual), stream channel and outwash fans composed of local basement and volcanic clasts

Older Alluvium (Qal), poorly sorted gravels and sands composed of local basement and volcanic clasts, being downcut by modern streams, 0 to 30 feet

Terrace Deposit (Qt), light red conglomerates composed of poorly sorted coarse sands and rounded cobbles of local basement and volcanic rock, 0 to 30 feet

Canebrake Conglomerate (Tcb), poorly sorted coarse sands and rounded cobbles composed of local basement and volcanic rock, 0 to 400 feet

Palm Spring Formation (Tps), interbedded buff poorly-sorted fine sands and silts with dark red well-sorted mudstones with fossil wood and reworked microfossils, 20 to 200 feet

Alverson Canyon Formation (Taf)

Plugs, Dikes, and Cinder Deposits (Tap), basaltic and hypersthene andesite intrusions, flows, and related cinder deposits

Upper Flows (Tafu), dark gray platey basalt flows, 30 to 120 feet

Volcanic Mudflows (Tal), massive chaotic lahars with volcanic clasts up to 15 feet diameter in a muddy baked matrix, 0 to 80 feet

Interflow sediments (Tas), poorly sorted volcanic conglomerates, fine sandstones, and well sorted volcanic mudstones with secondary evaporate minerals, 10 to 40 feet

Lower Flows (Tafl), light gray platey olivine basalt flows, 40 to 110 feet

Split Mountain Formation (Tsm), coarse sand and claystone ripup-clast-bearing cobble conglomerate composed of local basement and occasional volcanic clasts, 0 to 260 feet

Basement complex (Kgd), recrystallized limestone, mica schists, quartzites, fractured granodiorite, pegmatite dikes, intruded by massive quartz diorite

Rock Units from Fourt (1979), "Post-batholithic Geology of the Volcanic Hills and Vicinity, San Diego County, California," SDSU M.S. thesis.

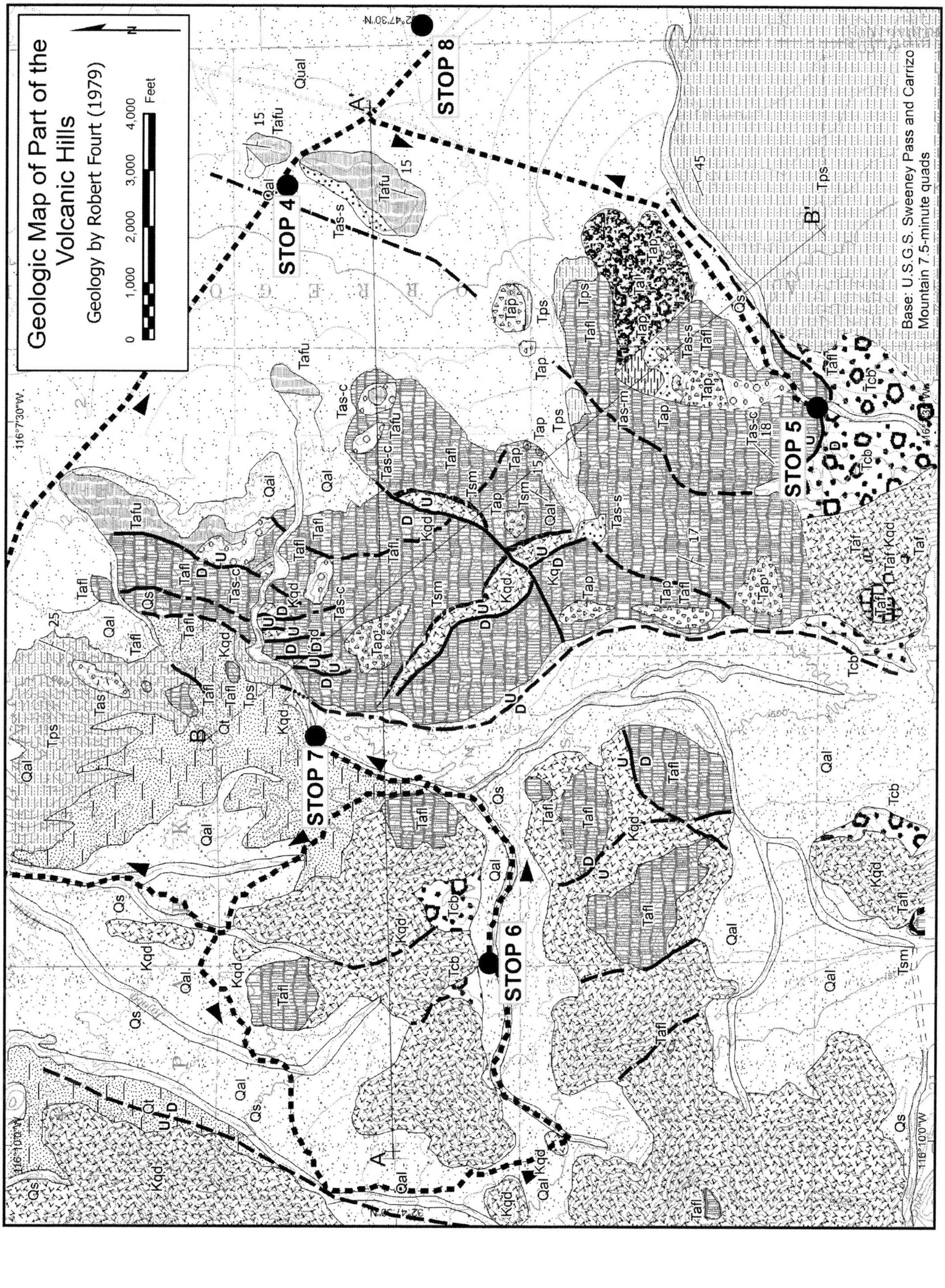
Geologic Map of Part of the Volcanic Hills
Geology by Robert Fourt (1979)
0 1,000 2,000 3,000 4,000 Feet
N
STOP 4
STOP 5
STOP 6
STOP 7
STOP 8
A
A'
B
B'
116°7'30"W
116°10'0"W
32°47'30"N
Base: U.S.G.S. Sweeney Pass and Carrizo Mountain 7.5-minute quads

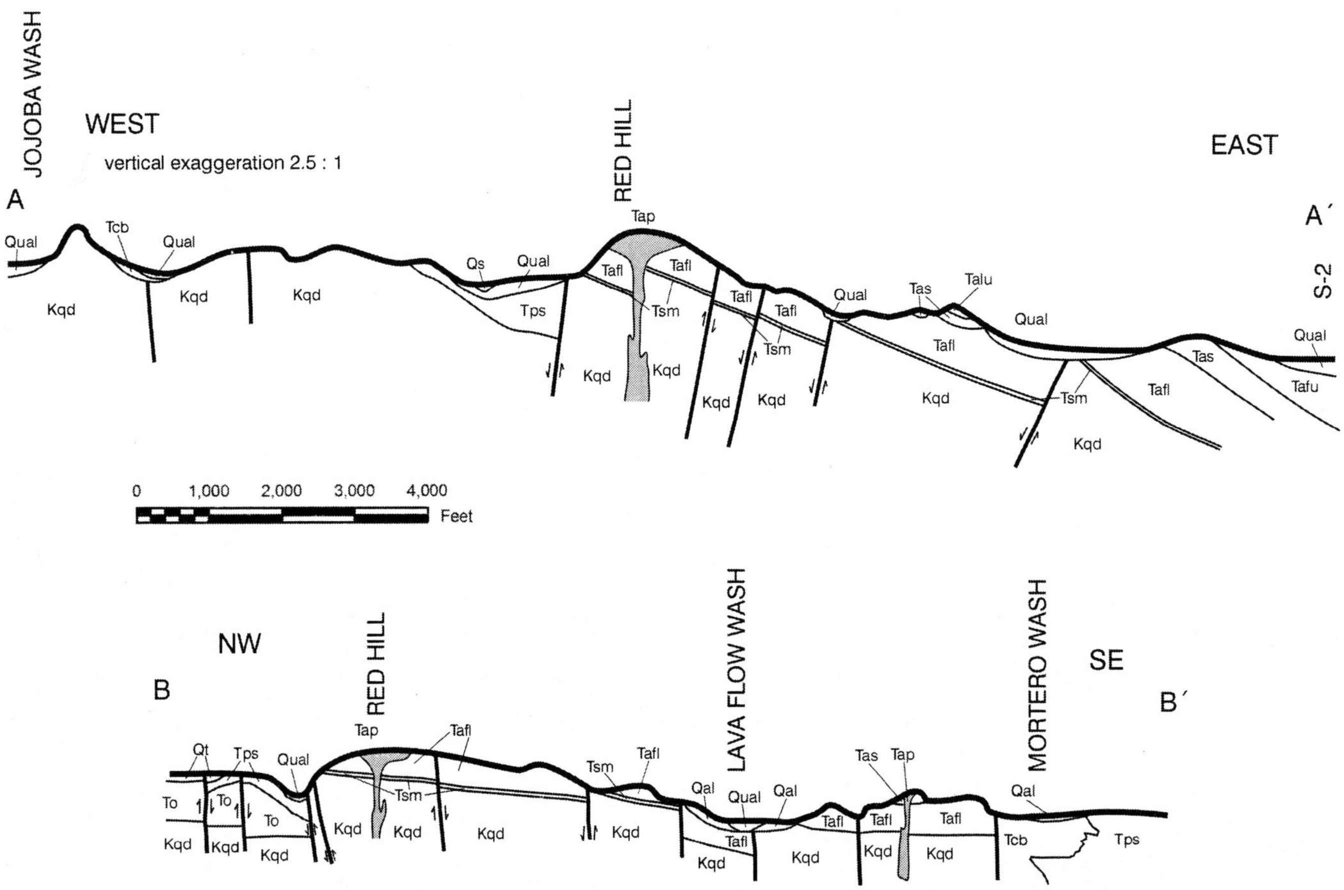

Cross Sections to accompany Fourt (1979) Volcanic Hills Map.

the base. Robert Fourt mapped this flow as a "lower flow" in his SDSU M.S. thesis of 1979.

16.9 Stop 4. Volcanic Hills. Park in the wide sandy shoulder before the low hill of basalt. This outlying hill of Alverson Basalt, upper flow (Tafu), is separated from the rest of the Volcanic Hills. Fourt (1979) distinguished upper and lower flows by an interflow sedimentary sequence. These deposits are mudstone, sandstone, and conglomerate. We will walk about 400 feet south along the base of the basalt to observe the contact of the basalt with the underlying sandstone facies of the interflow deposit. Notice the similarities and differences between similar contacts we saw yesterday. There is a prominent baked zone of volcanic breccia with underlying red coarse volcaniclastic sand. Ripups, angular clasts, and matrix supported gravel are among the evidence for a local high-energy environment of deposition.

Return to vehicles, and continue along Highway S2.

17.3 Mortero Canyon. Turn right. The large flat area is a good place to carpool. Soft sand and a severely rutted slope at Stops 6 and 7 will be difficult even for 4-wheel drive. We will return this way after Stop 7.

We are heading toward the eastern Volcanic Hills. Basalt flows and volcaniclastic deposits of the Alverson Canyon Formation (or Alverson Volcanics) are well represented here.

18.3 The rock on the right is a mudflow deposit. Large, up to 15-foot-diameter clasts of volcanic rock, make up this massive chaotic lahar, indicated as Tal on the accompanying map. The matrix is muddy and baked. On our left are fine sandstones and siltstones of the Palm Spring Group.

18.7 On the right is the contact between the lahar and lower basalt flow (Tafl), then a contact with the basalt and interflow sediments (Tas). The interflow deposits here consist of mudstone over sandstone over conglomerate. Fourt (1979) suggested the adjacent plug (Tap) was a possible source of the interflow deposits here. Fourt reported fossils in the interflow sediments, including fossil wood in the sandstone; fossil oogonia (female reproductive organs) of the green algae charaphyta, tentatively *Chara alversonus*; unidentified calcified shell fragments; and a small bone.

ANCIENT LOVE

Oogonia
Once loved in the Miocene
Longing for you now

—Rocky Sands, Jr.

19.0 We enter the canyon narrows. Basalt flows (lower) are present on both sides. On the left side you see drill holes near base level. There are at least three locations where rock samples were drilled to collect paleomagnetic data. Usually there are six to eight rock cores per paleomag sample point.

19.3 Stop 5. Park where the canyon widens. We turn around back to Highway S2 after this stop.

On the left is a clear contact between the basalt and Canebrake Conglomerate. Canebrake is the proximal facies of locally derived nonmarine sediment in the Palm Spring Group. There has been some attempts to correlate the Canebrake with the Hueso, Olla, and Arroyo Diablo formations of the Palm Spring Groups. However, because individual beds cannot be correlated across the basin, these efforts have been unsuccessful.

Stop 5 Speaker: Monte Marshall

Monte Marshall and his students collected paleomagnetic data which are compiled and synthesized for the first time in this volume. By incorporating data from here, the Coyote Mountains, the La Posta Pluton, and the west-

Angular basalt cobbles in fine sandy matrix below upper basalt flows.

ern Peninsular Ranges Batholith, Monte will speak to the implications for tectonic history far beyond these small hills. We will look at the problems associated with collected paleomag data from these locations. How do we measure original horizontal bedding? What was the original tectonic dip? These flows are dipping now, but how much dip was present at the time they were deposited? If these flows were viscous they might have a significant original dip. Paleomagnetic directions can be altered by hydrothermal processes. Some of these flows may not be fresh enough to obtain good data.

Return to the vehicles and turn around. Return to the highway.

21.3 Turn left at Highway S2, back to Jojoba Wash. Keep your light duty vehicles here and carpool. We will return after Stop 7.

22.1 Mile Post 55.

23.0 Road bends slightly to the left. The Coyote Mountains are on the right. The ruins of Dolomite Mine are visible at about 2 o'clock and one-fourth of the way up the side of the ridge.

23.7 Turn left to Jojoba Wash. The sign is small and easy to miss. (To the right is the road to West Dolomite Mine Trail.)

Soft sand and two rough sections make this route a hazard to passenger cars. Carpool with a 4WD or truck before proceeding.

Reset odometer to 0.0.

0.0 Jojoba Wash. We are heading into the western Volcanic Hills. Basalt flows and volcaniclastic deposits of the Alverson Canyon Formation are less prominent. We'll see more of the basement complex and alluvial deposits on this loop. Look for the pegmatite, schist, and gneiss in basement rocks. We will take special note of the terrace gravel deposits at Stop 6.

0.7 Cliff forming terrace deposit on the left. The road, which had been parallel to Highway S2, now turns south.

1.2 The low cliff-forming terrace ends.

1.5 Turn right at the fork.

1.9 On the left in the bedrock outcrop is one of many pegmatites with its distinctive line rock.

2.7 We have been getting closer to the steep hill composed of granitic intrusion into metamorphic rock, which contains mica schist, amphibolite, banded gneiss, and marble, higher on the hill. The granitic rock ("quartz diorite") is associated with the La Posta Pluton.

2.9 Hill of volcanic rock, on the left. The light patch on the hill in the distance across the alluvial fan is a marble deposit.

3.4 Pass through hills of basement complex (Kqd).

3.6 Turn sharply left (northeast). The older terrace deposits have an orange tint.

3.9 Terrace deposits on the left grade up into slope wash. These terrace deposits are quite extensive. They may represent a period of tectonic stability accompanied by higher rainfall than present conditions.

4.3 Keep left of the ravine cut into the terrace.

4.4 Stop 6. Park just off the road or stay on the road. Avoid driving on the fragile desert pavement. Avoid the dropoff into the caliche grotto at the head of a shallow ravine cutting into the older alluvial terrace.

Stalagmites of sand in small caves of eroded terrace alluvium.

Look over the edge at a curious phenomenon. The sand eroded out into small caves a few inches to a few feet across. Within these caves there appear to be stalagmites and stalactites of sand that has dripped and then been cemented. The source of the calcium may be the calcium-rich feldspars in the granitic rock of the La Posta Pluton. The basalts may also contribute significant calcium.

About 500 feet to the north up the small drainage is Canebrake Conglomerate (Tcb). In the bottom of this drainage is the nonconformity between the basement complex and Canebrake. A 10- to 15-foot-high older alluvial terrace deposit (Qal) overlies the Canebrake in an angular unconformity. The current channel is eroding the terrace, Canebrake, and basement.

In the basement observe marbles, pegmatites, and gneiss. This is the only stop where we see marbles in place.

Stop 6 Speaker:
Charlie Lough

These terrace gravels, sometimes called the Mesa Conglomerate, are extensive. They form a surface that laps partially up the lower portion of the slopes. This surface may represent a period of tectonic stability accompanied by higher rainfall.

4.5 Drop off into the wash below. This is the penultimate difficult part of our drive, impassible by low-clearance vehicles, and a serious impediment to return travel.

4.9 Canebrake Conglomerate on the left.

5.0 Turn left at the volcanic outcrop. Stay straight ahead. Hard right dead ends up the wash. In about 100 feet, a road to the right up the hill travels to Dos Cabezas. The road to the left up the side is our way back after Stop 7.

5.1 Hill of basalt on our left. These lower basalt flows of the Alverson rest on metamorphic basement. The rock is mostly gneiss cut by pegmatite dikes. There are a few small faults.

The terrace deposit (Qt) is on our left and older alluvium (Qal) is on our right.

5.4 The hill on the horizon to the right is Red Hill (elevation ~1,720 feet), a volcanic plug (Tap). Between the wash and Red Hill, Fourt mapped a thin unit of Split Mountain Formation (Tsm) over a sliver of basement complex.

5.7 End of the road. The wash narrows into a boulder pile, where you can go no farther by car. Park in the loop on the left. Deep loose sand is present on the right.

Stop 7. The wall of this narrow ravine reveals crystalline basement mostly made up of strongly folded gneiss with migmatite and pegmatite veins. The intrusions are mostly granitic in composition. A shear zone is present. In some locations just the mafic minerals and xenoliths appear to be stretched. In others, the whole rock appears strongly deformed. The irregular nonconformity between the basement and basalt indicates Miocene paleogeography.

Thick lower basalt flow on a 10-foot-thick base of volcanic breccia. Photo: Phil Farquharson.

We will hike about one-fourth to one-third mile into the gorge, then return to the vehicles.

The gorge provides good exposures of lower basalt flows, tuffs, volcanic breccia, baked zones on underlying volcaniclastic and sedimentary units, and additional contacts between the basement complex and volcanic rocks.

Some of the volcaniclastic units have granitic clasts that were either caught up in the melt before reaching the surface, or were entrained in the lahars after extrusion. Some of the sand-supported conglomerates appear to be deposits of debris flows.

Some of the basement includes cross-cutting pegmatite dikes with large feldspar, muscovite, and schorl. Some dikes express a distinct chilled margin. There is a foliated granitoid with red banded weathering apparently from the presence of hematite.

After the gorge widens, it narrows again as it bends to the right and the sandy bottom becomes bouldery. At this point return to the vehicles by backtracking up the wash.

6.3 Turn right (northwest) out of the drainage. This is the most difficult part of the drive, requiring 4WD and skill to negotiate the rutted slope without vehicle damage.

7.3 Turn right to rejoin the jeep trail we came in on.

8.8 Turn right onto Highway S2. The transpressive right-lateral Elsinore Fault cuts along the foot of the Coyote Mountains on our left. The recent streams have incised older alluvial terraces.

11.6 Stop 8. Imperial Highway Monument. In recent years the U.S. Border Patrol has used part of the pullout area as a checkpoint.

Aplitic dike in mica schist and strongly foliated gneiss. Photo: David Bloom.

We will contemplate the obelisk with the plaque:

"This is the desert.
There's nothing out here.
Nothing."
—N. Karavasiles

Bronze plaque on the Imperial Highway Monument, Anza-Borrego Desert State Park, at the San Diego-Imperial county line. Photo courtesy of California State Parks.

Name	Description	Position - WGS 84	Position - WGS 84	Altitude
		Lat/Lon hddd°mm.mmm'	UTM	
2006 Stop 1	Carrizo Badlands Overlook	N32 49.622 W116 09.866	11 S 578208 3632421	1249 ft
2006 Stop 2	Elsinore Fault in Canyon Sin Nombre	N32 49.946 W116 09.634	11 S 578565 3633023	970 ft
2006 Stop 3	Begin Hike to Alverson / Red Rock Contact	N32 51.055 W116 09.032	11 S 579488 3635079	708 ft
2006 Stop 3A	Alverson - Red Rock Contact	N32 50.186 W116 08.597	11 S 580179 3633479	1230 ft
2006 Stop 4	Outlier of Volcanics at Highway S-2	N32 47.799 W116 06.614	11 S 583310 3629094	948 ft
2006 Stop 5	Paleomag Data Talk - Monte Marshall	N32 46.322 W116 07.352	11 S 582181 3626356	1243 ft
2006 Stop 6	Caliche Grotto / Canebreak Conglomerate	N32 47.217 W116 09.375	11 S 579010 3627984	1491 ft
2006 Stop 7	West Mortero Wash	N32 47.752 W116 08.613	11 S 580191 3628982	1357 ft
2006 Stop 8	Imperial Monument	N32 47.373 W116 06.210	11 S 583947 3628313	901 ft

Place Names Associated with the Southern Anza-Borrego Desert and Coyote Mountains

For the SDAG 2006 Annual Field Trip

Diana Lindsay, *Sunbelt Publications, 1256 Fayette Street, El Cajon, California 92020*
DLindsay@SunbeltPub.com

AGUA CALIENTE

Agua Caliente is Spanish for "hot water." The phrase, as a place name, refers to the hot springs both at Warner Springs and at Agua Caliente in Vallecito Valley. The hot springs found at the foot of the Tierra Blanca Mountains at the east end of Vallecito Valley are today part of the 910-acre Aqua Caliente Regional Park, operated by the San Diego County Department of Parks and Recreation.

Part of the Pinyon Mountains detachment fault runs through Agua Caliente, providing fractured rock that allows heated ground water to percolate to the surface and exit as hot springs. The Elsinore Fault crosses the valley a short distance to the north.

The springs were originally home to the Haawii ("water in rock"), a band of Kumeyaay. According to Tom Lucas, the last full-blooded Kwaaymii who knew the history of Agua Caliente, the Haawi were driven from their homes about 300 years ago by a militant group known as Haakwapin who were not Kumeyaay or Cocopah. Possibly, they came from the north or Arizona. The Haawii moved to the eastern edge of Vallecito Valley in the area known today as the Vallecito County Park. The Haakwapin staged regular raids on the Haawii village until they retreated and merged with the Amat Haapshuu to the west of them. The Haakwapin then began raids on villages further to the west and into the mountains. "Finally, the combined desert and mountain Kumeyaay leaders met and decided the Haakwapin must be eliminated, and with their combined strength, they did just that." Lucas does not indicate if the Haawii returned to their original village.

State park lands adjacent to the county park are included in the California wilderness system. The Agua Caliente State Wilderness area is one of 12 wilderness units within Anza-Borrego Desert State Park (ABDSP). A rare name for Agua Caliente is Chicken Springs, according to Christopher Wray, author of *Historic Places in Eastern San Diego County*. This name referenced a rock feature found in the park.

Ref.: Cline 1990, 18; Lindsay 2001, 30-31; Lough 2006; Lindsay & Lindsay 1998, 173-4; Stein 1975, 1-2; and Wray 1999, 24.

ARROYO SECO DEL DIABLO

This dry mud-walled wash is the longest wash in the Carrizo Badlands. Translated from the Spanish as "Devil's Dry Wash," the walls show a variegated stratification with folding controlled by several antithetic left-lateral faults dissecting the Vallecito Creek-Fish Creek Basin (VCFCB). It is the type locale of the reddish Arroyo Diablo Formation, a widespread sandstone and mudstone unit derived from the Colorado River watershed.

Ref.: Dorsey 2006, 100; and Lindsay and Lindsay 1998, 156-7.

ARROYO TAPIADO

This "mud-walled canyon" is a literal description of its Spanish name, referring to the towering, eroded mud walls that have developed numerous caves, through the action of running water. It was named by park staff in the mid-1950s. Located in the Carrizo Badlands, north of Vallecito Creek, the canyon has sparse vegetation and sandstone concretions in the upper reaches of the wash. Historian Phil Brigandi found a 1930s reference for this canyon as "Wash of the Caves." The canyon is particularly known, in recent years, for its pseudokarst topography that has created numerous subterranean mud caves in the lacustrine Tapiado Formation.

Although the park was aware of the existence of some mud caves in the area before the mid-1970s, they were not aware of the extent or significance of the caves until Dwight Carey conducted a systematic scientific investigation of the area in 1975-76 for his research thesis. Carey discovered, mapped, and named most of the known caves today. He discovered that they were among the longest and largest pseudokarst caves yet reported. The San Francisco Bay Chapter of the National Speleological Society was the first organized caving group to visit the area in December 1982. Members of the Southern California Grotto first visited the area in March 1983.

Since that time, society members have had several mapping trips into the area, discovering some new caves and adding to the collection of colorful and graphic names. The best known caves are Cave Canyon or Big (Giant)

Mud Cave, Plunge Pool Cave and Carey's Big Mud Cave. Names for other caves as they appear from south to north include: First Falls, False Start, Dead End, Squeeze Pit or Tight Squeeze, Start Canyon, Bedding Plane, Dip Slope, Stairstep, Little Mud, Slump, Skylight, Oxbow, Chasm, Drainpit, E-Ticket, Schad's, Hidden, Corkscrew, Footprint, Falls, Slot Canyon, Meander, and Hard Rock.

Ref.: Brigandi 1998, 7; Lindsay and Lindsay 1998, 200-201; Lough 206; and Richards 1989, 37-38, 46-50.

BADLANDS

A badland is an intricately sculptured landscape that is a product of flash flood erosion in weak rock. It is either unprotected or insufficiently protected by vegetation. Badlands are characterized by a labyrinth of gullies separated either by round-crested or sharp-crested ridges. Wide, boulder-strewn valleys separate all of these low wrinkled irregular hills. Major badland areas of the Anza-Borrego Desert include Borrego, Carrizo, Vallecito, Fish Creek, and Coyote Creek.

The name "badland" (*malpais* in Spanish) comes from the supposition that these denuded hills have no commercial value. Traces of iron minerals give the landscape a colorful display of pink, red, yellow and green hues. Oxidation of iron results in pink, red and yellow colors. Green hues are the result of reduction of iron and may have sulfur mixed into them. The resultant multi-hued landscape led to the earlier name for the Borrego Badlands—the Painted Desert. Almost all of the badlands areas have important Miocene (mostly marine of the Imperial Group), Pliocene, and Pleistocene fossils.

Ref.: Chase 1919, 193; and Henderson 1938, 32.

CANEBRAKE

Canebrake literally refers to "a place overgrown with cane" or "a dense thicket of cane." The canyon and wash and the community that have grown at the mouth of the canyon take their name from the thick growth of reeds found in the canyon. The cane referenced is most probably *Phragmites australis*, commonly referred to as "carrizo" or "common reed." Canebrake has been inhabited since the late prehistoric times (ca. 500 BCE) by Kumeyaay Indians. San Diego County Supervisor James E. Jasper met an Indian at Vallecito in 1891 who said he lived "all alone in a wickiup at the cane brakes a few miles away."

The USGS first used the term Canebrake in 1908. John S. Brown referred to this area as "The Canebrakes" in his 1918 road log published for the US Department of the Interior. He said that one would find "a stream of good water...in a mountain canyon amongst a tangle of cane."

The Canebrake Conglomerate is a course-grained, grayish, proximal formation of the Palm Spring Group that has shed from steep mountain slopes and been lithified. It dates from the Pliocene-Pleistocene epochs, 0.9–4.2 MYA. While the term has been applied extensively to basin-margin deposits around the Salton Trough, paleontologists prefer a restricted definition of this unit where it can be positively correlated and thus maintain more consistent age controls for like fossils.

Ref.: Beauchamp 1986, 69; Brown 1920, 40; Dorsey 2006, 100 and Table 4, 365; Jasper 1934, v. 2, 301; Jefferson and Lindsay 2006; Leetch and Leetch 1992; Lindsay and Lindsay 1998, 175; Lough 2006; and Scott 1940, 30.

CANYON SIN NOMBRE

District Superintendent William Kenyon had asked Park Supervisor Carl Whitefield (1953–1956) the name of the canyon just east of Sweeney Pass. He was told that it had no name. That, Hispanicized, became its name—"the canyon with no name." Canyon Sin Nombre was cut by eroding floods, which have left high sheer canyon walls that expose the action of faulting and folding which has occurred in the layered sandstone walls. Guidebook author and State Park Commissioner Doc Parker aptly called the canyon "a geological hodge-podge," according to Resource Ecologist Mark Jorgensen. The mouth of this canyon also contains a healthy example of a smoke tree forest.

Ref.: Fairchild n.d.; Jorgensen 2000; and Lindsay and Lindsay 1998, 180–182.

CARRIZO

In Spanish, *carrizo* refers to a reed grass. Botanist Maximino Martínez, in his catalog of common names for Mexican plants, lists seven entries for carrizo for seven different genera. The carrizo found in the Anza-Borrego area is *Phragmites australis*—a tall, bamboo like plant found growing along moist arroyos, in cienegas, and at springs and waterholes. It is also the "cane" of Canebrake (see CANEBRAKE). Father Pedro Font first mentioned it on the 1775 Anza Expedition to California. Carrizo grows in the manner of desert palms—"feet in the water and head in the sun."

Indians used the hollow stem of carrizo for arrow shafts, pipe stems, and roofing, while the leaves were woven into mats. A cluster of three dozen arrows made of carrizo and arrowweed were discovered by Paul Johnson and Mark Jorgensen in the early 1980s. Carrizo roots were eaten raw, roasted, or boiled. Leaves were also cooked and eaten. Carrizo has also been used as a windbreak.

The name carrizo lends its name to a creek, a gorge,

a canyon, a marsh or cienega, a wash, a stage station, a palm oasis, a land grant, a route of travel, an impact area, a badlands area, a badlands overlook, a wilderness area, and a mountain.

The Carrizo Corridor was the route followed through the Carrizo Badlands by emigrants and the Butterfield Overland Mail between the Yuha Desert and Vallecito (see CARRIZO CORRIDOR and BADLANDS).

The Carrizo Badlands and Carrizo Canyon were designated state wilderness areas in 1982. The Carrizo Badlands contain world-class Mio-Plio-Pleistocene fossils. Carrizo Mountain is the highest point in the Coyote Mountains. It overlooks the Carrizo Badlands and the Carrizo Impact Area (CIA)—a World War II practice bombing area—to the north. It is just outside of the Coyote Mountains Wilderness Area administered by the BLM. The peak is now privately owned and is gravely threatened by mining. According to Park Superintendent Mark Jorgensen, "it is a vital bighorn lambing area, but the land was deeded by the BLM to private interests."

Ref.: Beauchamp 1986, 69; Henderson 1947, 19–22; Jorgensen 2000; Leetch and Leetch 1992, 96; Lindsay and Lindsay 1991, 144; *San Diego Union* 1939, 6; and Scott 1939, 24.

CARRIZO CORRIDOR

The Carrizo Corridor commonly applies to a 15-mile section of the original Southern Emigrant Trail between Carrizo Gap along Carrizo Wash and Vallecito—later a station stop for the Butterfield Overland Stage. Colonel Philip St. George Cooke, who led the Mormon Battalion through the Carrizo Corridor, said, "Altogether, it is the worst fifteen miles of road since we left the Rio Grande." It was more difficult to cross than open desert. Park Superintendent Mark Jorgensen quipped that it is noticeably more difficult traveling northwest in the sandy washes than it is going "downstream," whether riding bikes, walking, in a wagon, or in a four-wheel drive vehicle!

William P. Blake, with the Pacific Railway Survey, gave a vivid description in 1853. He said,

> The mountains on each side are entirely bald and free from soil. They present a most rugged and forbidding aspect, and reflect back into the valley the scorching rays of the sun, and keep off all the cooling breezes. The climate of the valley is thus exceedingly hot. It is even more uncomfortable than the open Desert. The sun's rays seem to have peculiar power and effect; and more animals die in this valley, after reaching the water, than on the open desert. Near to Vallecito, the valley grows narrow, and the ascent more rapid.

Former ABDSP naturalist Paul Johnson named the entrance to Vallecito Creek, from post mile marker 43 on Highway S2 (across from the entrance to the community of Canebrake), "Carrizo Corridor Overlook" in his *Weekender's Guide*. From here, one can get a clear view down the Carrizo Corridor.

Ref.: Edwards 1961, 29–30; Johnson 1987, 29; Jorgensen 2000; Lindsay 2001, 108-109; and Lindsay and Lindsay 1998, 197.

COYOTE MOUNTAINS

The Coyote Mountains, located south of the CIA, takes its name from the Aztec or Nahuatl word *(coyotl)* for the desert dog. Like the name Palm Spring (see PALM SPRING), Coyote Mountains sometimes is misspelled, and when it is, there can be confusion with Coyote "Mountain" north of Borrego Valley. The Coyote Mountains are included in a BLM designated wilderness area of 17,000 acres. The wilderness designation protects endangered and rare plants and animals found among the mountain ridges. The designation also provides extra protection for the rich fossil shellfish fauna found in the marine Imperial Group of the area (see DOMELANDS). Further protection applies to BLM Areas of Critical Environmental Concern (ACEC). An ACEC within this area is the Coyote Mountains Fossil Site.

The Coyote Mountains are a complex suite of almost every major rock type and geologic structure found throughout the Anza-Borrego region. The historic outline is briefly as follows, with dating approximate:

(1) 450 MYA—marine sediments in a shallow sea and organic limestones later metamorphosed into the marble stripes seen today high on mountain cliffs with large fragments found in washes, which have been eroded from the various strata found throughout the mountains.
(2) 100 MYA—an intrusion of granitic molten magma, which uplifted the marine sediments, metamorphosed the overlying sediments into schist, gneiss, marble, and dolomite. These metasediments are seen today as remnant roof pendants.
(3) 15–20 MYA—basaltic lava flows, related to initial rifting of the Gulf of California, cover approximately 100 square miles (see VOLCANIC).
(4) 4–5 MYA—Miocene-Pliocene marine advance

and retreat from the area, leaving the near-shore wind caves, offshore yellow muds, and delta-fan oyster shell hash "reefs" of the marine Imperial Group draped over earlier bedrock, like frosting on a cake.

Erosion has cut away many of the soft clay layers and left hard sandstone or oyster reefs projecting as isolated hills in badlands areas. Many of the reefs are composed almost entirely of organic calcite, chiefly oysters, firmly consolidated. In other places, there are large areas completely covered with fossil oyster shells, pectens and barnacles.

The first to record the occurrence of the fossil oysters along the washes to the southeast of the mountains was Father Pedro Font in 1775, an observant diarist on the second Anza expedition to California. He wrote:

> I have come to surmise that in olden times the sea spread over all this land, and that in some of the great recessions which histories tell us about, it left these salty and sandy wastes uncovered.... Indeed...one finds on the way many piles of oyster shells, mixed with the earth and half buried, and other shells and maritime signs. It is not possible that people should have made such mountains of shells by carrying them from the sea so great a distance merely to bury them in piles. All of which, although merely conjecture, has a high degree of probability.

Dr. Charles C. Parry, a botanist and geologist who traveled with the International Boundary Commission in 1849 also made note of the fossils.

Another feature of the area is "worm-eaten" limestone called rillensteine found in both Fossil Canyon and Painted Gorge. The channels in the limestone, called "wurmrillen," are caused when water containing carbon dioxide in solution dissolves the limestone and makes a solution of calcium bicarbonate. Rainwater contains some carbon dioxide. Rills are etched deeper and deeper with each rainfall.

In 1984, a hiker and backpacker named Andrew Gale discovered some fossil bone fragments in the Coyote Mountains and took them to Tom Deméré at the San Diego Natural History Museum. Investigation of the area led to the discovery of parts of a baleen whale, including its lower jaw. Other fossil animal evidence found includes a sea cow, sea turtle, toothed whale, barracuda, great white shark, and various invertebrates. A camel was found near the ancient shoreline. These large fossils date from the lower part of the Imperial Group, some 6–7 MYA.

Geologists have named many of the canyons and washes in the Coyote Mountains area, and these names appear in geological publications. Names originally used by Andrew D. Christensen in his 1957 master's thesis from UCLA include: Birdfoot, Vaca, Andrade, Garnet, Webb, Descanso, and Capote canyons from the northwest to the northeast side of the Coyote Mountains. Butaca Canyon is on the southeast side, and Cahuilla Canyon is on the south side. Chris I. Ruisaard named the canyon to the west of Fossil Canyon, Ocotillo Canyon, during his work on his master's thesis in 1979.

Patricia J. Bell provided more names while writing her master's thesis at UCLA in 1980. Gypsum Canyon is named for "a unique locality of bedded gypsum in the basal Imperial Formation, above the Alverson volcanics," according to Texaco geologist Charlie Winker. Bell called Butaca Canyon Lava Gorge for the volcanic rocks of the Alverson Canyon Formation. She named Tunnel Road off Fossil Canyon for an abandoned mineshaft at the top of the jeep trail. Chuckwalla Canyon and the Flatiron are located east of Fossil Canyon.

University of Chicago geologist Susan Kidwell has also contributed to the naming of features in the heart of the Coyote Mountains. They include Alpine Valley, Reba Ridge, Hidden Valley, and Marble Mountain. Alpine is "perched" on the Coyote range. Reba is to the east. Hidden is not visible from a distance, and Marble is obvious.

Ref.: Bell 1980; Bell-Countryman 1984, 45-70; Bolton 1930, v. 5, 123–124; Christensen 1957; Deméré 1993, 82–84, and 2000; Gudde 1969, 78; Hunter 1998, 13; Kidwell 2000; Laudermilk 1940, 10; Lindsay 2001, 126-128; Lindsay and Lindsay 1998, 184; Pickard 1996, 1–2; Remeika 1998, 59–85; Ruisaard 1979; San Diego Natural History Museum n.d.; and Winker 1996, 2000a, and 2000b.

DOLOMITE MINE

The largest and most accessible deposits of dolomitic limestone in San Diego County are found in quarries in a low group of hills near the Dos Cabezas railroad siding and on the southwest flank of the Coyote Mountains. Also called dolomite marble or magnesium calcium carbonate, it was used "mainly as sources of crushed and broken stone for use as roofing granules, poultry grit, and decorative stone." The Coyote Mountains deposit is the one referred to as the Dolomite Mine. It was trucked eight rugged miles to the Dos Cabezas siding.

Fred M. Elliot and James Elliot of San Diego owned the dolomite deposit, which was first claimed by the family in 1924. They leased the claim in 1951 to the Milroy Roofing Company of South Gate, who worked the deposit for about six months. The Elliots also worked the claim briefly during 1952-53. By 1957, only a small shack and a loading bin remained on the property.

The two major Dos Cabezas deposits of dolomite were named the Heathman and the Golden State. The Heathman Quarry, named for R.W. Heathman in 1924, was located only a quarter-mile from the Dos Cabezas siding. The claim remained idle until the 1950s when it was leased to the Campo Milling Corporation, which processed the dolomite limestone for roofing granules, chicken grit, and agricultural use. The property then remained idle from 1957 to 1960 when it was leased to Don Weaver of Jacumba who mined it for several years for decorative stone. It was worked from two locations, about 500 feet apart. Most of the dolomite was crushed, screened, and bagged adjacent to the railroad siding.

The Golden State deposit was located about a mile northeast of the Dos Cabezas siding. B.A. Sweet of the Pine Tree Portland Cement Company owned it. It was first prospected in the 1920s and worked in the 1950s.

Ref.: Lindsay and Lindsay 1998, 183; Parker 1979, 84; Walawender 2000, 19-21; and Weber Jr. 1963, 174-178.

DOMELANDS

In 1983, Skip Ruland wrote an article for the *San Diego Tribune* about an area on the western edge of Imperial County, north of Highway S2. He called the area the "Domelands," and the name stuck. The domes are eroded sandstone outcrops, similar to areas in southern Utah. Some of the caves are large enough to offer small groups shade with a spectacular view of the entrance to the Carrizo Corridor and the Carrizo Impact Area to the north. Domelands is within the BLM's Coyote Mountains Wilderness.

Ref.: Deméré 2006, 34; Ruland 1983, L7; and Schad 1986, 264.

ELSINORE FAULT

The Elsinore Fault Zone is part of the San Andreas Fault System. The Elsinore Fault and related erosional activity is primarily responsible for the drainage topography found along its path. The path of the Elsinore Fault crosses Mason Valley, Vallecito Valley, Mountain Palm Springs, and the Coyote Mountains from northwest to southeast. Basically, the route of Highway S2 follows the fault. Historically, the Southern Emigrant Trail and the Butterfield Overland Stage followed this natural trail through ABDSP (see SOUTHERN EMIGRANT).

Ref.: Bergen, Clifford, and Spear 1997, 123; Hunter 1998, 12–13; and Remeika and Lindsay 1992, 49–50.

HIGHWAY S2—IMPERIAL HIGHWAY

County Highway S2, connecting Ocotillo to Warner Springs and following much of the route of the old Butterfield Overland Mail, was once called the Imperial Highway and was the dream come true of the El Centro Chamber of Commerce. The chamber was the most enthusiastic supporter of its construction, viewing it as the "Gateway to Imperial County" as early as the late 1920s. They envisioned a "first class" roadway from El Segundo, on the coast, to El Centro, via Norwalk, Yorba Linda, Corona, Temecula, Warner Springs, and Mason Valley. It was their hope that the newly created state park would help draw visitors to their city.

In 1929, George Kellogg became secretary of the newly formed Imperial Highway Association. The name of the association told the road's destination—Imperial Valley. For years, the association urged the paving of the old Butterfield route, and the name Imperial Highway was noted on maps as a highway-to-be. Their dream moved one step toward reality when the Boyce Aten Post of the El Centro American Legion placed a plaque on the San Diego/ Imperial County line in 1933. The obelisk made of stone remains. The plaque was stolen in the 1970s, but a new graphite-rubbed plaster plaque was installed sometime in the late 1990s. This new plaque whimsically reads: "This is the Desert. There's nothing out here. Nothing." It is signed N. Karavasiles. Nina Karavasiles is the landscape artist who designed the San Diego Trolley stop at 70th Street and one of the unique sculptures found along Harbor Drive near the Maritime Museum. The original plaque displayed an American Legion medallion and read:

JANUARY 8, 1933
MEMBERS OF BOYCE ATEN
POST NO. 25
AMERICAN LEGION OF EL CENTRO
BLAZED THIS TRAIL
FOR THE BUILDERS
OF
THE IMPERIAL HIGHWAY

The California Development Company is credited with naming the newly reclaimed southern portion of the Colorado Desert at the turn of the last century. The developers felt the name Imperial would be more attractive to settlers than "Colorado Desert." They formed a subsidiary branch of the company, called the Imperial Land Company, and began colonizing the Imperial Valley. The earliest town was Imperial, established in 1901. Imperial County was formed in August 1907 when angry residents decided to break away from San Diego County.

Despite the long association with the name Imperial Highway, the San Diego Board of Supervisors officially changed the name of the highway in 1944 as a response to a petition submitted to them. The road from Scissors Crossing to the Imperial County line became the Great Southern Overland Stage Route of 1849.

Those who submitted the petition were of good heart to want to commemorate the Southern Emigrant Trail and the Butterfield Overland Mail route. The problem is that the Butterfield Overland Mail was only active from 1857 to 1861. The Southern Emigrant Route to California was the preferred all-weather route from 1848 to 1880. The Overland Mail route was never known as the "great," it was never "southern," and it did not exist in 1849. The name has never been corrected, and county maps still list it as the Great Southern Overland Stage Route of 1849. Imperial County, on the other hand, held on to the original name for many years with an exit originally on old Highway 80 and then on Interstate 8 for "Imperial Highway."

Ref.: Crouse 1952, 12–13; Gudde 1969, 150; Jorgensen 2000; Lindsay 1973, 96; Lindsay and Lindsay 1998, 184; Little 1982, 14; Mansfield 1944, B1; Mayne 1953, A18; San Diego County Department of Public Works 2000; Weight 1952, 12–14; and Wray 1999, 100.

JOJOBA WASH

The abundant presence of jojoba or goatnut, *Simmondsia chinensis*, in a wash south of Carrizo Badlands Overlook off Highway S2, accounts for its name. The road in the wash loops through the Volcanic Hills via Jojoba and West Mortero Washes to return to Highway S2. The jojoba is dioecious—having male and female reproductive organs on separate plants. The female plant bears inch-long nuts noted for their oil content. At one time, it was believed that the production of commercial jojoba would save the sperm whale because of its oil content. The problem was that it took almost seven years for the plants to produce nuts, and not until then could farmers determine which were the female plants. Therefore, half of the crop could not be harvested.

Jojoba oil is used for cosmetics, as a precision lubricant, for cooking, car wax, and, after oil removal, as livestock feed. The nuts were a food source for Native Americans and were used in making a coffee-like drink. The seeds were eaten fresh or ground into powder to make the drink. Seeds were gathered in May and June. Bighorn sheep and deer eat the plant. Park Superintendent Mark Jorgensen reports that bighorn feed so heavily on some jojobas that "they are cropped down like bonsai plants."

Ref.: Jorgensen 2000; Lindsay and Lindsay 2006, 211.

MOONLIGHT CANYON

Campers who enjoyed taking a moonlight hike to Inner Pasture from Agua Caliente Hot Springs may have named Moonlight Canyon. The canyon is located south of the hot springs, in the Tierra Blanca Mountains, and is within Agua Caliente Regional Park. Moonlight streaming down the narrow canyon would have adequately lit the route (see AGUA CALIENTE and TIERRA).

Ref.: Lindsay and Lindsay 1998, 174.

MORTERO WASH

Morteros are deep grinding holes in granitic rocks and boulders made by Indians while pounding and grinding seeds and desert plants. An elongated rock or pestle was used to pound the vegetative matter in the bedrock mortar or mortero. Often in association with morteros are metates or slicks, which were used with a small smooth rock or mano to grind seeds into flour with a roller-pin action. Metates are smooth or slick areas found on rock surfaces. Morteros and metates are commonly found throughout the park where Indians encamped.

Mortero Wash and its tributaries, all off Highway S2 near the San Diego and Imperial County line, take their names from their proximity to many Indian sites found at Indian Hill and Piedras Grandes. West Mortero Wash is the primary drainage for the western half of the Volcanic Hills and offers a superb geological hike from the Jojoba Wash loop, past Red Hill, and out to Highway S2 at mile 55.

Ref: Lindsay and Lindsay 1998, 192, 204, 208; and Welch 1956.

PALM SPRING

The California fan palm (*Washingtonia filifera*) lends its name to many sites in the desert. Palm Spring, located just north of Vallecito Creek in the Carrizo Corridor, was the site of a Butterfield Stage Station, midway between Carrizo and Vallecito, built next to an all-year freshwater spring. Before the station was built, a grove of tall palms grew around the spring, and the oasis was a refreshing stop for soldiers, early emigrants and mountainmen. However, after gold was discovered in California, the impact was immediately felt at Palm Spring. By the time the stage station was built in 1858, all of the palm trees had been chopped down by emigrants to feed their campfires. Park personnel planted the palms now found there.

Water has existed at this mesquite-surrounded watering

spot since Indian times, as is shown by the scattering of shards and the remains of cooking fires and camp middens. A monument now marks the site of the old way station. The site has been known by various names, including Palmitas, Palmitos, Pamitas, Pometo, Palmetto, Palmetta, Sulphur Spring, and Palm Springs. The USGS lists the site as Palm Spring. A common error even in current geologic studies is to refer to the site and/or the formation as Palm Springs.

Palm Spring also refers to the gray to gray-and-tan terrestrial sandstone and claystone sediments found throughout the badlands of Anza-Borrego, above the yellowish sediments of the Imperial Group. The Palm Spring Group includes the proximal basin-margin Canebrake Conglomerate, the basinward Hueso and Olla formations, the lacustrine Tapiado Claystone, and distally the Arroyo Diablo Formation of Colorado River provenance. The Arroyo Diablo Formation at the base of the Palm Spring Group dates from the Pliocene epoch, some 3 to 4 MYA, and contains fragments of silicified wood, which is common throughout the fluvial formation. It is similar to the lacustrine Borrego Formation. Former Ranger-geologist Paul Remeika did a study on the wood fragments found in this lower Palm Spring Group. His study indicated that the wood was mainly willow, cottonwood, laurel, and California walnut.

Ref.: Edwards 1961, 3, 9, 34–41, 49; Jasper 1934, v. 2. 389; Jorgensen 2000; Knaak 1995 Lindsay and Lindsay 1998, 53–54, 59, 111, 177, 184–185, 199; Remeika 2006; Von Till Warren and Roske 1981; Welch 1955 and 1956; and Wray 1999.

PENINSULAR RANGES PROVINCE

Even though ABDSP is part of the Colorado Desert, much of the park falls within the Peninsular Ranges (geomorphic) Province rather than the Salton Trough (geomorphic) Province. The surface features characteristic of the Salton Trough Province abut and partially overlap the underlying bedrock of the Peninsular Ranges Province. (Geomorphic provinces are a geographic unit characterized by a distinctive geologic history and similar relief features throughout.)

The Peninsular Ranges Province is a mountainous plateau that has been uplifted steeply on the eastern side and titled westward, creating a distinct eastern escarpment like that of the Sierra Nevada. The highest elevations in the province are found along this rugged eastern side or scarp face.

The backbone of the Peninsular Ranges is igneous rock of the Peninsular Ranges Batholith that dates to the Cretaceous and/or Jurassic Period, some 95 to 155 MYA. This batholithic material consists of granitic rocks and contains deposits of gold, tungsten, gem minerals, feldspar, quartz and molybdenum.

Ref.: Bergen, Clifford, and Spear 1997, 30; Dibblee Jr. 1954; Remeika and Lindsay 1992, 15, 29, 31–36, 58; and Norris and Webb 1990, 257–261, 277.

SOUTHERN EMIGRANT TRAIL

The all-weather road to California, followed by literally thousands of emigrants after the discovery of the California gold fields, connected New Mexico and Arizona to California via the Gila Trail and a route across the Colorado River and the Imperial sand dunes to the Carrizo Corridor. From there, the trail went through Box Canyon and up San Felipe Valley to Warners Ranch. The Southern Emigrant Trail split at Warners Ranch with one branch leading southwest across and down the mountains to San Diego by way of Santa Ysabel. The other branch continued northwest over Sunshine Summit to Temecula, San Gabriel and finally Los Angeles.

The Southern Emigrant Trail was the American version of the old Sonora Road used by Mexico to dispatch mail to their California outposts. Of all the roads to California, this southern route proved to be the only all-weather route, even though it remained extremely challenging during the dry, hot summer months. In the months following the announced gold rush, the trail was literally covered with the carcasses and bones of animals that had succumbed to the desert heat.

The Southern Emigrant Trail was the route later followed by the San Antonio and San Diego Mail Line and the Butterfield Overland Mail. Highway S2 follows much of the original route.

Ref.: Dixon 1965; and Lindsay 1973, 57, 68–69.

TIERRA BLANCA MOUNTAINS

The Tierra Blancas, Spanish for "white land," probably refer to the mineralized white hillsides surrounding Agua Caliente Hot Springs, which are part of the Tierra Blanca Mountains (see AGUA CALIENTE).

Ref.: Lindsay and Lindsay 1998, 173.

VOLCANIC HILLS

The most accessible feature of volcanic activity in the Anza-Borrego area is found in the Volcanic Hills and Red Hill area, south of Highway S2, west of the San Diego/Imperial County line (see HIGHWAY S2). The massive lava flows, consisting of basalt, andesite, and airborne ash, solidified here some 16-18 MYA during the Miocene epoch. These volcanic rocks are found in the Carrizo Impact Area,

Coyote Mountains, Dos Cabezas, Jacumba Valley, and the Table Mountain area (see COYOTE MOUNTAINS). This unit is generally known as the Alverson volcanics (or andesite) and has been considered for Alverson (Canyon) Formation status.

The Volcanic Hills lava flows are compositionally the same as that found on Table Mountain, 10 miles south and almost 3,000 feet higher. Table Mountain, however, was subjected to the Laguna escarpment uplift. This volcanism indicates activity and faulting along the Elsinore Fault Zone coincidentally with the earliest stages of the rifting of peninsular California from the Mexican mainland (see ELSINORE). Names in the area reflect this activity: Volcanic Hills, Red Hill, and Lava Flow Wash.

Ref.: Cassiliano 2006, 129; Lindsay and Lindsay 1998, 183, 204, 212.

WIND CAVES

Most landforms in the desert are sculptured by a combination of weathering, mass wasting—landslides and mudflows, running water, and wind. Running water is the most effective agent of desert erosion because there is insufficient vegetation to protect slopes from rapid runoff in the aftermath of desert cloudbursts. Wind is also effective as small sand particles become aerial abrasives that etch and sharpen desert features.

Good examples of water and wind erosion are seen by the occurrence of various caves found in the park area. Three areas specifically have places called Wind Caves, even though the major force responsible for their formation is water. The Borrego Mountain Wind Caves are found near the Borrego Mountain Dropoff. The tilted and twisted strata of Arroyo Diablo Formation sandstones found here are pierced like Swiss cheese by water and, later, wind erosion. The Split Mountain Wind Caves are carved in marine beds of the Imperial Group as are the Coyote Mountains Wind Caves.

Ref.: Lindsay 1973, 1; Lindsay and Lindsay 1998, 126, 153–154, 184; and Rightmer 1996, 167.

(The convention for Earth Science terms in this paper is Jefferson and Lindsay, cited below.)

SELECTED REFERENCES

Beauchamp, R. Mitchel. 1986. *A Flora of San Diego County, California*. National City, California: Sweetwater River Press.

Bell, Patricia J. 1980. "Environments of Deposition, Pliocene Imperial Formation, Coyote Mountains, Imperial County, California." Master's thesis. University of California, Los Angeles.

Bell-Countryman, Patricia J. 1984. "Environments of Deposition of the Pliocene Imperial Formation, Coyote Mountains, Southwest Salton Trough." In: C.A. Rigsby, ed. *The Imperial Basin—Tectonics, Sedimentation and Thermal Aspects*. Fullerton, California: Society of Economic Paleontologists and Mineralogists, Pacific Section, 1984.

Bergen, Frederick W., Harold J. Clifford, and Steven G. Spear. 1997. *Geology of San Diego County: Legacy of the Land*. Diane M. Burns, ed. San Diego: Sunbelt Publications.

Bolton, Herbert Eugene. 1930. Translator and editor. *Anza's California Expeditions*. 5 vols. Berkeley: University of California Press.

Brigandi, Phil. 1998. "Caves, canyons and drop-offs." *Borrego Sun*, 6 August, 7.

Brown, John S. 1920. "Routes to Desert Watering Places in the Salton Sea Region, California." *Water-Supply Paper #490-A*. US Department of the Interior. Geological Survey. Washington: Government Printing Office.

Cassiliano, Michael. 2006. "Mammalian Biostratigraphy in the Vallecito Creek-Fish Creek Basin." In: George T. Jefferson and Lowell Lindsay, eds. *Fossil Treasures of the Anza-Borrego Desert*. San Diego: Sunbelt Publications.

Chase, J. Smeaton. 1919. *California Desert Trails*. New York: Houghton Mifflin Co.

Christensen, Andrew Dougan. 1957. "Part of the Geology of the Coyote Mountain Area, Imperial County, California." Master's thesis. University of California, Los Angeles.

Cline, Lora L. 1979. "The Kwaaymii: Reflections on a Lost Culture." Imperial Valley College Museum Society, El Centro, California. Occasional Paper No. 5.

Crouse, Harriet. 1952. "History and Hot Springs." *Westways Magazine* 44 (October): 12–13.

Deméré, Thomas A. 1993. "Fossil Mammals from the Imperial Formation (upper Miocene–lower Pliocene), Coyote Mountains, Imperial County, California." In: Robert E. Reynolds and Jennifer Reynolds, eds. *Ashes, Faults and Basins*. Special Publications 93–1. Redlands, California: San Bernardino County Museum Association, 82-85.

______. 2000. Telephone conversation with the author, 22 March.

______. 2006. "The Imperial Sea: Marine Geology and Paleontology." In: George T. Jefferson and Lowell Lindsay, eds. *Fossil Treasures of the Anza-Borrego Desert.* San Diego: Sunbelt Publications.

Dibblee Jr., T.W. 1954. "Geology of the Imperial Valley Region, California." *Geology of Southern California.* Edited by Richard H. Jahns. Bulletin 170. San Francisco: California Division of Mines.

Dixon, Ben F. 1965. *Camp Salvation: First Citizens of Calexico, An Oasis on the Gold Rush Trail.* The Benjamin Franklin Junior Historical Series No. 10. San Diego: Squibob Chapter, E Clampus Vitus.

Dorsey, Rebecca. 2006. "Stratigraphy, Tectonics, and Basin Evolution in the Anza-Borrego Desert Region." In: Jefferson and Lindsay, eds.

Edwards, E.I. 1961. *Lost Oases Along the Carrizo.* Los Angeles: Westernlore Press.

Fairchild, Frank L. n.d. "Places and Features Names." Copy in author's files.

Gudde, Erwin G. 1969. *California Place Names: The Origin and Etymology of Current Geographical Names.* 3rd edition. Berkeley and Los Angeles: University of California Press.

Henderson, Randall. 1938. "Just Between You and Me." *Desert Magazine* 1 (March): 32.

______. 1947. "Palms of the Carrizo Country." *Desert Magazine* 10 (April): 19–22.

Hunter, Christy. 1998. "Roadside Geology of Imperial Valley." In: Lindsay and Hample 1998, 5–17.

Jasper, James A. 1934. "Trail-Breakers and History-Makers of Julian, Ballena, Mesa Grande, Oak Grove, Warner Ranch, Banner, Cuyamaca in San Diego County, California: History, Biography, Reminiscences." 2 vols. Unpublished manuscript. San Diego Historical Society Research Archives.

Jefferson, George T. and Lowell Lindsay, eds. 2006. *Fossil Treasures of the Anza-Borrego Desert.* San Diego: Sunbelt Publications.

Johnson, Paul. 1987. *Weekenders Guide: Point of Interest and Walks Along the Paved Roads of Anza-Borrego Desert State Park.* Borrego Springs, California: Anza-Borrego Desert Natural History Association.

Jorgensen, Mark. 2000. Letter and resource data sent to the author, 2 February.

Kidwell, Susan. 2000. E-mail correspondence to the author, 4 May.

Knaak, Manfred. 1995. "Palmitas, Vanished Little Palm Oasis." ABDNHA *Sand Paper*, November.

Laudermilk, J.D. 1940. "The Rillensteine Case." *Desert Magazine* 4 (December): 10–12.

Leetch, George and Jean Leetch. 1992. *A Collected History of Canebrake, California.* Julian: Canebrake Improvement Association.

Lindsay, Diana. 1973. *Our Historic Desert: The Story of the Anza-Borrego Desert.* San Diego Union-Tribune Publishing Co.

______. 2001. *Anza-Borrego A to Z: People, Places, and Things.* San Diego: Sunbelt Publications.

Lindsay, Lowell and Diana Lindsay. 1991. *The Anza-Borrego Desert Region.* 3rd edition. Berkeley: Wilderness Press.

______ 1998. *The Anza-Borrego Desert Region.* 4th edition. Berkeley: Wilderness Press.

______ 2006. *The Anza-Borrego Desert Region.* 5th edition. Berkeley: Wilderness Press.

Little, Ida. 1982. *Imperial Valley Historical Markers.* Holtville, California: Imperial Valley Pioneers.

Lough, Charles F. 2006. Personal communication with the author, 8/13.

Mansfield, Richard. 1944. "Good Afternoon." *San Diego Tribune-Sun*, 16 November, B1.

Mayne, Mel. 1953. "Vallecito Road Improvements Asked." *San Diego Union*, 1 March, A18.

Norris, Robert M. and Robert W. Webb. 1990. *Geology of California.* 2nd edition. New York: John Wiley & Sons, Inc.

Parker, Horace. 1979. *Anza-Borrego Desert Guide Book.* Revised by George and Jean Leetch. Borrego Springs, California: Anza-Borrego Desert Natural History Association.

Pickard, Greg. 1996. "Coyote: A Lesson in Survival." Colorado Desert District *Tracks*, December, 1–2.

Remeika, Paul. 1998. "Marine Invertebrate Paleontology and Stratigraphy of the Vallecito-Fish Creek Basin: a Historic Review, Synthesis and Revision." In: Lowell Lindsay and William G. Hample, eds. *Geology and Geothermal Resources of the Imperial and Mexicali Valleys.* San Diego: San Diego Association of Geologists.

_______ 2006. "Ancestral Woodlands of the Colorado Desert Plain." In: Jefferson and Lindsay, eds.

______ and Lowell Lindsay. 1992. *Geology of Anza-Borrego: Edge of Creation*. San Diego: Sunbelt Publications.

Richards, Bob. 1989 "The Arroyo Tapiado Mud Caves." *California Caver* 39 (Summer): 37–38, 46–50.

Rightmer, David A. and Patrick L. Abbott. 1966. "The Pliocene Fish Creek Sturzstrom, Anza-Borrego Desert State Park, Southern California." In: Abbott and Seymour 1996, 165-184.

Ruisaard, Chris Ivan. 1979. "Stratigraphy of the Miocene Alverson Formation, Imperial County, California." Master's thesis. San Diego State University.

Ruland, Skip. 1983. "Desert backpack lets you walk an ocean floor." *San Diego Tribune*, 27 December, Leisure 7.

San Diego County Department of Public Works, Road Registry. 2000. Telephone conversation with Eileen Lindquist, Road Register Administrator, 30 March.

San Diego Natural History Museum. n.d. "The Geology of Coyote Mountain." Copy in author's files.

San Diego Union. 1939. 7 May, 6.

Schad, Jerry. 1986. 1st edition. *Afoot and Afield in San Diego County*. Berkeley: Wilderness Press.

Scott, Tracey M. 1939. "Desert Place Names." *Desert Magazine* 2 (April): 24.

______. 1940. "Desert Place Names." *Desert Magazine* 3 (February): 30.

Stein, Lou. 1975. *San Diego County Place-Names*. San Diego: Tofua Press.

Von Till Warren, Elizabeth and Ralph J. Roske. 1981. "Cultural Resources of the California Desert, 1776–1980: Historic Trails and Wagon Roads." Bureau of Land Management, California Desert District. Russell L. Kaldenberg, Series Editor and Cultural Resources Program Manager. Copy in General Plan Archives, ABDSP.

Walawender, Michael J. 2000. *The Peninsular Ranges: A Geological Guide to San Diego's Back Country.* Dubuque, Iowa: Kendall/Hunt Publishing Co.

Weber Jr., F. Harold. 1963. "Mines and Mineral Resources of San Diego County, California." County Report 3. San Francisco: California Division of Mines and Geology.

Weight, Harold O. 1952. "Petrified Wood Along the New Butterfield Trail." *Desert Magazine* 15 (June): 10–15.

Welch, Jack. 1955. Ranger patrol notes, 17–18 March. Copy in author's files.

______. 1956. Ranger patrol notes, 17 April. Copy in author's files.

Winker, Charles D. 1996. "Stratigraphy of a Marine Rift Basin: Neogene of the Western Salton Trough, California." In: Patrick L. Abbott and John D. Cooper, eds. *Field Conference Guide 1996*. Annual meeting of American Association of Petroleum Geologists. Book 80. Fullerton, California: Society of Economic Paleontologists and Mineralogists.

______. 2000a. Telephone conversation with the author, 3 May.

______. 2000b. E-mail correspondence to the author, 7 May, and 8 May.

Wray, Christopher. 1999. "Historic Places in Eastern San Diego County and Imperial County." Unpublished manuscript submitted to Sunbelt Publications, El

---

Diana Lindsay is the author of *Anza-Borrego A to Z: People, Places, and Things* and the co-author of the *Anza-Borrego Desert Region: A Guide to the State Park and Adjacent Areas of the Western Colorado Desert*. She received her M.A. from San Diego State University in history and geography. She also was a contributor to *Geology of the Elsinore Fault Zone San Diego Region*, SDAG/SCGS Volume 31 – 2003. She is president of the Anza-Borrego Foundation and Institute.

Part II. Papers

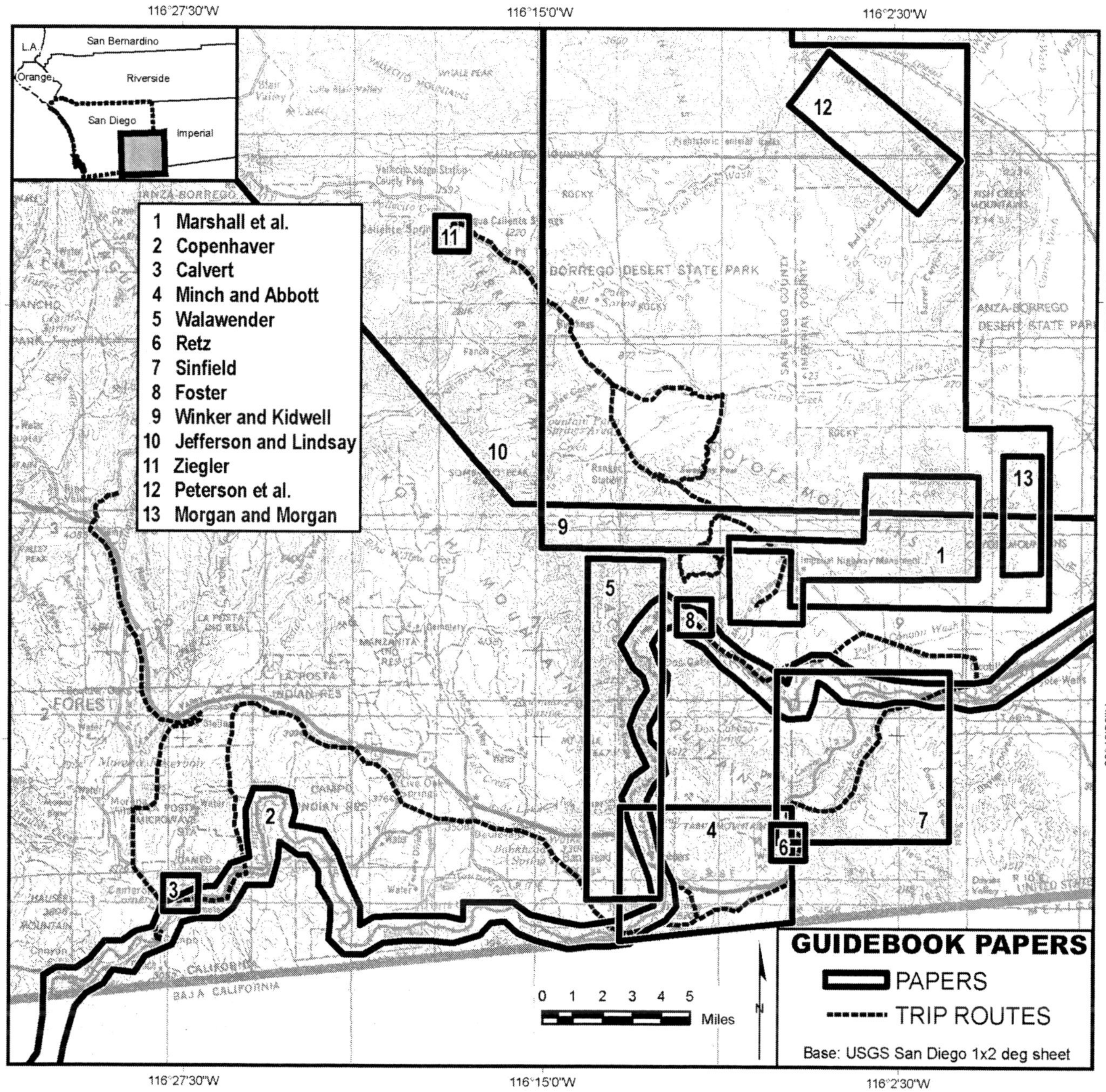

What Regional and Local Paleomagnetic Data Tell Us about the Tectonics of the Coyote Mountains, Southwest Imperial Valley, Southern California

Monte Marshall, Neal Mace, Bradley Erskine, and Edward Jansen
Department of Geologic Sciences, San Diego State University, San Diego, California 92182
Monte Marshall: mmarshall@geology.sdsu.edu

INTRODUCTION

This paper will document some of the contributions of paleomagnetic studies to the better understanding of the movement of both large and small blocks of the earth's crust in San Diego and Imperial Counties. Paleomagnetism, or the study of the magnetization that was acquired by many rocks when they formed, has been one of the key sciences in discovering plate tectonics, mapping the paleopositions of the continents, and measuring smaller scale tectonic activity. In fact, not only did the interpretation of the magnetic anomalies recorded in the oceanic crust when it was created at spreading centers lead to the discovery of seafloor spreading and continental drift, but the study of the linear magnetic anomalies recorded in the pillow basalts of the oceanic crust of the east Pacific led to much of our understanding of the geologic history of the western margin of both North and South America.

To better understand the geologic context of the 2006 SDAG field trip to the Coyote Mountains, we need to look at the plate tectonic history of this part of California. Probably nowhere in the world is the present and past interaction between plates better recorded and understood than the interaction between the Pacific (PP), Farallon (FP), and North American plates (NAP) along the western margin of North America. We understand much of this interaction because of Tanya Atwater's (1970) pioneering application of the linear marine magnetic anomaly pattern recorded on the northeast Pacific seafloor to decipher the tectonic history of the western North American continental crust. The magnetic anomaly pattern recorded in the oceanic crust of the Pacific Plate showed that the intermediate Farallon Plate had been spreading away from the Pacific Plate and moving easterly to be subducted under the western edge of the North American Plate for more than 100 million years. Then, about 27 Ma, the Farallon Plate together with the FP-PP spreading center were totally subducted near the latitude of San Diego, leaving the Pacific Plate in direct contact with the North American Plate. From this time more and more of the Farallon Plate was totally subducted north and south along the coast of California, thus lengthening the suture/transform boundary between the Pacific Plate and the North American Plate as one triple junction moved north and the other moved south. A triple junction is a place on the earth's surface where three plate boundaries intersect at a point. Until about 17 Ma, this strike-slip plate boundary lay underwater, between the base of the continental slope and high on the continental shelf. With time more and more of the shear moved on land. The southern San Andreas transform—right-lateral strike-slip fault system moved to its present position on dry land only about 6 Ma (Atwater, 1989). This happened when the southern triple junction had migrated as far south as Mazatlan. At that moment the triple junction consisted of 1) a northwest-southeast trending transform fault in the continental borderland to the north of Mazatlan, 2) the East Pacific Rise (the spreading center between the Pacific Plate and the southern remnants of the Farallon Plate) which trended northeast-southwest and ran into the base of the continental slope at Mazatlan at almost a right angle, and 3) a northwest-southeast trending subduction zone extending south of Mazatlan, where the subduction of the Cocos Plate (a remnant of the Farallon Plate) was causing the Mid-America Trench as it went beneath Mexico. Apparently the stresses of this geometry and/or a north-south zone of crustal weakness that formed the shallow proto-gulf allowed the East Pacific Rise to migrate north as a series of northeast-southwest spreading centers alternating with northwest-southeast transform faults, forming the present deep water Gulf of California. Simultaneously, this tore the Baja California Peninsula block away from mainland Mexico and made it part of the Pacific Plate. The northernmost spreading center now underlies and causes the Salton Trough, and the present, so called southern San Andreas fault system, with its almost 300 km of right-lateral strike-slip, begins near the southeast shore of the Salton Sea and is merely the northernmost and longest of this string of transform faults!

In 1980, Bruce Luyendyk and his students at UCSB began to discover, through detailed paleomagnetic sampling in the Los Angeles area that large crustal blocks, including the Transverse Ranges and the northern Channel Islands, had been uniformly rotated 30 to 90 degrees clockwise. This rotation around a near-vertical axis began about 16 Ma, at the same time that the right lateral shear zone of the PP–NAP boundary moved on land in this area (Hornafius et al., 1986; Carter et al., 1987). Even though the exact mechanics of how crustal blocks can rotate around almost vertical axes in zones of right and left lateral shear is still poorly understood, both clockwise and counterclockwise rotations of crustal blocks have been observed in strike-slip fault zones all over the world.

RESULTS OF PALEOMAGNETIC STUDIES IN SAN DIEGO COUNTY AND WESTERN IMPERIAL COUNTY

There are six areas in southernmost California whose paleomagnetic data is of interest to our study of the Coyote Mountains (Table 1). In three of the areas, including the Coyote Mountains, the paleomagnetic data is taken from oriented cores collected from basalt flows of the Alverson Volcanics. One study measured the paleomagnetic directions in the sedimentary rocks of the Imperial and Palm Spring Groups in the Vallecito-Fish Creek basin immediately north of the Coyote Mountains. The other two studies used cores collected from plutonic rocks. One sampled the La Posta Pluton in the eastern half of the Peninsular Ranges Batholith (PRB). The other sampled gabbros and granitic plutons in the western half of the batholith.

Coyote Mountains

A paleomagnetic study was made by Mace (1981) of 19 basalt flows having ages between 15 and 20 Ma at 4 outcrops in the south Coyote Mountains (Figures 1 and 3). Figure 2 shows the directions of the magnetization acquired by these flows as the magnetite crystals cooled through their blocking temperature. Using alternating field demagnetization, the cores were cleaned of any secondary components acquired since the basalts cooled. The cleaned directions were corrected for bedding tilt. The directions are plotted on a stereonet, exactly the way trend and plunge of geologic lines are plotted in structural geology (remember?!). Two additional conventions are seen on paleomagnetic stereonet plots. Since the directions of magnetization are unit vectors and can plunge down or up, the former are plotted as SOLID points (•) on the lower hemisphere. These directions were acquired during time of normal polarity, like today, when the geomagnetic field points north and down. The latter are plotted as OPEN circles (o) on the upper hemisphere. These rocks acquired their magnetization in a time of reversed polarity, that is, when the geomagnetic field pointed south and up. The second convention in paleomagnetic studies is to show alpha 95 circles, that is, circles whose radii reflect the amount of between-sample scatter in directions. These circles on the stereonet define the size of a cone that contains the true magnetization direction with a statistical confidence of 95 percent.

> The interpretation of paleomagnetic stereonet plots is as follows: any large difference between the latitude where the magnetization formed, i.e., the paleolatitude (PL) of the sample, and its present latitude suggests a possible N-S translation of the rock. For all its complexity the shape of the geomagnetic field is that of a simple dipole and the magnetic plunge or inclination (I) varies from zero at the equator to 90 degrees at the magnetic poles. Although the geomagnetic pole is currently about 13 degrees away from the geographic pole, the two poles have been found to coincide if the gyrations of the geomagnetic pole are averaged over several tens of thousands of years. The accuracy of the paleomagnetically determined paleolatitude obviously increases with increased sampling of the paleogeomagnetic field directions at a place over time. When this is done the paleolatitude of a locality is given by the formula tan(PL) = 2tan(I). This simple formula is the basis for much of our paleolatitude information and paleogeographic reconstructions. Given the symmetry of the geomagnetic field, there is, alas, *no* way to calculate *paleolongitudes*.
>
> Any significant deviation of the magnetic azimuth or declination from north or south suggests possible clockwise or counterclockwise rotations around a mostly vertical axis of the area where the samples were taken. There is some ambiguity in interpreting the sense of the rotation. A paleomagnetic direction pointing down and northeast can suggest either a 45-degree clockwise rotation or a 315-degree counterclockwise rotation. In the absence of additional constraints, Luyendyk's Razor, or the principle of 'least astonishment' is used.

The paleomagnetic directions found in the south Coyote Mountains range from south and up, to west-northwest, and up, to northeast and up (Figures 1 through 3, and

Table 1. Paleomagnetic data from basalts in the Alverson Volcanics, sedimentary rocks in the Imperial and Palm Spring Groups, and plutonic rocks of the Peninsular Range Batholith, southern California.

AREA	NO. OF FLOWS	POINTS IN TIME	AGE (Ma)	Δ LATITUDE (degrees)	ROTATION (degrees)	REFERENCE
Jacumba, CA	4	3	19 ± 3[H]	10N	20CW	E.J. Jansen, Senior Thesis, SDSU (1978)
Mortero Canyon, Volcanic Hills	5	4	18 ± 2	10N	16CW	M. Hickle, Master's Thesis, SDSU (incomplete, 1981)
SOUTHERN END, COYOTE MOUNTAINS						N.W. Mace, Master's Thesis, SDSU (1981)
Fossil Canyon, south (FCA & B)[a]	5	2	19 ± 0.8[R]	0	0	
Fossil Canyon, north (FCN)	4	1	20 ± 0.6[R]	8N	110CW	
Butaca Canyon (BC)	6	4	17 ± 0.5[R]	8N	61CW	
Painted Gorge (PG)	4	2	15 ± 0.5[R]	10N	43CW (2), 130CCW (2)	
Vallecito-Fish Creek Basin, Imperial and Palm Spring Groups	NA	107	1.0 to 4.3	12N	35CW	N.M. Johnson et al. (1983)
La Posta Pluton, Jacumba Mountains (eastern batholith)	NA	20	94	0	0	D.T. Symons et al. (2003)
Multiple plutons (western batholith)	NA	13	125 to 100	11N	42CW	B. Erskine, Master's Thesis, SDSU (1982)

Notes:
CW = Clockwise
CCW = Counter-clockwise
NA = Not applicable

a = Sample identification (see Figures 1 and 2)
H = Hawkins (1970)
R = Ruissard (1979)

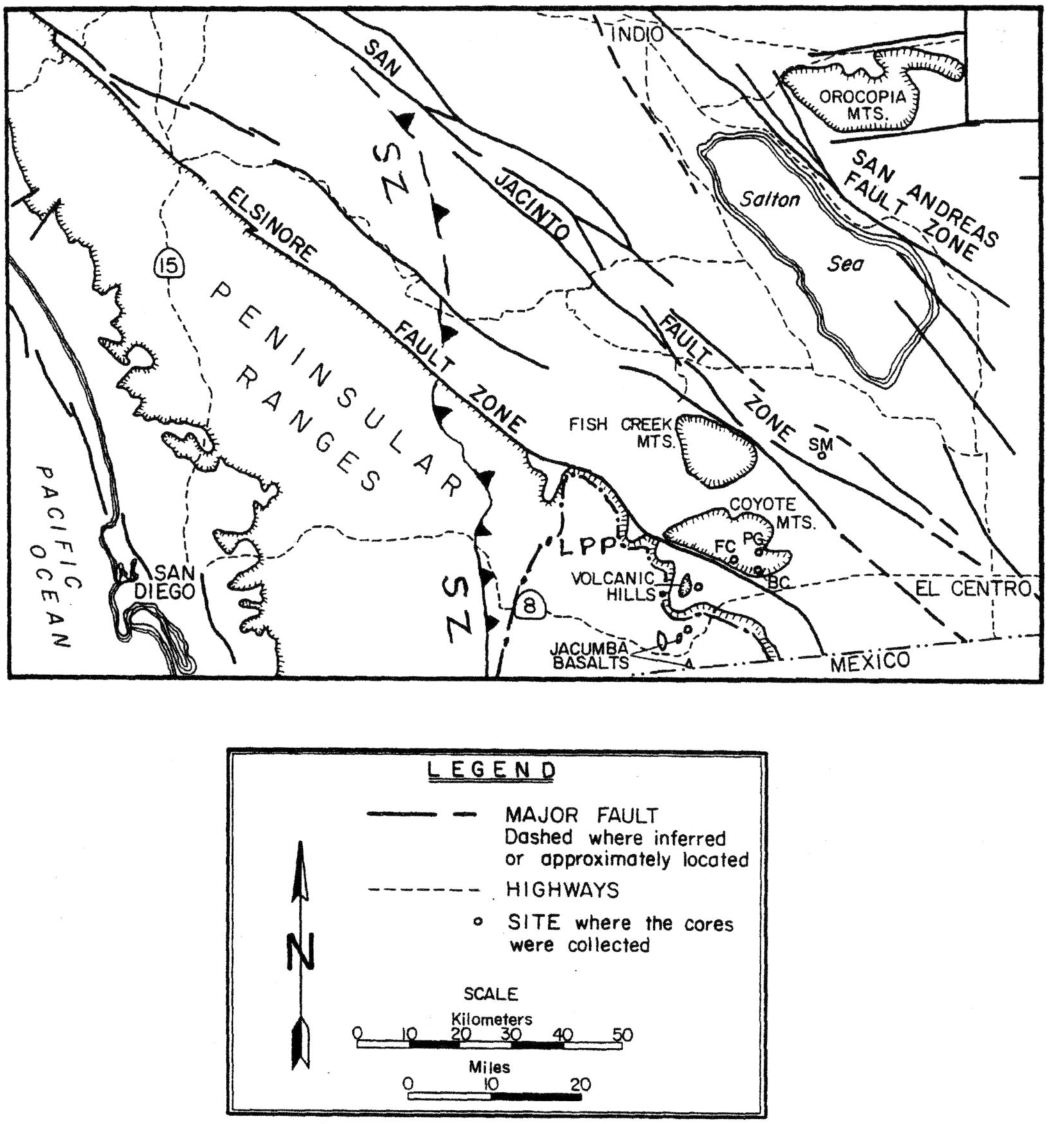

Figure 1. Map showing the paleomagnetic study sites in the Volcanic Hills, Coyote Mountains, and Jacumba basalts. This map also shows the location of the possible suture zone (line with barbs labeled SZ, dashed where inferred or approximately located) between the eastern and western parts of the Peninsular Ranges Batholith. In general, the magnetite-ilmenite line and the other geochemical and geophysical lines lie fairly close to the suture line. The location of the La Posta Pluton is shown by dash-dot lines labeled LPP on either side of Jacumba. The three major fault zones crossing the Peninsular Ranges and Salton Trough are shown schematically.

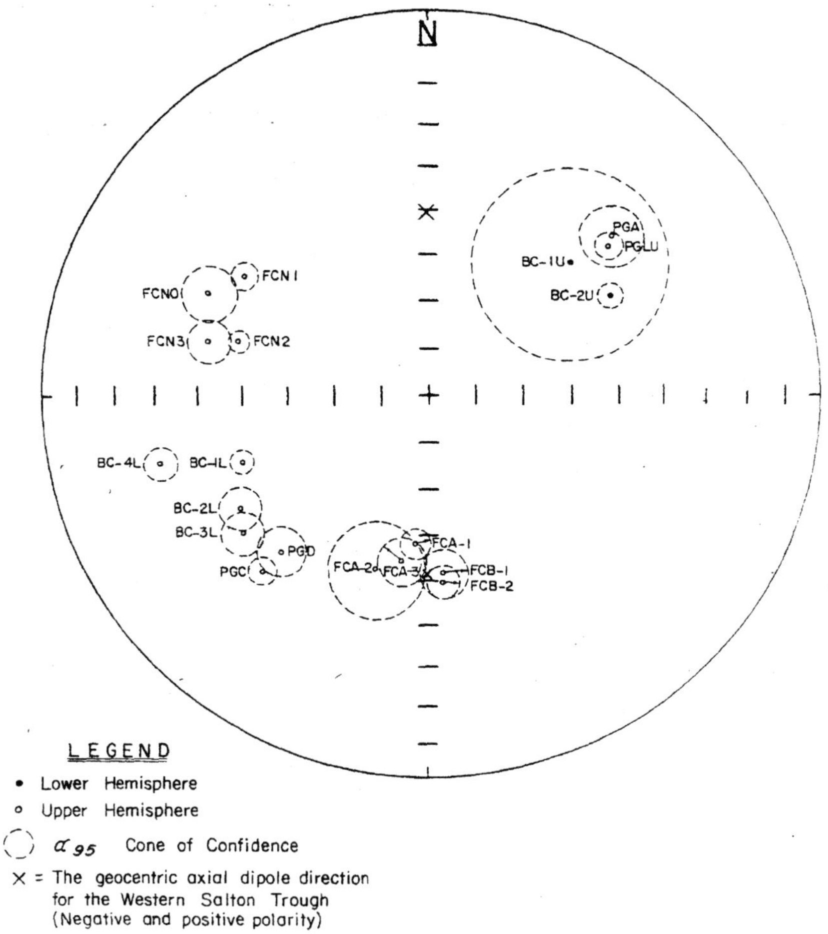

Figure 2. Stereonet plot of the average magnetization directions of 19 basalt flows outcropping at the south end of the Coyote Mountains—after alternating field cleaning and bedding tilt corrections have been made (Mace, 1981).

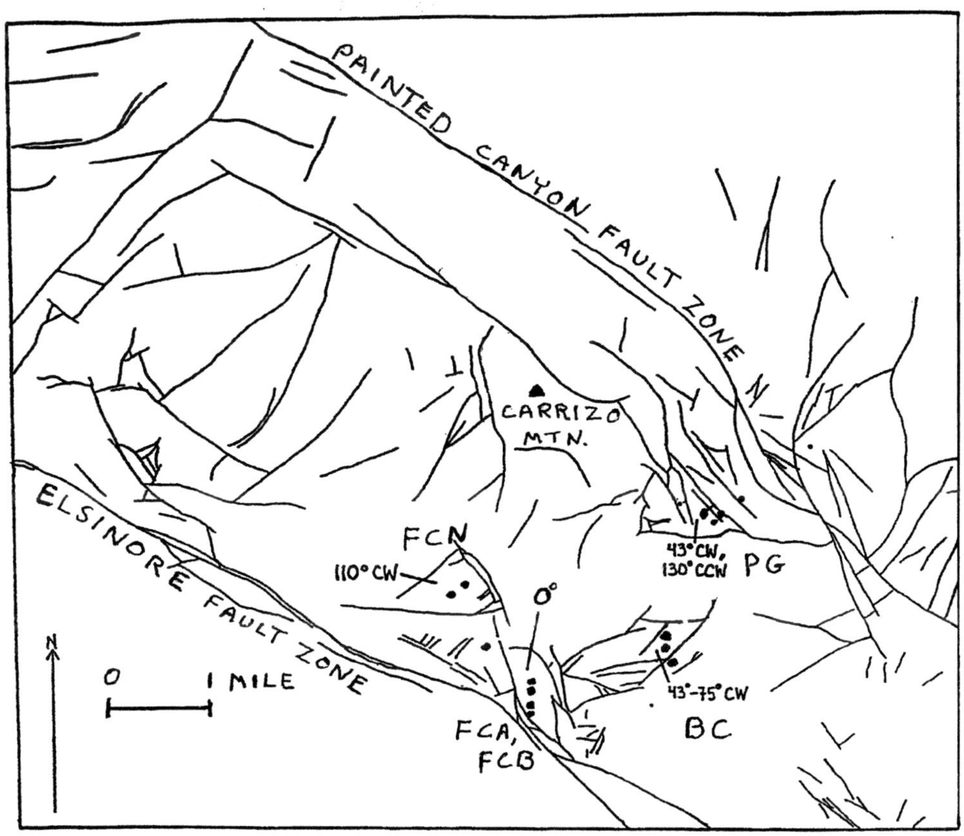

Figure 3. Map showing the dense and complicated fault network in the southern Coyote Mountains. Groups of dots are the four paleomagnetic sampling sites. The sites are labeled as in Figure 2 and in the text, and the amount and sense of paleomagnetically suggested rotation is given for each site. The fault network is from Christensen (1957), the paleomagnetic data is from Mace (1981). Figure courtesy of B. Luyendyk.

Table 1). All the inclinations are fairly close to that expected for this latitude. Most of the sites show a possible, small north translation of about 10 degrees. Some of this translation may reflect the 2 to 3 degrees of actual northward translation on this side of the San Andreas that has occurred since the opening of the Gulf at 6 Ma. The declinations, however, suggest that the basalts at the mouth of Fossil Canyon (FCA, FCB) are unrotated, whereas those 2 km up the canyon (FCN) have been rotated 110 degrees clockwise, and those 2 km east of the mouth of Fossil Canyon, in Butaca Canyon (BC), have been rotated about 60 degrees clockwise. Note here that two of the six BC flows are normally magnetized, four are reversely magnetized, and that the average directions of the two sets are almost antipodal—a good test of the stability of the magnetization (Figures 2 and 3, and Table 1). The flows of Painted Gorge (PG), 2 km northeast of site BC, yield puzzling results. The paleomagnetic directions in two flows near the canyon mouth suggest a 43-degree clockwise rotation, whereas a flow 400 m farther up the canyon to the northwest (PGA) and one (PGLU) 200 m northeast of PGC and PGD appear to have been rotated 130 degrees counterclockwise!

Mortero Canyon, Volcanic Hills

A paleomagnetic study of five basalt flows in the Alverson Volcanics was made in the Volcanic Hills, about 15 km west of the Elsinore Fault and the Coyote Mountains (Figure 1; Hickle, 1981). The paleomagnetic directions show no significant translation and a possible, small clockwise rotation of about 15 degrees (Figures 2 and 3, and Table 1).

Jacumba

Ed Jansen (1978) made the first paleomagnetic study in southernmost California and sampled four flows of the Alverson Volcanics just north and east of the small community of Jacumba (Figure 1, Table 1). This area is about 15 km due south of the Volcanic Hills and about 20 km south-southwest of the Coyote Mountains. One of the flows rests on whitish sandstones of the Split Mountain Group, the other three were apparently deposited directly on the granitic basement of the La Posta Pluton. The paleomagnetic directions are almost identical to those found in the Volcanic Hills basalts.

Vallecito-Fish Creek Basin

In the only study using sedimentary rocks in the area of concern, Johnson et al. (1983) measured the paleomagnetic directions at 107 places over a 5 km-thick section of the Imperial and Palm Spring Groups in the deep basin immediately to the north of the Coyote Mountains (Figure 1, Table 1). These are the youngest rocks used in this paper and were deposited between about 1 and 4 Ma, and their paleomagnetic directions are somewhat scattered. The study average direction shows a small northward translation and a 35-degree clockwise rotation of a large area in the basin. Most, if not all, of this rotation occurred after 1 Ma.

The La Posta Pluton

This paleomagnetic study by Symons et al. (2003) of the 94 Ma La Posta Pluton near Jacumba used a modern cryogenic magnetometer to tease paleomagnetic directions from rocks that failed to yield stable results in an earlier (Erskine, 1982) study reported next (Table 1). It is the only, apparently successful, paleomagnetic study in the eastern half of the Peninsular Ranges Batholith, the plutonic rocks east of Gastil's (1990) magnetite-ilmenite and suture line. Although this eastern part of the batholith supposedly contains nonmagnetic ilmenite ($FeTiO_3$) and no magnetite (Fe_3O_4), their paleomagnetic directions are probably due to a small amount of tiny magnetite crystals included in silicate grains. Twenty sites along a 45 km stretch of I-8 yielded an average paleomagnetic direction which is so close to the expected Cretaceous geomagnetic field direction for this area that they concluded that this part, at least, of the eastern PRB has not been translated or rotated.

The Western Peninsular Ranges Batholith

Erskine (1982) measured the paleomagnetic directions at 29 sites in the Peninsular Ranges Batholth, ranging from Mission Gorge, San Diego, and east to the La Posta Pluton, and from the US-Mexico border north to Escondido (Table 1). The plutons are mostly early Cretaceous and range in age from about 125 to 100 Ma. The paleomagnetic directions of the 12 sites that had stable magnetizations cluster within a cone whose radius is about 10 degrees. None of the samples from the eastern half of the PRB had stable magnetizations, and their magnetizations were so weak that he was led to determine the magnetic mineralogy. The western samples contained magnetite and the eastern samples had essentially nonmagnetic ilmenite as their main iron-titanium oxide. He was thus the first person to define a magnetite-ilmenite line running northwest-southeast along the center of the PRB (Erskine and Marshall, 1980). Unlike the La Posta Pluton, the average paleomagnetic direction from the western half of the PRB lies far from that expected in this region in the early Cretaceous and he concluded that this half had been translated 11 degrees north and rotated about 40 degrees clockwise. As in the studies above the 11 degrees of northward translation may be mostly noise, but the 40-degree clockwise rotation *is* large enough to probably be real.

SUMMARY AND CONCLUSIONS

Even though the 2006 SDAG field trip concentrates on the Coyote Mountains, paleomagnetic data taken farther west even than the 2005 SDAG field trip have been presented in order to show the plate tectonic context, the "Big Picture," that the relatively small Coyote range fits into. Let's begin with a discussion of both the Vallecitos-Fish Creek basin and Coyote Mountains taken together. Both areas are on the western edge of the Salton Trough, a region that has been undergoing crustal extension/rifting for at least the past 6 million years. The Moho depth here is only about 25 km, much less than under the PRB just to the west (Wallace, 1990). The crust here probably consists of down-dropped blocks of the eastern PRB (the La Posta Pluton?) sitting on a gabbroic lower crust formed by mafic magmas coming up from the mantle during rifting. The Vallecito-Fish Creek area subsided to form a basin that accumulated up to 5 km of marine and fresh water sediments (Johnson et al., 1983). The Coyote Mountains which still have some of the same, older sedimentary rocks left on them was probably part of this basin. When right-lateral motion on the Elsinore fault began about 1 Ma, the shear folded, uplifted, and rotated both crustal blocks. Since the mountains are adjacent to a transpressional bend in the Elsinore fault, this block was uplifted more than the basin block to the north, bringing the more erosion-resistant granitic basement up to the surface. The higher compressional stresses on the mountain block possibly caused it to be more highly faulted and broken into small blocks that rotated both clockwise and counterclockwise.

The lack of significant rotations at the Jacumba and Volcanic Hills sites may be due to their being farther from the Elsinore fault zone and also sit on thicker granitic crust. Since the Volcanic Hills are down on the desert floor, about 1000 m lower in elevation than Jacumba at the crest of the PRB, they are probably separated from the Jacumba site up by one or more of the lystric normal or detachment faults that formed the Salton Trough.

Discovering that the PRB actually is apparently two batholiths is one of the most interesting and geologically significant parts of this summary. After the existence and location of the magnetite-ilmenite line had been worked out, mostly by Gordon Gastil (1990), various studies began to show that this line also marked abrupt changes in many other rock properties, such as isotopic ratios, rare earth element patterns, kinds of metamorphic inclusions, and possibly marked a fundamental suture between two separate batholiths (Symons et al., 2003). Since batholiths are usually formed over subduction zones, eastern San Diego and western Imperial Counties have experienced two episodes of subduction, one about 90 Ma and an earlier one about 120 Ma. The western plutons are intruded into andesites and geophysical studies suggest that the lower crust of the western batholith is more mafic than that of the eastern batholith and consists of island arc or oceanic basaltic and gabbroic rocks (Wallace, 1990). The plutons of the eastern batholith apparently intruded first the Proterozoic granitic and metamorphic rocks that formed the continental slope and shelf and then the mostly shallow water clastic and calcareous sediments resting on them, along what would have been the coast of western Arizona or Sonora in the late Cretaceous.

Trying to tie all this together, the western half of the PRB seems to have formed as an island arc on the Farallon Plate (or on one of the others plates in the northeast Pacific in the early Cretaceous) somewhere off the coast of western North America. In late Cretaceous time subduction of the Farallon Plate beneath the west coast of North America produced an Andean-type magmatic and volcanic arc. Ninety million years of erosion have since removed the upper 10 to 15 km, including all the volcanoes and upper parts of the plutons. The end of this arc magmatism could have been caused by the arrival of the western half of the batholith arc at the subduction zone or by a change in the dip angle of the subducting slab. The time at which the island arc arrived at the subduction zone and became sutured to the eastern batholith is not known. If the southern end of the arc collided with the eastern batholith first, the northern end should have swung to the east before colliding with the eastern batholith, thereby causing the 40 degrees of clockwise rotation found in the paleomagnetic study of the western batholith.

ACKNOWLEDGMENTS

I (MM) want to acknowledge my former senior thesis and grad students for cheerfully generating much of the paleomagnetic data presented in this paper and for that reason well deserve to be co-authors of this paper. I feel constrained to admit that this paper would never have been written by me in my supposed golden and carefree years without the gentle (?) and interminable pushing of one David Bloom!

REFERENCES CITED

Atwater, T., 1970, Implications of plate tectonics for the Cenozoic tectonic evolution of western North America: GSA Bulletin, v. 81, pp. 3518-3536.

Atwater, T., 1989, Plate tectonic history of the northeast Pacific and western North America, *in* Winterer, E., Hussong, D., and Decker, R., eds., The Eastern Pacific Ocean and Hawaii: Boulder, Colorado, GSA, The Geology of North America, v. *N*.

Carter, J., Luyendyk, B., and Terres, R., 1987, Neogene clockwise tectonic rotation of the eastern Transverse Ranges, California, suggested by paleomagnetic vectors: GSA Bulletin, v. 98, pp. 199-206.

Christensen, A., 1957, Part of the geology of the Coyote Mountain area, Imperial county, California: Master's Thesis, University of California, Los Angeles.

Erskine, B., and Marshall, M., 1980, Magnetite- and ilmenite-series sub-belts in the northern Peninsular Ranges batholith, southern California: Their association with magnetic stability and implications on source regions: Eos (Transactions, American Geophysical Union) v. 62, pp. 849.

Erskine, B., 1982, A paleomagnetic, rock magnetic, and magnetic mineralogic investigation of the northern Peninsular Ranges batholith, southern California, Master's Thesis, San Diego State University.

Gastil, G., Diamond, J., Knaack, C., Walawender, M., Marshall, M., Boyles-Reaber, C., Chadwick, B., and Erskine, B., 1990, The problem of the magnetite/ilmenite boundary in southern and Baja California, *in* Anderson, J. L., ed., The nature and origin of Cordilleran magmatism: Boulder, Colorado, Geological Society of America Memoir 174.

Hawkins, J., 1970, Petrology and possible tectonic significance of late Cenozoic volcanic rocks, southern California and Baja California: GSA Bulletin, v. 81, pp. 3323-3338.

Hickle, M., 1981, A paleomagnetic study of the Volcanic Hills, western Salton Trough, California: unfinished master's thesis, San Diego State University.

Hornafius, S., Luyendyk, B., Terres, R., and Kamerling, M., 1986, Timing and extent of Neogene tectonic rotation in the western Transverse Ranges, California: GSA Bulletin, v. 97 pp. 1476-1487.

Jansen, E., 1978, A paleomagnetic investigation of the Jacumba basalts: Senior Thesis, San Diego State University.

Johnson, N., Opdyke, N., Woodard, G., Zeitler, P., and Lindsay, E., 1983, Rates of late Cenozoic tectonism in the Vallecito-Fish Creek basin, western Imperial Valley, California: Geology, v. 11, pp. 664-667.

Mace, N., 1981, A paleomagnetic study of the Miocene Alverson Volcanics of the Coyote Mountains, western Salton Trough, California: Master's Thesis, San Diego State University.

Ruissard, C., 1979, The stratigraphy of the Alverson Formation, Salton Trough, California: Master's Thesis, San Diego State University.

Symons, D., Walawender, M., Smith, T., Molnar, S., Harris, M., and Blackburn, W., 2003, Paleomagnetism and geobarometry of the La Posta Pluton, California, *in* Johnson, S., Paterson, S., Fletcher, J., Girty, G., Kimbrough, D., and Martin-Barajas, A., eds., Tectonic evolution of northwestern Mexico and the southwestern USA: Boulder, Colorado, GSA Special Paper 374, pp. 135-155.

Todd, V., Erskine, B., and Morton, D., 1988, Metamorphic and tectonic evolution of the northern Peninsular Ranges batholith, southern California, *in* Ernst, G., ed., Metamorphic and crustal evolution of the western United States (Rubey Volume VII): Englewood Cliffs, New Jersey, Prentice-Hall, pp. 894-937.

Todd, V., Shaw, S., and Hammarstrom, J., 3003, Cretaceous plutons of the Peninsular Ranges batholith, San Diego and westernmost Imperial Counties, California: Intrusion across a Late Jurassic continental margin, *in* Johnson, S., Paterson, S., Fletcher, J., Girty, G., Kimbrough, D., and Martin-Barajas, A., eds., Tectonic evolution of northwestern Mexico and the southwestern USA: Boulder, Colorado, GSA Special Paper 374, pp. 185-235.

Wallace, R., 1990, The San Andreas Fault System, California: USGS Professional Paper 1515.

Walawender, M., Girty, G., Lombardi, M., Kimbrough, D., Girty, M., and Anderson, C., 1991, A synthesis of recent work in the Peninsular Ranges batholith, *in* Walawender, M., and Hanan, B., eds, Geologic excursions in southern California and Mexico: Boulder, Colorado, GSA Guidebook for the 1991 Annual Meeting, San Diego, California.

Main Stem: Geology of the San Diego & Arizona Eastern Railway, San Diego and Imperial Counties, California, U.S.A. and Baja California, Mexico

George C. Copenhaver, Jr., C.E.G., Project Engineering Geologist
Geocon, Incorporated, San Diego, California
copenhaver@geoconinc.com

ABSTRACT

In 2003 and 2004, a reconnaissance study of geologic, historic, geographic and cultural settings was conducted across the Peninsular Range following the 139-mile-long alignment of the San Diego & Arizona Eastern Railway. The study extended southward from San Diego, California into northern Baja California, Mexico, re-entered the United States near the crest of the Peninsular Range mountains, then descended eastward through the Carrizo Gorge into the Salton Basin of the Colorado Desert.

Mapping of geology along the railway alignment reveals a cross-section of Paleozoic through Early Cretaceous metamorphic and igneous rocks intruded by Jurassic through Late Cretaceous plutons of the Peninsular Range Batholith. Later Tertiary through Quaternary rocks overlap the edges of the central Peninsular massif.

More detailed studies, especially within metasedimentary rocks of the Carrizo Gorge segment, allowed some preliminary assessments of structural geology and potential geotechnical constraints along the reactivated railway. Faults, landslides, rockfalls and potentially adverse slope conditions in weak rocks along the railway and within tunnel openings are approximately delineated.

Particular attention is paid to Tunnel No. 7 for future reopening and landslide buttressing, Tunnel No. 8 groundwater seepage for drainage and grouting, the unique wooden trestle at Goat Canyon for landslide stabilization and trestle scour-protection, and Tunnel No. 16 for east portal slope-stabilization.

A review of ongoing operations of the railway and recommendations for future improvements indicates fulfillment of a 150 year-old dream of San Diegans for an eastern "main stem" railway that connects this important corner of the United States to a multinational rail network.

INTRODUCTION

The primary purpose of this paper is to present the geology of the San Diego & Arizona Eastern (SD&AE) Railway alignment which was reopened in 2004. Although not intended to be an exhaustive research paper, the text does provide a summary of readily accessible published references regarding: geology, geography, botany, indigenous cultures, and highlights of the U.S. Westward Expansion.

Sprinkled throughout the text are personal experiences, recollections, and interesting and different points of view, as well as heretofore little-known oral history and anecdotes that bring life to the story of this historic railroad.

In short, this paper provides a benchmark for future work to broaden technical knowledge of this poorly explored corridor through the Peninsular Ranges of Southern California and Northern Baja California.

After more than twenty years of silent rails, it is now possible for full-size trains, with both freight and passenger capability, to take advantage of the southernmost rail route in the United States of America, a route that connects to a nationwide and international transportation network. Although the San Diego & Arizona Eastern Railway has historically operated primarily as a freight carrier, it may now be destined to afford passengers the delight and enjoyment of taking this unique and picturesque journey.

Crossing deep canyons, passing through dark tunnels and over numerous side-hill trestles, the railroad climbs east from the Pacific Ocean to the crest of the Peninsular Ranges Mountains, and finally back down to sea-level and below, in the Colorado Desert. To make this journey, the railroad enters Mexico at Tijuana, Baja California, and returns to the United States in mountainous terrain about half way between Tecate and Campo.

While currently operating a limited, but growing freight business, Carrizo Gorge Railway, Inc. (CZRY) has not excluded the possibility of expanding their operations to passenger and possibly commuter service. In the author's opinion, population growth in coastal Southern California, as well as in nearby mountain and desert communities will make passenger service inevitable.

Geologic Reconaissance Sheet

STATION	ELEV.	DISTANCE FROM SAN DIEGO	STATION	ELEV.	DISTANCE
San Diego	00	00 Miles	Matanuco	350	27.0 Miles
National City	0 0	4.5 •	Eduardo	570	31.8 •
Chula Vista	2 8	8.6 •	Redondo	766	36.1 •
Palm City	3 5	10.6 •	Loma	1000	39.7 •
San Ysidro	9 0	14.5 •	Rita	1325	44.4 •
Tia Juana	9 5	15.1 •	La Puerta	1575	47.9 •
Tijuana	100	15.7 •	Tecate	1685	52.3 •
Arguello	145	19.7 •	Lindero	2139	59.1 •
Garcia	220	23.5 •	Int'l Boundary	2160	59.4 •

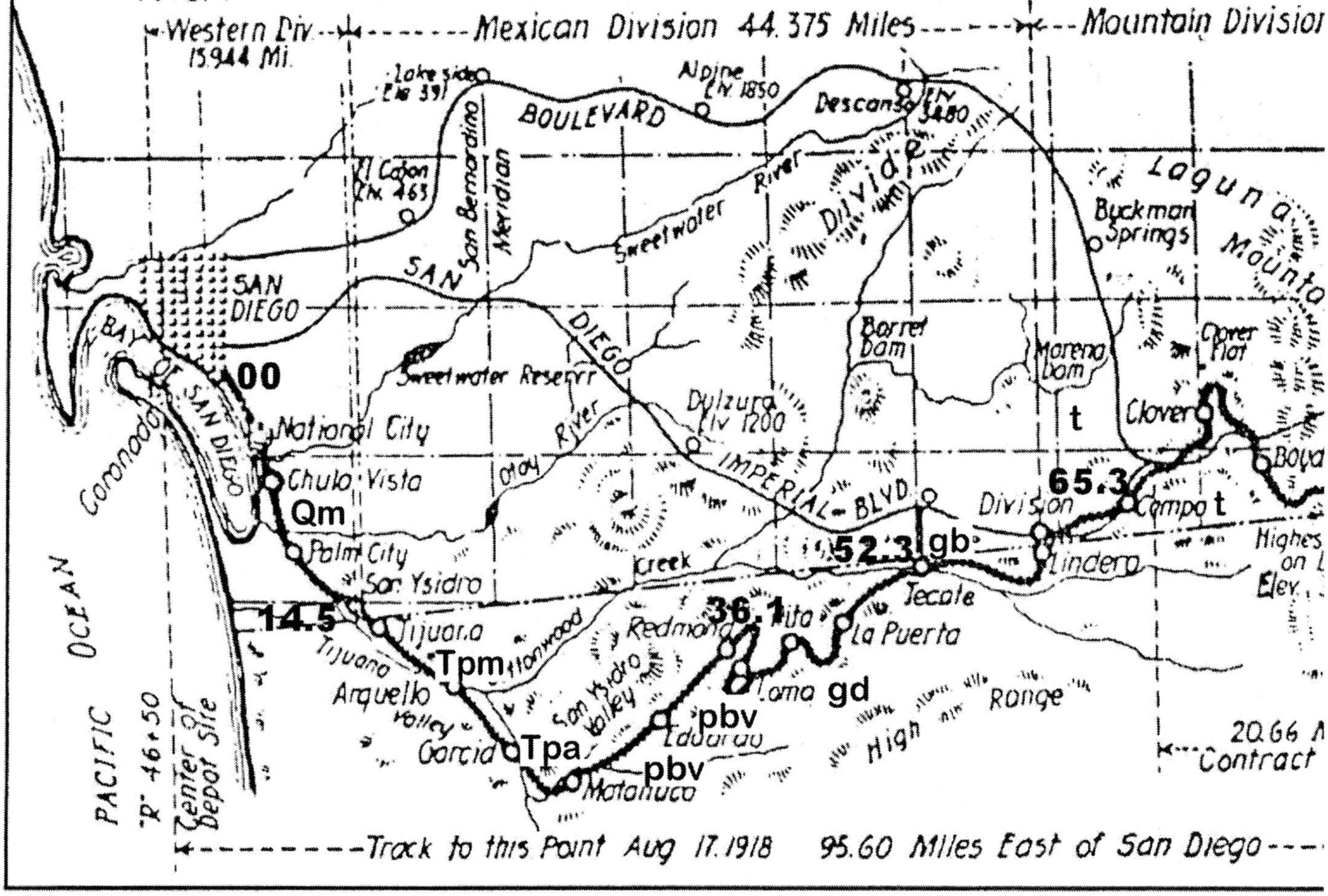

Figure 1. Geologic reconnaisance sheet, showing route through Carrizo Gorge (dotted right-of-way) subsequently completed in November 1919.

Note stations later renamed or abandoned:

Boyd became Pueblo
Carlos abandoned with Sugarloaf fill pit
Yuha and Marea replaced by Plaster City

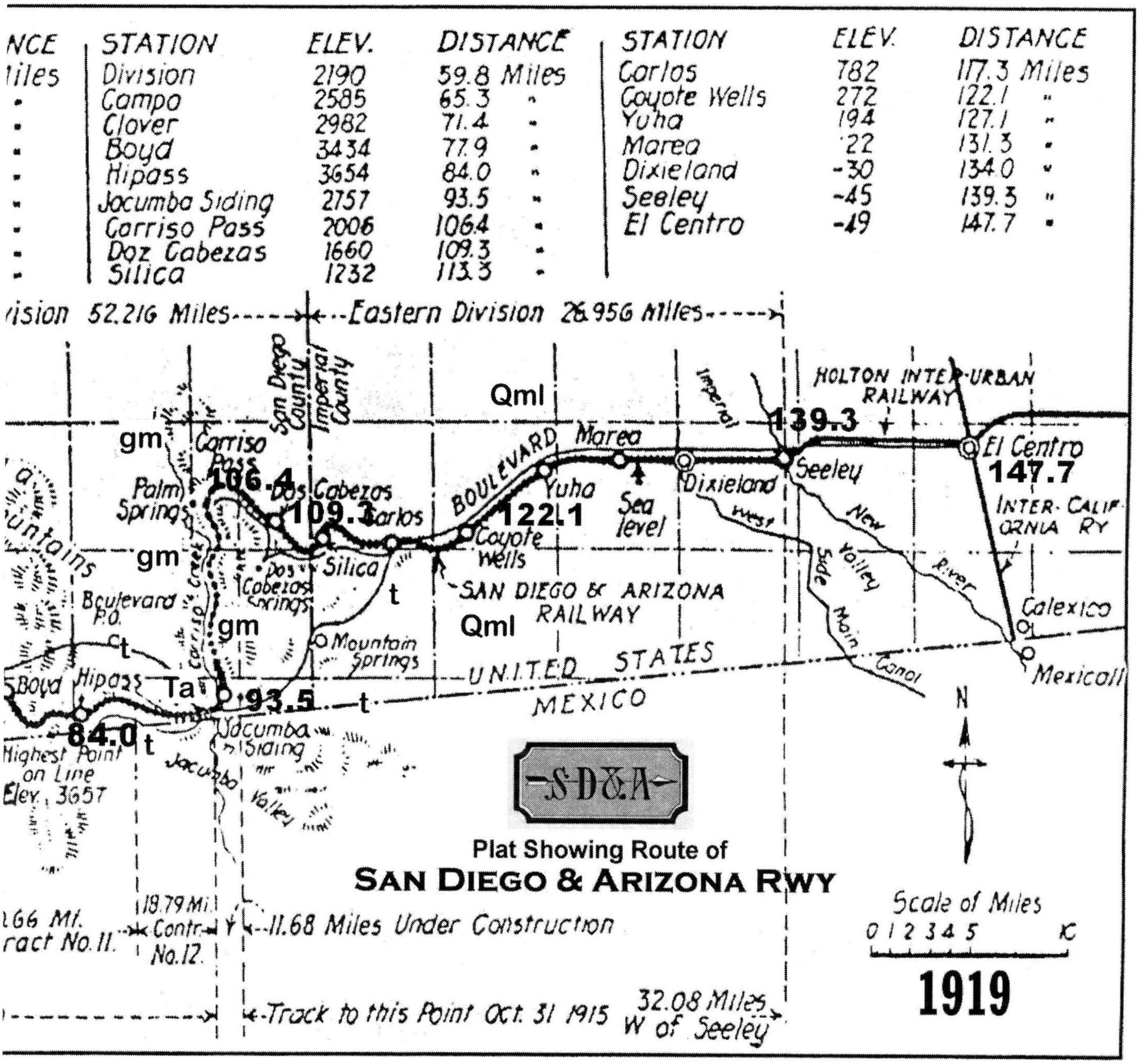

Base Map Scan courtesy Border Rail Institute

Geologic Explanation

(From Geol. Soc. America Mem. 140 and Calif. Geol. Survey, San Diego Sheet)

Qm	marine terraces
Qml	marine and lacustrine
Tpm	Pliocene marine
Tpa	Pliocene andesitic volcanics
gd-gb	older batholithic granitoids and gabbro
pbv	pre-batholith volcanics
gm	pre-batholith meta-sediments and pegmatite
t	La Posta tonalite pluton

The Carrizo Gorge is Open!

Welcome news for railroad buffs and historians, alike, "THE CARRIZO GORGE IS OPEN!" — was the news received by the author in a short electronic message, later confirmed by a telephone call to Carrizo Gorge Railway, Inc. On February 19, 2004, the first train since 1983 was taken down (railroad east) through the formerly "abandoned" Carrizo Gorge Segment of the San Diego & Arizona Eastern Railway. In addition, a test train, provided by Union Pacific Railroad, was then taken up (railroad west) through the Carrizo Gorge; it consisted of three locomotives, one tank car, one flat car and two passenger cars. This was the first train in over 20 years to pass through the scenic Gorge and across the legendary wooden Goat Canyon trestle (Personal communication, Charles Coval, Parsons and CZRY, 2004).

Although not present to witness the February 19, 2004, history-making event, the author was there in January, 2004, just prior to opening the last tunnel to be cleared of thousands of tons of rock and debris. The feelings experienced by the author, while climbing through the first small opening linking the rails at Tunnel No. 16, were likely akin to what railway pioneers must have felt while witnessing the linking of the first transcontinental rails at Promontory Summit, Utah, in 1869! As a result of this experience, the author became avidly interested in personally mapping the geology along the San Diego & Arizona Eastern Railway alignment, the results of which are summarized herein.

Housekeeping

For discussion purposes, the railroad flows from west to east, that is, from it beginnings in San Diego on the Pacific Coast, to Seeley in the desert. Therefore, this guidebook is written with the assumption that rail passengers will be traveling in an easterly direction, from San Diego toward Seeley — and perhaps beyond.

As such, in the text that follows, regardless of the rail's compass directions, which twist and turn, east will refer to traveling eastward along the rails and west will refer to traveling westward along the rails. For example, the west portal of each tunnel is the portal entered first when traveling eastward from San Diego. The east portal is on the Seeley side of the tunnel.

Geographic directions, for the sake of clarity, will generally be given in parentheses. For example,...when the train leaves the east (northerly) portal of Tunnel No. 4, it has crossed back into the United States from Mexico...

References to the right and left sides of the train assume that rail passengers are traveling from San Diego to Seeley. In the case of photographic captions, the camera view will be referenced to true geographic compass directions. Elevations are given relative to Mean Sea Level, abbreviated, MSL.

Helpful Hint: Bring a couple of County and/or State road maps (including Baja California) along for the train ride. These will be useful in determining where the train is along the alignment relative to nearby geographic points of interest.

Official Surveys

During James Buchanan's Presidency (1857-61), his Secretary of War, Jefferson Davis, received funding to do "official" surveys for the first transcontinental railroad route.

By the late 1850s, Jefferson Davis and his "army" of scientists and United States Topographical Engineers, under the command of Major W.H. Emory and Lieutenant R.S. Williamson, completed a 12-volume report detailing the feasibility of five separate routes for the first transcontinental railroad to the West Coast.

As it turned out, Jefferson Davis and his surveyors, along with most Southerners who had settled/invested in the southwestern border regions, including San Diego, favored the southernmost route. Also Andrew B. Gray's choice for a transcontinental railroad route, this southern border alignment would connect the port of New Orleans, on the Gulf Coast, with the Port of San Diego, on the Pacific Coast (Hanft, R.N., 1988; Brands, R.W., 2002; Bain, 2000).

After the Mexican War (1846-48), via the Gadsden Purchase in 1853, northern territories in Mexico (particularly the Mesilla Valley along the Rio Grande in south-central New Mexico) were specifically acquired by the United States Government to provide a convenient corridor for the much anticipated southern transcontinental railway route to the Pacific Coast. It was to be the "Main Stem" of western railways (Gray, 1856).

Unfortunately for the southernmost route, with its terminus in San Diego, Jefferson Davis' surveys, while having elaborate maps and voluminous scientific descriptions, did not contain elevations, profiles, and other engineering data needed to prepare cost estimates for bidding and construction.

THE ROUTE, GENERAL GEOLOGY AND GEOGRAPHY OF MAJOR SEGMENTS

The following summary of geology and other items of interest will cover major segments of the route that were highlighted in an engineering study in 1995. The San

Diego Association of Governments (SANDAG) reviewed the feasibility of reopening the approximately 139 miles of railway (San Diego to Seeley) for regular train service in a study titled San Diego & Arizona Eastern Railway, Element No. 1 — Engineering Study, prepared in March, 1995, by the Barrett Consulting Group. Inspections of all bridges and tunnels between San Ysidro and Plaster City, a distance in excess of 120 miles, are detailed in Chapters 3 and 4 of the SANDAG study.

Almost 150 years after the National Boundary and Railroad Surveys (approximately 1848-56), in what may be a sesquicentennial event, the San Diego Association of Governments (SANDAG), in 2003, approved funding a business plan for reopening the San Diego & Arizona Eastern Railway. During that same year, successful preliminary tunnel repairs, preparatory to reopening the line, were largely completed by the current operator of the line, San Diego-based Carrizo Gorge Railway, Incorporated (CZRY). In February, 2004, CZRY accomplished the feat of reopening the Carrizo Gorge Segment to a full-size test-train; provided by rail pioneer Union Pacific, the locomotive and rail cars passed through all tunnels and over all bridges without incident.

The 15.5-mile San Diego to San Ysidro Segment, not covered in the SANDAG engineering study, will be described next, to provide an overall cross-section of the entire route.

THE SAN DIEGO TO SAN YSIDRO SEGMENT — MILEPOSTS 0.0 to 15.5

The Railway Alignment

Starting from downtown San Diego and building the railway southward toward Baja California, Mexico, the alignment is reminiscent of one of the original "southern route" options considered in the 1850s by the Davis and other surveys; in fact, it closely follows some of the mapped southern reconnaissance routes.

From San Diego, south to the International Border at Tijuana, elevations range from about 10 to 100 feet msl, respectively.

There were, no doubt, records of an old trail that was successfully used by indigenous peoples and early California explorers/settlers, since the route selected through Northern Baja California closely follows what was originally an old mule-and-wagon track.

This trail was well-known by early Mexican Ranchero Grantees, such as the likes of Juan Bandini. It served the Bandini Rancho at Tecate, and was later used by United States Army couriers as early as 1848. Still later, the same path was effectively used up to, and after, the American Civil War; it also provided a route for military mail service (sometimes affectionately known as the "Jackass Mail") between San Diego, Camp Yuma (later known as Fort Yuma), and points east (Pourade, 1966).

Spreckels most likely made the railroad alignment decision he did, because he and his chief engineer, E.J. Kallright, were aware of previous surveys for a proposed route in Northern Baja California by Kallright's former employer, the Southern Pacific Railroad. Years before Spreckels' announcement of the San Diego & Arizona Railway project, Southern Pacific had obtained a concession from the Mexican Government that would have precluded Spreckels' use of the route. But, E.L. Harriman, head of Southern Pacific, decided to secretly joint venture the project with Spreckels, supplying $3,000,000 to begin work (Kirchner, 1988).

Why? Serendipity and pure politics may be why. At that time, San Diegans still resented Southern Pacific for their parent's ("Big Four") blockade of an eastern route from San Diego. And the novel, The Octopus (Norris, 1901), among other so-called muckraker books of the day, demonized the rail monopoly's image along with other "robber barons" of the day.

In stark contrast, however, John D. Spreckels' image was excellent, considering his previous and continuing contributions to San Diego. His family was also well-known for their benevolence in contributing to victims of the catastrophic, 1906 San Francisco Earthquake. Following the secret E.L. Harriman deal with Southern Pacific, he fully reimbursed investors for a previously failed railway project, adding to his already favorable reputation. How clever of Southern Pacific, considering growing public distrust of rail monopolies, to invest in a popular civic leader's plans!

Following the 1906 earthquake and fires, Spreckels moved from San Francisco to San Diego. The stage was now set for 1907, when Spreckels would appear to defy the big railroad monopolies, by starting construction of the San Diego & Arizona Railway (later renamed San Diego & Arizona Eastern Railway).

Geology

Geologically, the San Diego to San Ysidro Segment was a good place to begin railroad construction. The relatively flat coastal plains and low hills south of San Diego are underlain by the nearly horizontally bedded, clastic marine and non-marine Pleistocene Lindavista and Bay Point formations (Kennedy, 1975). These strata, capped by late Quaternary gravelly river terrace deposits required a relatively light excavating effort during grading operations, but provided a firm base upon which to lay rail after ballast was applied. Similar geologic conditions extend southward, well in to Northern Baja California.

Several major stream drainages, such as Chollas Creek, and the Sweetwater, Otay, and Tijuana Rivers, were crossed on combinations of embankment fills and bridges supported by driven piles.

The San Ysidro Landslide

Just east of San Ysidro (formerly Tijuana, USA, and currently a community within the City of San Diego), approximately between Mileposts 14 and 15.5, the railroad traverses the western edge of bluffs that form the toe of a huge landslide complex. These bluffs are underlain by the late Oligocene/early Miocene Otay Formation (Deméré, 1985). Bentonite clay lenses commonly occur within these clastic sedimentary deposits.

Known as the San Ysidro Landslide, historical movement is unknown. Recent drilling and geologic surface mapping suggest that it was probably active thousands of years before present (Geocon, 2003, 2004).

During the Ice Ages, when sea level was about 100 feet lower than it is today, the Tijuana River cut deeply into the toe of the bluffs east of the railroad, exposing "slippery" layers of bentonite clay. Unsupported, the bluffs failed, creating the very large landslide mass that we see today.

After the last Ice Age ended, about 11,000 years before present, glacial ice melted and sea level rose. Base level rose along with sea level. Unable to cut below the now higher base level, the Tijuana River deposited its bed load against the toe of the San Ysidro Landslide. Today, therefore, the slide is effectively buttressed by geologically young river deposits of silt, sand, and gravel.

Faulting and Seismicity

An interesting geologic finding about the San Diego to San Ysidro Segment is that the original railway alignment fortuitously avoided the Rose Canyon fault, the only active fault in the immediate San Diego metropolitan area. John D. Spreckels' groundbreaking ceremony in 1907 was located just south of this fault, which in recent years was found to extend offshore into San Diego Bay (Kennedy and Welday, 1980).

The main San Diego rail terminal, the Santa Fe Depot (currently the AMTRACK Station), is situated adjacent to tracks now controlled by Burlington Northern/Santa Fe (BNSF). It is located approximately two miles north of the 1907 San Diego & Arizona Railway groundbreaking. This two-mile-long section of track crosses the "Downtown Fault Zone," officially designated as an Alquist-Priolo Special Studies Zone.

Prior to any new construction (public or private) in these Special Studies Zones, a fault-location investigation is mandated by the State of California and the City of San Diego to verify the presence or absence of active faults (California Division of Mines and Geology, 1991; City of San Diego Development Services Department, 2003; USGS Point Loma 7.5' topographic quadrangle map). Considering, however, the short length (about two miles) of alignment impacted by known earthquake faults and historical earthquake epicenters, the San Diego & Arizona Eastern Railway appears to be at relatively low risk for ground rupture during major seismic events.

Further Reading

The "Rise and Fall of San Diego" (and of Sea Level with its effects on the San Diego landscape) is well-described in a book on San Diego geology by Dr. Patrick Abbott, Professor of Geology at San Diego State University (Abbott, 1999). A former undergraduate classmate of the author, Dr. Abbott has written many technical papers and several books on San Diego geology.

The Alignment in Years to Come

One of several suggestions for future improvements to the urban portion of the San Diego & Arizona Eastern Railway would be to excavate a series of deep trenches for the alignment beneath surface streets, thus improving railroad safety and mitigating noise. Engineering testing and geological inspection of these trenches will be necessary to insure cut slope stability.

Geological and paleontological examinations of trench excavations could lead to the discovery of valuable fossil and lithologic (stratotypes) data that would improve our understanding of San Diego's geologic history (Walsh, 2005). The San Diego Natural History Museum has documented a large cluster of fossil occurrences in late-Tertiary and Quaternary formations near the area where the original groundbreaking ceremonies for the railway were held (Personal Communication, Hugh Gardner and Steven Walsh, 2004).

In addition, archeological examinations during construction excavations often document the occurrence, or creation, of human artifacts; these are often in the form of rock chips left behind from toolmaking (known as debitage). Even nondescript items, such as small chips of rock, can potentially add to our knowledge of indigenous people's cultural history.

SAN YSIDRO TO DIVISION, THE "MEXICO SEGMENT" — MILEPOSTS 15.5 to 60.0

The Railroad

Proceeding southeasterly into Baja California, the alignment in Mexico recently came under the control of a privatized company, Ferrocarriles Peninsulares del Noroeste (FPN); this route, however, is still known as Ferrocarril Tijuana y Tecate (T&T). The privatized company, FPN, cooperates with Carrizo Gorge Railway, Inc., based in San Diego, California, and is involved in the overall effort to reopen the San Diego & Arizona Eastern Railway to regular freight and passenger service.

From west to east, elevations along this 44.5-mile-long segment vary from approximately 100 to 2,200 feet, respectively.

Geology

A north-northwest trending linear fault, approximately 45 to perhaps 60 miles long, has been mapped by Gastil and others (1971), principally through pre-batholithic basement rock in northwestern Baja California. Known as the Vallecitos fault (Falla de Vallecitos), it projects approximately toward the similarly northwest trending Tijuana River Valley. Although there has been speculation over the years about a possible connection between these two linear trends, the jury is still out, waiting on defendable evidence. Little historic seismic activity has been associated with this fault; as well, it does not appear on "official" Mexican maps.

Hillsides of Tijuana abound with active and inactive landslides. Unlike the hills east of San Ysidro, bluff tops, mesas and side-hill slopes along the Tijuana River and its tributaries are intensely developed with homes and businesses; these slopes are underlain by shallow layers of weak bentonitic clay that are more fractured and disturbed by faulting than their counterparts north of the Border.

Fortunately, however, over the years the Mexican portion of the Spreckels railroad (Tijuana y Tecate) through the Tijuana River Valley has shown little, if any, evidence of distress resulting from landslide movement and/or settlement (Personal communication with geotechnical coworker, R. Garcia, 2003).

As the alignment turns to a northeasterly direction, ascending into canyons above the main drainage valleys, and past the towns of La Mesa (The Table) and La Presa (The Dam), the rails pass through Tunnel No. 1 (at Kilometer 16.7) and Tunnel No. 2 (at Kilometer 17.36). Both of these short tunnels are bored through fractured, but strong, prebatholithic metavolcanic rock.

Tunnel No. 3, several kilometers further on, was cut through poorly consolidated, fluvial Eocene sandstone and conglomerate (Gastil and others, 1971). Since the 1930s, Tunnel No. 3 has been plagued by fires and cave-ins. Finally, in recent years, a small American contractor, that would eventually become the Carrizo Gorge Railway, Inc., solved the problem. The roof of the tunnel was removed, thus creating a deep open trench in its place.

As the rails continue to climb, higher still, they enter Valle Redondo (Round Valley). At this point, near Kilometer 33.9 (Milepost 36.6), the route leaves behind a short stretch of metavolcanic rock, and crosses over onto granitic rock. Here, to gain altitude, the alignment is built as two horseshoe, or "balloon curves." That is, the rails turn 180° back on themselves for a short distance and then turn 180° again to continue their northeasterly journey toward Tecate.

While twice doubling back to gain altitude, the alignment passes through hillsides underlain by granodiorite and tonalite, intrusive members of the Cretaceous Peninsular Ranges Batholith. Throughout this section, substantial cuts and fills in post-batholithic volcanic rock as well as in batholithic intrusive rock (Gastil and others, 1971) were required for the first time as construction proceeded from west to east (Hanft, 1984).

As the route continues in an easterly direction, past the brewery town of Tecate, Kilometer 60.0 (Milepost 37), passengers will likely several see new construction projects, including a Toyota truck factory.

Gabbroic and granodioritic intrusive rocks of the Peninsular Ranges Batholith, perhaps Jurassic to early Cretaceous in age (Gastil and others, 1971), are exposed southerly and easterly from Tecate. Highly fractured, less resistant to surficial weathering, and millions of years older than nearby younger siliceous granitic rocks, these mafic rocks often develop a deep clayey grus, or regolith of clayey soil.

These deeply weathered zones are mined for reddish clay that has been used for years in local tile and brick making operations. Brick-making continues to this day, within view of the railway alignment. "Beehive kilns," or "furnaces," are used to "fire" sun-dried clay. Bricks-to-be are stacked along the walls of the kiln; after firing the kiln and its contents are dismantled and everything is marketed.

Beginning a few kilometers southwesterly from Tecate, and continuing on easterly to where the rails cross back into the United States inside Tunnel No. 4, the railroad alignment begins to cross, back and forth, onto the western edges of the central core of the Peninsular Ranges

Batholith. This enormous body of granitic rock forms the crest of the Peninsular Ranges. Elongate in a north-northwest to south-southeast direction, this thumb of hard dense rock, mostly of tonalitic composition, extends for many miles from Southern California well into Baja California. (Strand, 1962; Gastil and others, 1971; Walawender, 2000.)

Geologic Maps

Recent geological research and mapping in Northern Baja California, by San Diego State University and Universities in Mexico, have resulted in more systematic and coordinated geologic studies south of the International Border than to the north, in San Diego and Imperial counties (Harris, Mary E., 1986). Over half of the Baja California Peninsula was geologically mapped at a scale of 1:250,000 by students and staff from San Diego State University and *Universidad Autonoma De Baja California* (Gastil, and others, 1971).

A continuous series of official 1:50,000-scale, colored geologic maps of Northern Baja California, was produced by the Mexican government, Secretaria de la Presidencia and Secretaria de Programacion y Presupuesto, in the early 1970s.

In 2004, the U.S. Geological Survey released the 1:100,000-scale, Preliminary Geologic Map of the El Cajon 30'x60' Quadrangle, Southern California, covering areas of San Diego and Imperial counties that had not been previously officially mapped (Todd, 2004).

For purposes of this summary paper, the general geology of the San Diego & Arizona Eastern Railway alignment through Baja California (Tijuana y Tecate) is excerpted from Gastil's (1971) work. The four 1:50,000 official Mexican Map Sheets used herein are: Tijuana (Map No. I 11C69, out of print), Murua (Map No. I 11D61), La Presa (Map No. I 11D71), and Tecate (Map No. I 11D62). Like United States Geological Survey quadrangle maps, these maps represent the "official" geological survey of Mexico.

DIVISION TO JACUMBA AND DUBBER SPUR, THE "CAMPO SEGMENT" — MILEPOSTS 60.0 TO 96.0

From west to east, from Division, up to Hipass, and then down to Dubber Spur, elevations across the Campo Segment vary from about 2,200 to 3,660 to 2,675 feet msl, respectively. The high point along the San Diego & Arizona Eastern Railroad alignment occurs at Hipass, Milepost 84.5.

Geology

Soon, the rails turn to a northerly direction and re-enter the United States inside Tunnel No. 4, at Milepost 60.0, near the old Division siding. This 598-foot-long tunnel cuts through massive granitic rock, named the Japatul Valley Tonalite (Todd, 2004), and requires no shoring except near the more easily eroded portals at each end.

The presence of tonalitic rock in Tunnel No. 4, in contrast to gabbroic and dioritic plutons back in the Tecate area, is a last reminder that the rails are passing through the transition from older pre-batholithic rocks and earliest batholithic intrusions, to younger batholithic intrusions in the core of the Peninsular Ranges. After exiting Tunnel No. 4, and crossing the bridge over Campo Creek, located near the old Division siding, the route turns to a more easterly direction. Here, the alignment parallels Campo Creek another 3 to 4 miles before cutting into the massive and concentrically zoned La Posta Tonalite pluton. This distinctive pluton is perhaps the largest single granitic body (or *massif*) in this part of the Peninsular Ranges Batholith (Walawender, 2000; Todd, 2004).

Continuing east and entering the grassy valley around Campo, the San Diego & Arizona Eastern Railway climbs ever upward along the Campo Creek drainage, passing the old wooden Camp Lockett barracks, the Campo Railway Depot (Milepost 65.8), and the Pacific Mill.

The rails then follow Clover Creek drainage, ascending onto granitic hills, shaped by an orthogonal pattern of fractures and joints. Everywhere, the outcrops of rock are composed of tonalite, containing calcium and sodium feldspars that weather more easily than the potassium feldspars of a true granite (Larsen, 1948). Thousands of years of chemical and mechanical weathering, combined with a crosscutting joint pattern, has produced surficial rock that is oftentimes decomposed to depths in excess of 50 feet.

Groundwater

In some low-lying tributary-drainages to Campo Creek, groundwater is found by drilling until hard, impermeable, tonalite is reached. Depending upon the surrounding topography, these surficial accumulations of decomposed tonalite can oftentimes be drilled and produced as unconfined groundwater aquifers. Deeper wells can also be productive when they encounter ancient fault zones with water-saturated fractures.

Gold Processing

A short distance past the Division siding and bridge over Campo Creek, the remains of a semi-modern gold processing facility adjacent to Campo Creek still stand.

Likely last used in the late 1980s, this mill is thought to have been used to process gold ore from small mines when the price soared over $800 per ounce. The mines were probably in the San Ysidro Mountains and/or the Pine Valley areas, located to the west and to the north, respectively.

Fire and Railroad Security

In the Fall of 2004, a wildfire started in Mexico and burned northward, crossing the International Border over the ridge above Tunnel No. 4 and ignited wooden east-portal shoring- timbers. In addition, immediately east (northerly) of Tunnel No. 4, wooden railroad ties on the steel bridge over Campo Creek burned. Rails and ties on the west (southerly) bridge approach and at the east (northerly) end of Tunnel No. 4 were subsequently repaired and/or replaced.

This is a good example of how over 20 years of inactivity, accompanied by severe drought conditions, can lead to railroad alignment deterioration. To add insult to injury, a nearby high-voltage electric transmission line prevented aerial firefighting tankers from effectively dropping fire retardants.

Firefighting equipment is carried on all rolling stock operated by Carrizo Gorge Railway, Inc. Often, firefighting rail-patrols follow each train that passes through fire-hazard areas. For additional fire safety, the company maintains a zone of cleared weeds and brush along the entire alignment.

Security patrols and fire prevention practices along the alignment are being increased (personal communication, Carrizo Gorge Railway, Inc.). Regarding security in the Border zone, a story in the San Diego Union-Tribune cites a statement by Roger Maier of the United States Customs and Border Protection Service, that it "...has the staffing necessary to address any sort of freight traffic at the crossing near Campo." (December 31, 2004).

Campo

Campo is the site of the United States Cavalry Camp Lockett. This was the last cavalry post built in the United States. Substantial remnants of Camp buildings, some still in use, are present on the hills immediately south of the Campo Depot (Milepost 65.8). During World War II, Camp Lockett served both as a convalescent hospital and as a detention facility for German and Italian Prisoners of War. Current efforts funded by the County of San Diego are underway to preserve the old camp by filing a nomination for Historic Landscape and a Historic District. Other goals, in part, are to create a Management Plan as a guide for its preservation and the development of a County park (Personal communication from Richard Borstadt, 2005).

The Author's Personal Stories and Memories of Campo

For many years after World War II, the Camp barracks, auditorium and gymnasium served as school facilities for the Mountain Empire High School District. I graduated with the MEHS class of 1958. At quiet moments, in those days, Morse Code could be heard clear across the valley, clicking away through open windows at the Campo Depot on the SD&AE Railway.

While a student at Mountain Empire High School, I cultivated an interest in science. Delving too deeply into chemistry, I nearly destroyed the school's chem-lab, more than once, with explosive experiments! In an unusual chem-lab experiment, with a familiar flavor, I introduced my wide-eyed classmates to chicken-fried snake. A huge diamondback rattlesnake's life was shortened near some boulders in a popular school lunch area — known as Senior Park. Shortly thereafter, I prepared a special chem-lab lunch with my mother's deep-fryer. Advertised as a special recipe for chicken, all who partook were more than pleased with the tasty feast, but wondered why only "neck sections" were served.

During long hot summers at Campo, a few hardy students worked part-time as "tunnel guards" for the San Diego & Arizona Eastern Railway along the Carrizo Gorge alignment. They watched for rockfalls, fires started by sparks from passing trains, and other disturbances of an unusual or mystical nature; it is interesting what a day in summer heat can do to one's brain cells. Some brought back tales of frying eggs on the railroad tracks, as well as hair-raising tales of their summer "ordeal" — survival in the Carrizo Gorge.

Ralph Goff, a high school classmate of mine, is a past Tribal Chairman of the Campo band of Kumeyaay (pronounced Coo-me-eye) Indians, and is a current Vice Chairman. I remember Ralph as a quiet, strong (actually very strong) virtuoso of the Mountain Empire High School wrestling team! As a wrestling teammate of his, my recollection is that virtually no one could pin this superb wrestler. It is my firm belief that if the Ralph Goff, of the Mountain Empire High School wrestling team of 1958, had qualified for the 2004 International Olympics, he could have out-wrestled the equivalent of Rulon Gardner! Needless to say, I didn't have a ghost of a chance in a bout with Ralph.

Kumeyaay Indians

Since World War II, the Campo Band of Kumeyaay Indians have made numerous economic advances with tribal lands that are beneficial to them as well as to the community in general. Chairman Goff's ability with quiet diplomacy over the years has been a steady force, leading the Campo Band toward sovereignty. Some of his activity is described in a book entitled, *The Campo Indian Landfill War* (McGovern, 1995).

The introduction of gaming in 2001, as well as planned developments for tribal lands, continues to improve business and cultural activities in the Tecate Divide-Interstate Freeway 8 corridor.

Recently, the Campo Tribe received the honor of being one of only eight tribes (chosen from over 900 Western Hemisphere entries) to open exhibits at the first National Museum of the American Indian at the National Mall in Washington, D.C. The Campo exhibit is part of the "Our Lives" gallery that emphasizes examples of Native American Culture in modern times.

Unique in all of the exhibits, Indian Gaming at the Golden Acorn Casino shows how gaming has profoundly and profitably affected Indian lives at home. Youth Programs, Historic Cultural Traditions, Environmental Activities, and Cross-Border Relations with Baja California, Mexico, are also highlighted in the exhibit.

Golden Acorn Casino

As the rails climb eastward, past Cameron Corners and Miller Creek, they cross reservation land belonging to the Campo Band of Kumeyaay Indians. With the exception of the Golden Acorn Casino, located a few miles north near Interstate Freeway 8, these lands are mostly undeveloped.

The rolling terrain, with mixed chaparral cover, consists of small valleys, grasslands, and springs shaded by groves of stout Valley Oak (*Quercus agrifolia*). Yellow acorn seeds from this oak species were a major food source for early members of the Campo Band of Kumeyaay Indians. Hence the name, Golden Acorn Casino, for the gaming facility that provides jobs and increased revenues for the Campo Band, as well as for residents in the surrounding region.

On long hikes, exploring San Diego County mountain backcountry and geology, I have frequently found ancient Kumeyaay camps where solid granitic rock outcrops are dimpled with *morteros* or mortars, shallow holes in the rock surface used for grinding seeds and acorns.

Mining at Campo

Living in the Campo area during my formative years sparked an enduring interest in rocks and minerals — long before I received my formal education in geology. It helped, of course, that Campo was a center for mining feldspar, silica, and small deposits of molybdenum and gold (Weber, 1963).

Feldspar and Silica

By far, the most important mining operation in the Campo area was for feldspar from pegmatite dikes. Used for ceramics and fluxes, the largest single deposit was at the Pacific Mine, located about five miles northwest of Campo (Weber, 1963, Pl. 1, Location 50a, p. 76). Once extracted, feldspar was transported by trucks from the mine to the Pacific Mill for processing.

A small tonnage of quartz (silica) was also produced from the Pacific Mine.

Heading east out of Campo, and a little east of Cameron Corners, the tall old corrugated metal mill building still stands on the north side the San Diego & Arizona Eastern Railway tracks, just before the State Highway 94 grade-crossing at about Milepost 67.

Molybdenum and Gold

Sub-economic deposits of molybdenite (MoS2), a soft metallic gray sulfide of molybdenum, occur in northwest-striking aplite dikes located in, and just north of a cut made for the San Diego & Arizona Eastern Railway alignment in 1914. Located at about Milepost 65, these dikes cut through granitic rock about 2,000 feet west of the old stone Campo store (Weber, 1963, Pl. 1, Locality Nos. 321 and 325; p. 195-198).

A small deposit of gold was discovered less than one-half mile north-northwest of the above-noted molybdenum deposits, possibly along an extension of the same dike system that contains the molybdenite. Gold was reported to have been mined from a "porphyry" deposit in a shaft that subsequently caved-in (Weber, 1963, Pl. 1, Location No 159, p. 138).

The porphyritic texture observed in this gold deposit is similar to the coarse-grained, salt-and pepper texture typical of crystalline granitic rocks, a texture oftentimes associated with the mineral molybdenite. Although further exploration will be required, there is always the possibility that both gold and molybdenum could occur close to each other in these deposits. Molybdenum commonly occurs with gold, and sometimes increases in occurrence along gold-bearing veins that extend into granitic wall-rock (Boyle, 1980).

Ranching and the Campo Gunfight

Ranching has historically been an important part of Campo's economy. In the 1860s and 1870s, the Gaskill brothers pioneered and operated a successful ranching business near the old stone Campo store.

During the American Civil War (1861-65) and the years immediately following, San Diego County experienced prolonged drought as well as financial depression; in San Diego, the population dwindled to less than 200. Compared to hard times in the City of San Diego, the Gaskills were relatively successful with their ranching operation, and so, became an attractive target for bandits.

In the mid-1800s, Campo was a lawless region where rustlers and bandits roamed back and forth across the International Border. In 1875, a notorious gang of bandits targeted the Gaskill Ranch and the old stone Campo store, resulting in the deadliest gunfight in San Diego County history (Pourade, 1964). At least 8 people were killed, including a former territorial governor of Baja California, Don Antonio Sosa. Although the ranchers won the battle, lawlessness only ended after United States Military Patrols were established along the U.S.A./Mexico Border.

Border Problems and Security

Now, over a hundred years after the end of the "wild and wooly" Frontier Era, and American Civil War concerns with rebel insurrectionists, the United States/Mexico Border region is, again, all but an "armed camp." This time, the "arms" race results from concerns over illegal immigration and heightened security issues growing out of the September 11, 2001, terrorist attacks on New York and the Pentagon. Not since World War II, or at any other time since, has this Border region been so heavily patrolled.

Despite immigration problems, the Carrizo Gorge Railway, Inc., along with historical and cultural cross-border organizations such as the International Border Rail Institute, are actively promoting positive educational and benevolent programs. During these particularly difficult times along the Border, these programs are proceeding, hand in hand with the establishment of additional and more effective security measures.

The Zoned Batholith, Campo to Jacumba — Mileposts 66.0 to 92.9

La Posta Pluton

Unique mineralogical changes in granitic rocks underlying the terrain being crossed between about Mileposts 66 and 92.9, will most likely be invisible to the casual train passenger as well as to most geologists. These rocks, known as the La Posta Tonalite, are the largest plutonic intrusive rocks within the Peninsular Ranges Batholith.

This huge body of rock straddles the United States/Mexico Border and is exposed over an area of approximately 1,400 square kilometers (Walawender, and others, 2003). Like a north-to-south elongated bull's-eye, this gigantic tonalite core is concentrically zoned into four bands, varying in width from approximately 4 to 10 miles.

These four distinct concentric mineralogical zones are distinguishable one from the other by the presence of small percentages of mostly dark-colored accessory minerals. From the outermost ring to the center of the "bull's-eye," they are: 1) sphene-hornblende, 2) large-biotite, 3) small-biotite, and 4) muscovite-biotite, respectively. These subtle differences in accessory mineral content, readily apparent with high magnification, can sometimes be distinguished megascopically, that is, with the unaided eye.

From west to east, along the San Diego & Arizona Eastern Railway alignment, the outermost sphene-hornblende zone is first encountered just beyond the Campo Railroad Depot, at approximately Milepost 66. From here, the tracks traverse the sphene-hornblende zone as they climb eastward out of Campo Valley and past Clover Flat. At about Milepost 73, before reaching Miller Creek, the rails cross the transition from the outermost sphene-hornblende zone into the next zone — the large-biotite zone.

Between about Mileposts 73 and 81, the rails traverse the relatively narrow, large-biotite zone. En route to Milepost No. 81, the tracks cross over Miller Creek, as well as over Stony Canyon via the steel trestle over State Highway 94/Campo Creek (Upper Campo Creek Viaduct).

Between about Mileposts 81 and 91, the rails cross over the approximately ten-mile-wide, small-biotite zone, the most extensive zone traversed by the San Diego & Arizona Eastern Railway alignment. Immediately east of the Tecate Divide, between about Mileposts 85 to 86, the small-biotite zone is briefly interrupted by a lens of two-mica leucogranite, locally known as Boulevard Monzogranite.

Hipass, also known as the Tecate Divide, is at Milepost 84.5. At an elevation of 3,660 feet MSL, it is the drainage divide at the crest of the Peninsular Ranges. West of the divide, water flows toward the Pacific Ocean; east of the divide, water flows to the Salton Trough — structural extension of the Gulf of California. The Salton Sea, a shallow landlocked body of water, covers a portion of the Salton Trough.

Between about Mileposts 91 and 92.9, the San Diego & Arizona Eastern Railway alignment passes over a short section of pre-Cretaceous hybrid gneiss, granodiorite, and quartz diorite (Weber, 1963).

At about Milepost No. 92.9, the tracks cross an unnamed, north-northwest trending fault. With an apparent down-to-the-east sense of movement, this fault separates metasediments to the west, from the Table Mountain Formation to the east. For the most part, the Miocene Table Mountain Formation is a collection of volcaniclastic sediments and lava flows. Hot springs in the community of Jacumba are apparently associated with this fault.

Although not exposed along the San Diego & Arizona Eastern Railway alignment, zone four, the innermost muscovite-biotite zone of the La Posta Tonalite, is reported to crop out a short distance northwesterly from the community of Jacumba (Walawender, 2003). Additionally, it is thought that the muscovite-biotite zone may be present at depth, buried under the younger Table Mountain Formation.

The four La Posta Tonalite zones, from outermost to innermost, 1) sphene-hornblende, 2) large-biotite, 3) small-biotite, and 4) muscovite-biotite, are generally thought to reflect melt-crystallization temperature-gradients for the La Posta pluton. Highest temperatures are thought to have occurred in the outer sphene-hornblende zone, while the lowest temperatures occurred in the innermost muscovite-biotite zone. Continuing research into the nature of these concentric zones may lead to explanations for localized occurrences of economic deposits of gem minerals and metals in this region of the Peninsular Ranges.

Physiography, Erosion, and Slope Stability

Backing up a bit, to Milepost 77.27, the railroad alignment crosses over a deep canyon on the Stony Canyon Bridge. Unique at the time, this high steel bridge had the longest steel bridge-spans in the world at the time of construction (Railway Age, 1915). This bridge, also known as the Upper Campo Creek Viaduct, separates the railroad grade from State Highway 94 and Campo Creek, below. At this location, Campo Creek has eroded deeply into the large-biotite zone of the La Posta tonalite pluton.

The view from this high bridge may give cause to the casual train traveler, not to mention the professional geologist, to ponder how eons of erosion might have reduced the summit of the Peninsular Ranges to a broad crest of seemingly featureless rolling topography. For one thing, for countless thousands of years, Campo Creek (and others like it) have transported weathered and decomposed rock downstream. Following joints and fractures, it eventually carved the present-day canyon, exposing solid La Posta Tonalite bedrock.

When compared to alpine mountain ranges, such as the Rocky Mountains of Colorado or the Sierra Nevada of Central California, with their glacially sculpted pinnacles and arêttes, the Peninsular Ranges in eastern San Diego County and Northern Baja California fall strikingly short. Instead, the low relief, hummocky summit is thought to be all that is left of an old erosion surface, or *peneplain*. Viewed from a distance, the entire top of the Peninsular Ranges appears remarkably flat.

Conventional wisdom holds that, starting over 50 million years ago in late Cretaceous or early Tertiary time, this portion of the Peninsular Ranges was worn-down by ancient east-to-west flowing rivers (Miller, 1935; Minch and Abbott, 1973). Subsequently tectonic forces pushed the Peninsular Ranges up to somewhere near their present elevation. In isolated valleys, near the western foothills of the Ranges, remnants of ancient river channels containing truck-sized, water-rounded boulders, are preserved in paleo-channel deposits scoured into the ancient batholith surface.

An alternative explanation for the seemingly featureless crest of this area of the Peninsular Ranges is that it may, with a relatively small amount of erosion, approximate the rounded top of the huge La Posta pluton as it was emplaced in Cretaceous time (Walawender, personal communication, 2005). Subsequently, the pluton may have been buried by a large volume of softer sedimentary and volcanic rock that has long since been eroded away.

Moving eastward along the tracks to the Tecate Divide at Hipass (Milepost 84.5, 3,660 feet msl) the surrounding terrain changes character. It becomes noticeably more scenic, with large stacks of bizarre "large weathering boulders," punctuated with occasional sets of balanced rocks.

Steeply dipping (inclined) joints and fractures can be easily viewed by the casual train passenger in unsupported excavations into La Posta tonalite. These planes of weakness, enlarged by thousands of years of chemical and mechanical weathering, separate and isolate large blocks of tonalite which will one day become the next generation of "large weathering boulders" at the ground surface.

Over the years, it has become apparent that some railroad cut slopes, more than others, have had a tendency to slough and/or to be subject to rockfalls. The results of this natural geologic process, collectively known as *mass wasting*, require constant vigilance and periodic cleanup along the alignment. Future mitigation of problem slopes, among other solutions, could include: laying back slopes to a gentler inclination, tiebacks (soil-nails) to hold slope materials in place, scaling (removing) loose rock, and/or concrete grouting ("shotcrete") on slope faces to stabilize loose materials and mitigate erosion.

Problem slopes can be delineated by geologic observation and measurement of joint/fracture configurations, followed by engineering calculations to determine the apparent safety-factor for suspect slopes. Because of the reconnaissance nature of this paper, slopes that could become subjects for future investigation are only casually mentioned herein.

Geology, Gems, Gold, and Feldspar in the Jacumba Region

Geology

Jacumba Valley is a nearly closed, graben and horst structural basin bounded by northwesterly trending faults with apparent dip-slip separation.

In the Jacumba Valley, these weaknesses have provided pathways through the upper crust for molten magma to reach the ground surface — in this case, in the form of andesitic volcanoes with their associated ash flows and lahar breccias. Additionally, these fractured zones have provided conduits for deposition of metallic ores and gem minerals, as well as for thermal springs.

The succession of sandstones, conglomerates, volcaniclastic deposits, and interbedded andesitic/basaltic flows are considered one of the most complete stratigraphic sections preserved within the uplifted and deeply eroded Peninsular Ranges batholith.

The oldest units in this Jacumba section were originally considered to be the same age as the Cretaceous Cabrillo Formation, present in the La Jolla and Point Loma areas of San Diego (Minch, J.A. and Abbott, P.L., 1973). These strata have since been shown to be Miocene.

Gems

A number of economic mineral occurrences are clustered in the hills and valleys north of Jacumba. Gem pegmatite minerals, including tourmaline, garnet, and beryl, estimated at values approaching half a million dollars, are rumored to have been recovered from a few "pockets" in the Packrat Mine (Weber, 1963, Pl. 1, Mine No. 74; Walawender, 2000).

Scattered occurrences of tungsten, wollastonite, garnet, and other skarn-associated metals and minerals have also been reported along the northern, eastern and western peripheries of Jacumba Valley (Weber, 1963, Pl. 1).

Gold

Although not officially reported, gold is known to occur in the Jacumba area, both in bedrock veins and in alluvial (placer) deposits.

One anecdotal story tells of a prospector, or trainman, who panned gold from drainage culverts in the Carrizo Gorge Segment of the San Diego & Arizona Eastern Railway alignment (personal communication, Charles Coval, 2004).

Over 15 years ago, in the Jacumba Mountains, located several miles north-northeast of the community of Jacumba, I discovered a small vein of quartz within Triassic metasediments which had been intruded by the La Posta tonalite. Being the curious geologist, I sampled some rust-stained portions of an exposed narrow quartz vein in the old mineshaft (prospect) workings. The sample was submitted to an independent chemical laboratory (Clarkson Laboratory) where it was determined that it contained over one-half ounce of gold per ton.

This occurrence lines up with a number of other metallic and gem-mineral localities (including the Packrat Mine) along a linear west-northwest trend. This trend may also cross the small-biotite and muscovite-biotite zones in the surrounding La Posta pluton.

For the modern prospector this would represent a technical success, but a practical failure! In a broader sense, however, it is marginally successful in that it shows that an individual can still find buried treasure by digging in the ground. It also shows that, long long ago, some persistent old prospector, likely without benefit of an education in geology, was the first to discover this quartz vein and its gold.

Back then, mining equipment was generally limited to what a prospector and his burro could easily carry, likely not much more than a pick, shovel, and steel gold-pan, or *Battea*. The latter was used to extract tiny specks of heavy gold from the ore.

Following traces of "color" up rugged arroyos, our intrepid prospector eventually found the vein of quartz, laced with traces of precious yellow metal. Digging an inclined shaft more than 30 feet into the ground, following the richest portions of the vein, he likely stripped the best ore, packed up, and moved on — always looking for the ever elusive Big Bonanza.

For me, at today's gold prices (over $400 per ounce) this claim was a potentially significant find.

Most modern gold-mining technologies can turn a profit mining rock assaying only fractions of an ounce of gold per ton. Beyond just the price of gold, however, there must also be sufficient quantities of gold-bearing reserves to justify and support a full fledged mining operation, not to mention surviving economic uncertainty.

With today's stringent environmental regulations in force, especially on public lands, it is likely that only the smallest of mining operations might be approved by regulatory agencies. To make things even more difficult, this mine and surrounding land was declared Wilderness not long after I rediscovered this long-forgotten prospect.

There is no official record of this gold mine — only my unofficial description noted in the paragraphs above.

Thousands of explorers, prospectors, and immigrants, attracted to California by the Gold Rush of 1849, dug uncounted exploratory pits like the one described above in search of buried riches.

An authority on western mining history concludes:

"Most of the mineral strikes were tempting at first. They attracted millions of [investor] dollars from Europe and the Northeast, but returned comparatively few in dividends. Nevertheless, the money was not wasted. It was expended in payrolls, transportation, cost of materials, and construction of which most was locally purveyed and from which the general locality benefited. No matter [how rich] it was…, a working mine was doing its share toward western development." (Young, 1970).

In other words, the freedom to explore for valuable and/or strategic resources on public lands, as well as establishing efficient transportation networks, were key factors in the creation our world-class California economy.

Feldspar

On the way down into Jacumba Valley, keen-eyed train passengers will likely spot scattered white rock outcroppings and exploration pits, artifacts of long-abandoned feldspar prospects and mines. Small tonnages of feldspar, produced from pegmatite dikes, were trucked to the Pacific Mill at Campo for processing.

Jacumba — Trivia and Memories

Local Art and Gravity

A local impressionist's cycloramic painting of Jacumba Valley decorates the restaurant ceiling at the Jacumba Hot Springs Spa. Individual mesas and volcanic necks are recognizable in this surreal, domed 360-degree view of surrounding terrain.

My interpretation of this representation is that it is a "fish-eye" view of the Jacumba Valley by someone lying on their back, staring upward at stars and constellations between sunset and moonrise. And, just perhaps, could it be that an exotic mixture of trace chemicals from the local hot springs provided inspiration for this abstract image?

In a chance meeting with the cyclorama artist, I was asked if the extremely low Jacumba Valley gravity anomaly could have adverse physical effects on humans. Unaware of who was asking the question, and being neither a geophysicist nor bio-geophysicist, I wisely, and thankfully, dodged the question.

Later, however, I did verify that a gravity low (ranging from about -80 to -90 milligals) does, in fact, occur between Miller Creek and Jacumba (Biehler, 1979). One simplistic geologic interpretation may be that the density of the La Posta tonalite is significantly lower (hence, less pull by gravity) than older, denser rock intruded by the huge pluton. In the emplacement process, molten rock "floated" and "stoped" its way upward as the high-temperature magma-melt cooled and solidified into the massive, concentrically zoned Peninsular Ranges tonalite core.

As to any known adverse health effects, I suppose the jury is still out; and I'm still dodging the question!

The Old Erosion Surface, Geologic Processes, and Darwin

As noted in a previous section, (The Zoned Batholith) the "old erosion surface", or peneplain (Miller, 1935), must be very old, so old that millions of years were necessary for erosion to have removed the top of a mountain range about two miles higher than the existing surface (Walawender, 2000). The ancestral Peninsular Ranges may have at one time been equivalent in height to the Sierra Nevada Mountains in central California, or to the Rocky Mountains in Colorado. Although not as spectacular as these rugged, glacially carved, and saw-tooth capped summits, the gradual processes that reduced the Peninsular Ranges still represents a colossal event.

The "old erosion surface" may be a good example of a premise in geology called *Uniformitarianism*. This principle assumes that faulting, erosion, climate change, and other natural phenomena proceed at mostly "uniform" rates, and that catastrophic geologic events are the exception, not the rule. For example, given millions of years, the gradual processes of weathering and erosion (mass wasting) can remove entire mountain ranges, and/or carve huge gorges like the Grand Canyon.

Reality, however, may be a combination of *Uniformitarianism*, interspersed with catastrophic events that wreck regular havoc, such as, volcanic eruptions, earthquakes, severe weather (tornadoes and hurricanes), flooding, and climatic changes.

For example, the submarine subduction earthquake of December 26, 2004, was a truly catastrophic event — especially for humans living along the coasts of the Indian Ocean. In written human history, this event was similar to the 1883 volcanic eruption at Krakatoa and subsequent tsunamis.

On the scale of geologic time, the interval between the 2004 and 1883 events was less than one-millionth of the approximately 248 million years of San Diego County geologic history, from Triassic to present! If catastrophic events such as these typically occur about every 200 or so years, this suggests that more than a million similarly catastrophic events have likely occurred since the age of dinosaurs, the Mesozoic Era.

With the exception of a few hardy species, like sharks, cockroaches and ants, what does this say for the concept of Earth's so-called "fragile environment"? It may simply confirm what geologists have known from the stratigraphic record all along, even before Darwin published his research on evolution, that the environment and most species in it have always been repeatedly disturbed, destroyed and/or recycled. This may also explain the acceptance by some geologists of an older premise called *Catastrophism*. For them, a better title for Darwin's 1859 book, "The Origin of Species," might have been The SURVIVAL of Species!

Another take, might be that catastrophic events tend to occur at relatively regular intervals over geologic time — hence the appearance of uniformity when blended and "air-brushed" over time spans measured in millions of years. For a number of reasons (mostly scientific), I decline to argue the merits of another "...ism," and most especially, those that suggest: 1) a supernatural origin of species, 2) ethereal explanations for fossils, and/or 3) other demonstrative evidence for the age of the Earth.

But, I am willing to *listen* to different positions on geologic phenomena, provided I am allowed to weigh merit, based on 40 years of geologic field experience!

The Jacumba Valley

Jacumba Valley Drainage

As the San Diego & Arizona Eastern Railway tracks head northeasterly and then northerly across Jacumba Valley they parallel the mouth of Boundary Creek (near the Border with Mexico), and then the headwaters of Carrizo Creek (adjacent to Titus Siding), respectively.

Jacumba Valley Farming

On the east side of the Valley, on the site of a former dairy, organic vegetables are grown in rich volcanic soil. An area of limited year-round rainfall, irrigation is required to produce row crops in this high "desert" environment.

Clothes Optional

After the rails pass Round Mountain (a volcanic cone, or "neck" on the left side of the train), and pass under Interstate Freeway 8 near the north end of Jacumba Valley, a large billboard (on a volcanic hillside on the right side of the train) advertises De Anza Cove Resort. At first, this sign may seem out of place — advertising a "clothes-optional" dress-code. As the train passes by the resort, however, the billboard's meaning may becomes abundantly clear, especially if Resort guests are taking an illegal "birthday-suit-only" stroll along the tracks!

Jacumba Valley Aggregate

Recent and older alluvium in the Carrizo Creek channel, as well as Lower Basalt flows between Titus Siding (Milepost 94) and Dubber Spur (Milepost 96) contain sand and gravel deposits that may be of suitable quality for construction materials and/or for crushed aggregate. Within railroad property itself, economic deposits of sand and gravel may occur near the old railroad construction camp at Dubber Spur (Milepost 96), as well as upstream from the bridge crossing an unnamed tributary to Carrizo Creek (Milepost 95.7).

Additionally, large outcrops of granitic rock within railway property may be a good source of crushed stone — depending, of course, on the thickness of non-economic weathered overburden.

Jacumba Valley Springs

Nearby mountain springs were relied upon by railroad construction workers at Dubber Spur. From south to north, these necessities of life received colorful names such as, Arsenic Spring, Lone Tiger Spring, and Camp Sizzle Spring.

Railroad Construction in Jacumba Valley

Powder magazines, one of the few totally intact remnants of the old railroad construction camp at Dubber Spur, were cleverly made to fit into wind-sculpted caves in large granitic outcrops in the hills surrounding the camp.

A "good luck" decoration on the threshold of one small magazine (perhaps used to store detonators[?]) consists of a horseshoe-shaped impression inlaid with bits of quartz and colorful stones.

THE CARRIZO GORGE — AN OVERVIEW

Geology of the Carrizo Gorge (and the Desert Line in general) is unique when compared with other scenic railways. With seventeen tunnels in the Gorge alone, and excellent exposures of geologically complex rocks, the Gorge Segment of the Railroad provides the train traveler with superb views of tens of square miles of some of the oldest rocks in Southern California.

Like the Royal Gorge and Glenwood Canyon short lines in Colorado, and other scenic railroads in the Canadian Rockies, the Alaska Range, and Copper Canyon in Mexico, the San Diego & Arizona Eastern Railway, especially through the Carrizo Gorge, provides its own share of unique geological and scenic vistas.

An array of variegated stone, with an artist's palette of red, brown, gray, and white literally surround the traveler along about 10 miles of track. Early San Diego & Arizona Eastern Railway passengers were WOWED by the ever changing and contrasting desert terrain. The steep-sided Carrizo Gorge, rugged side canyons, and colorful rock formations, were witnessed as fleeting glimpses, while the train passed in-and-out of tunnels and over wooden trestles.

Petrology

Views now plunge nearly 1000 feet (and almost as far upward) into the chasm occupied by Carrizo Creek, revealing a jigsaw puzzle of folded, faulted and metamorphosed, as well as intrusive and extrusive rocks.

Virtually a *mélange*, or mixture, the Carrizo Gorge provides unparalleled cross sectional views of some of Southern California's most spectacular geologic scenery. These hard, dense, metamorphosed rocks began as soft calcareous-muds and/or silty fine-grained sands on a broad, widespread sea floor. Millions of years ago, during the Triassic and Jurassic Periods, these seas extended over much of what is now Western North America. They represent approximately 248 million years of geologic events, including the Mesozoic age-of-dinosaurs.

Fossils have rarely been found in these older rocks. The few fossils discovered were shallow-marine mollusks such as ammonites (Hudson, 1922; Merriam, 1958). These marine sediments were first deposited in widespread flat-lying, pancake-like layers, intruded by volcanoes along island-arcs, buried, and finally heated and compressed to form metamorphic rocks. In the Carrizo Gorge, the train traveler is afforded a front-row-center "screening" of the aftermath of eons of intense heat and pressure, faulting and folding, and uplift and erosion, resulting in the colorful present-day display of intensely folded, faulted, and fractured rock.

Structure — Faulting and Folding

Major, as well as minor faults mapped by the author exhibit at least two intersecting sets of shears that trend northwest and northeast. Geologic maps prepared by the United States Geological Survey (Hoggatt, 1979), in areas immediately north of the Carrizo Gorge, show similarly oriented, intersecting shear-sets.

Northwest-striking fault-sets were found to be more likely to displace formations as young as Pleistocene, whereas northeast-striking faults (optimal trends approximating N.30°E.) did not cut rocks younger than Cretaceous. This finding may also be applicable to northeast-striking faults near Tunnel No. 5, and/or to fault intersections mapped by the author in the northern portions of the Gorge Segment, starting at the east (northerly) end of Tunnel 16.

Along the westerly side of the Carrizo Gorge, starkly contrasting features, like half-mile-long, white, straight-edge lines of pegmatite dike-walls, enclosed by brown to dark gray metasedimentary rock, are visible from the left (westerly) side of the train. On the right (easterly) side, however, more folded and faulted rocks of the same variety are exposed in cuts and tunnels. At least in part, this contrast is due to variations in ancient heating and compressive forces at depth that, like "play-dough putty," have locally distorted both the metasedimentary rock and pegmatite dikes on the right (easterly) side of the rails.

Landslides

Ancient as well as recent landslides that have failed along weak and/or faulted foliation planes and/or dike-contacts are ubiquitous along the Carrizo Gorge Segment of the San Diego & Arizona Eastern Railway alignment. Most slides seem to have resulted, at least in part, from the presence of *adversely* inclined (out-of-slope) dipping surfaces and rapid erosional undercutting of steep natural slopes. Eventual hillside collapse, possibly accelerated by seismic events and/or by undercutting during original railroad construction, will likely require mitigating measures from time to time.

In the 1920s and early 1930s two major railroad tunnels, Tunnel Nos. 7 and 15 in the Carrizo Gorge Segment, were damaged and/or closed by landslides. Tunnel No. 7 was abandoned and bypassed with daylight track in early 1933. Tunnel No. 15 was closed by a large bedrock landslide in 1932; this destroyed tunnel was quickly bypassed with shoofly track in the form of Tunnel No. 15 1/2 and the 636-foot-long, 185-foot-high Goat Canyon trestle (Copenhaver, 1988).

Seventeen tunnels were required to construct the Carrizo Gorge segment of the San Diego & Arizona Eastern Railway, more than the thirteen tunnels needed by the Central Pacific to cross the Sierra Nevada Mountains, at Donner Summit, during construction of the first transcontinental railroad.

The Carrizo Gorge segment, because of the number of tunnels, steep cuts, serpentine rail-curves and geological hazards has been, and continues to be, an important factor in overall railway operations, including both maintenance and repair.

Geomorphology

For the geomorphologist, it is important to note that many of the side canyons, wind-gaps and ridge lines, on both sides of the Carrizo Gorge, have borrowed their trend and form from underlying geologic planes of weakness, for example, foliation and intersecting faults.

DUBBER SPUR TO DOS CABEZAS, THE "CARRIZO GORGE SEGMENT" — MILEPOSTS 96.0 to 109.7

The Carrizo Gorge Segment represents the scenic, historic, and geological highlight of a trip on the San Diego & Arizona Eastern Railway. It is only about 13 miles long, but it cost about one-fourth the amount spent to complete the entire railroad (Hanft, 1984; Kirchner, 1988). Of the 21 tunnels needed to complete the railroad from San Diego to Seeley, 17 are in the Carrizo Gorge.

Traveling from west (southerly) to east (northerly), this segment of the railroad alignment descends about 955 feet. From Dubber Spur (approximate elevation 2,675 feet) the rails wind precariously down through the Carrizo Gorge to Dos Cabezas (approximate elevation 1,720 feet).

Petrology

After crossing the trestle over a tributary creek to Carrizo Creek at Dubber Siding (Milepost 96.0), the train passenger will notice a terrain change, to rugged, bouldery, light gray outcrops of Boulevard monzogranite (Walawender, 2003). (This batholithic rock is probably the same as the "leucomonzogranite" differentiate of the La Posta pluton described by Todd, 2004). It may represent one example of several smaller plutons (another being the Indian Hill monzogranite) that intrude the central zone of the much more massive La Posta tonalite, as well as Triassic metasedimentary rocks. Boulevard monzogranite contacts are irregular, and have xenolithic inclusions of both La Posta tonalite and metasediments.

Landslides

Cut slopes along the Carrizo Gorge segment of the San Diego & Arizona Eastern Railroad frequently expose fracture and joint-planes that are inclined (dip) out-of slope, leading to sloughing and rockfalls.

A good example of a small rockfall, approximately 30 feet wide, can be seen by the observant train traveler on the right side, in an alignment cut at approximately Milepost 96.4.

Many small rockfalls can be relatively easily mitigated by scaling (also known as "dental work," or "barring") and/or by excavation that would lay slopes back to reduce slope steepness.

Tunnel No. 5 — Mileposts 96.6 to 96.7

Tunnel No. 5 generally represents the "gateway" to the scenic and historic Carrizo Gorge. Past Milepost 96.7, the tracks follow a more sinuous northerly path, still in granitic rock (most likely La Posta tonalite).

On the left (westerly) side of the railroad alignment, Carrizo Creek changes gradient from a gently inclined desert wash as it crosses Jacumba Valley, to a shallow arroyo with increased gradient adjacent to Tunnel No. 5, and finally into a steep-sided deep canyon (the Carrizo Gorge).

Train passengers heading east (northerly) down into the Carrizo Gorge can now see the locomotive (engine) more often than before, since the train is going around more numerous and sharper curves. Only the left (westerly) side of the train has a good view of the Carrizo Gorge; passengers riding on the right side of the train will quickly shift their attention — and/or their seats. At this point, the tracks are hugging the steep face of the arroyo that will soon become the rugged, bouldery, and barren Carrizo Gorge.

Faulting at Tunnel No. 5

A short distance further, approximately between Mileposts 96.5 and 96.7 the railroad alignment crosses a significant zone of cross faults. Previous studies show only a single east-northeast striking fault truncating the northwesterly striking faults that are responsible for the Jacumba Valley structural basin (Minch and Abbott, 1973; Gastil and Bushee, 1961; Brooks and Roberts, 1954).

At least three sub-parallel, north-northeast trending faults were mapped by the author.

The first (strike N.80°E., dip 80° north), was exposed at approximately Milepost 96.5, in a cut on the right (easterly) side of the tracks.

The second (strike N.80°E., dip 55° south), was exposed just outside the west (southerly) portal of Tunnel No. 5, at approximately Milepost 96.6.

The third (strike N.80°E., dip 30° north), was exposed just outside the east (northerly) portal of Tunnel No. 5, at approximately Milepost 96.7.

The southernmost of these three cross faults, is all in granitic rock. It has the most developed gouge, with about 2 inches of clay on the hanging-wall plus a brecciated footwall, 5 to 10 feet thick. This fault is very similar to, and may be an extension of, a fault mapped approximately two miles to the east by Minch and Abbott (1973). To the west, this fault parallels a linear drainage called Walker Canyon.

The middle fault has the least gouge, less than about one inch in thickness.

The northern fault appears to be a gouge-breccia zone, about two feet in thickness.

The west (southerly) portal of Tunnel No. 5 was excavated through metasedimentary gneiss, while the east (northerly) portal encountered tonalite (possibly La Posta tonalite); the contact between these two vastly different rock types occurs somewhere in Tunnel No. 5. This zone of faulting may mark the northern boundary of the Boulevard monzogranite (Walawender, 2003).

Although the age of these cross faults is not known, it is most likely pre-Holocene (more than 11,000 years since the last movement). Fault gouge in the middle fault appears to be strongly oxidized, perhaps hydrothermally altered, and without evidence that iron-oxide mineral coatings along the gouge zone have been disturbed. In arid climates, mineral coatings such as these are thought to take more than 10,000 years to form (Birkeland, 1984).

Slope Stability at Tunnel No. 5

Previous inspections of Tunnel No. 5 have noted shoring distress caused by fractured rock "squeezing" against the left (westerly) wall (SANDAG, 1995). This report concluded, however, that the tunnel is in generally good condition and recommended additional shotcrete to stabilize areas of loose rock. Considering the intensity of fracturing caused by prehistoric faulting, another option might be to "daylight" the tunnel. That is, remove the tunnel's roof and create an open trench through the relatively low hill. This might also mitigate rockfalls like those that occurred adjacent to the east (northerly) portal fault in January, 2005.

Trails

At about Milepost 96.9, the track approaches a clockwise bend around a hill (underlain by tonalite) that allows an unobstructed view across the now rapidly deepening Carrizo Gorge. With luck, train passengers will be able to spot a steeply inclined trail along the west wall of the Gorge; also descending into the Gorge, it is at about the same elevation as the train tracks. Locals know it as the McCain Trail.

Buck McCain, artist-sculptor-rancher, made use of this little-known trail after a train derailment with a Coors beer shipment into the gorge. He spent three days with pack horses to haul out the brew, a favorite for thirsty cowboys (McCain, personal communication, 2006).

This ranching McCain family was the first to settle in the Jacumba Mountains in 1868 (Lindsay, 2001). It is the only known trail, other than those created during railroad construction in the early 1900s, to enter and exit the upper reaches of the Carrizo Gorge. Local lore suggests that it was constructed by the McCain's to access water for their cattle.

It is possible, however, that the trail is older, and was simply maintained and/or improved by early ranching families. Pre-ranching use is alluded to by reports describing the trail as first being used by parties of immigrants in 1849, and as a primitive branch trail used by United States Boundary Commission topographical engineers before that (Pourade, 1966).

Lieutenant Cave J. Couts established Fort Calhoun in 1849 as a base for United States Boundary Commission surveyors. Fort Calhoun was later renamed Fort Yuma in the 1850s (Pourade, 1966).

Accompanied by Company A of the First Dragoons, commanded by Lt. Cave J. Couts, this same trail was apparently explored by Andrew B. Gray, engineer with the United States Boundary Commission. Upon receiving Couts's request, Gray temporarily left his surveying duties in San Diego to guide an immigrant train (known as the "Collier" party) through the Carrizo Gorge.

Use of the "McCain Trail" by surveyors and immigrants is an indication that the rugged Carrizo Gorge was explored early-on as a possible first transcontinental railroad route.

Geology, Geologic Hazards, and Railroad Construction

By approximately Milepost 97.1, the rails have gone around the first clockwise half-circle that cuts into massive granitic rock, resembling La Posta tonalite. Instead of bouldery outcrops of previously encountered Boulevard monzogranite, the tonalite exhibits increasingly irregular contacts with xenolithic inclusions and roof-pendants of older rock. A variety of weathered and foliated, brown Triassic metamorphic rock, early Cretaceous gabbro, and an increased frequency of resistant light gray to white pegmatite dikes cut across the tonalite country rock.

Just after rounding the bend, the traveler may see the first of many wooden side-hill trestles (ahead, at about Milepost 97.4), clinging precariously to the dizzyingly steep-sided Carrizo Gorge. The asymmetrical design (long downhill side versus short uphill side) of these "Billy-Goat-like" structures necessitated the use of wood, instead of steel. According to railroad specialists, wood was preferred over steel as it is less affected (dimensionally) by the extremes of temperature typically experienced in the Gorge.

At about Milepost 97.4, Carrizo Creek is over two hundred feet below the railway alignment. Here it has carved itself deeply into fractured, weathered, and foliated Triassic schist and migmatite-gneiss of the Indian Hill-Tule Mountain Roof Pendant (Walawender, 2003). Most tributary canyons, as well as the majority of the south-to-north draining central gorge of Carrizo Creek, are eroded into intensely folded and faulted metamorphic rock. Many of these slopes exceed 100% (greater than 45 degrees down from horizontal) inclinations, foretelling the potential for extreme erosion and susceptibility to rockfalls.

As the rails approach the first Gorge trestle, at about Milepost 97.4, slopes above the alignment, west (southerly) and east (northerly) from the bridge, for at least 100 feet in either direction, are underlain by an ancient rockfall. Because of the presence of over-steepened slopes and adversely dipping (out-of-slope) foliation planes and parallel fault shears, this slope will likely require stabilization fills and/or buttressing at some future time. Rockfall potential, however, could be alleviated by building gabion walls, a barrier to earth movement and rockfalls created by stacking cubical steel-wire baskets, filled with heavy rock (no shortage of rock in this area).

Beyond approximately Milepost 97.7, to about Milepost 104.4 (a distance of almost 7 miles), most of the terrain crossed by the railway alignment is underlain by Triassic metasedimentary rock cut by numerous pegmatite dikes that were intruded along foliation and/or fault planes. Faults, joints, and contacts tend to be linear to curvilinear in character, often exhibiting differentially eroded notches in the walls of the rapidly deepening Carrizo Gorge.

From time to time, the observant traveler may spot small, irregularly shaped masses of gray to light brown tonalite, or other granitoid rock. These sparse intrusive masses represent isolated invasions into the Indian Hill-Tule Mountain roof pendants and older batholithic plutons.

Mining and Mineralization

The track rounds another bend at about Mileposts 97.5 to 97.6, roughly opposite the confluence of Tule Creek with Carrizo Creek. Here, railway cuts expose large pegmatite dikes cutting across Triassic metasediments composed of amphibolite, coarsely crystalline light gray limestone (marble), and skarn.

After passing over a bridge, at about Mileposts 97.6 to 97.7, an approximately 10- to 15-foot-thick bed of limestone (marble) is exposed in an alignment cut on the right side of the tracks.

These units occur in an approximately one-half-mile-wide irregular zone of limestone-bearing metasedimentary rocks that also contain tungsten and other skarn mineralization, including wollastonite and garnet. Good outcrops of massive garnet can be found adjacent to the contact-zone with the above-noted marble beds.

Shortly after passing the Tule Creek-Carrizo Creek confluence, the railroad alignment crosses a zone of erratically distributed metallic mineralization that approximately follows the contact between batholithic rock and the intruded metasedimentary rock.

Intrepid and persistent, early California prospectors combed these mountains and valleys in search of valuable metals and minerals. Abandoned prospects, shafts and adits, attesting to the presence of those nameless explorers, have been documented by the California Division of Mines and Geology (Weber, 1963, Pl. 1; Prospect/Mine Numbers 448 and 470 - tungsten; 477 - vermiculite [altered biotite]; and 479 - wollastonite, Ca[SiO3].

This belt of mineralization may also include a cluster of gem-pegmatite occurrences around Sacatone Spring and the Packrat Mine (Weber, 1963, Pl. 1; Prospect/Mine Number 74), located approximately three miles northwesterly, beyond the west rim of the Carrizo Gorge. These mineral localities are oriented along a northwesterly striking linear trend that, if extended far enough to the southeast, would come close to the previously described gold prospect found by the author.

During intrusion of the La Posta pluton, magmatic heating is likely to have mobilized an assortment of exotic metals and rare-earth elements into favorable hosts, such as *infiltration skarns*, resulting from recrystallization and silica alteration of calcareous units (Einaudi, 1975). It may also be possible that later gem pegmatite dike occurrences are more frequent where proximal to skarn zones, especially where skarns have formed deep in metasedimentary roof pendants (Walawender, 2004, presentation during the San Diego Association of Geologists annual fieldtrip).

Landslide at Tunnel No. 7, Repair Discussion — Milepost 98.7

Landslide mitigation at the west (southerly) portal of abandoned Tunnel No. 7 is a good example of a useful landslide repair by excavation and placement of stability fills. This slide, in metasedimentary rock, has blocked the west (southerly) portal of Tunnel No. 7 since the early 1920s. Spreckels set off a blast of over 50 tons of black powder in the middle of the landslide, thinking the entire block could be displaced into the gorge (like a snow avalanche). It was the largest black power explosion known at that time (see Figure 2).

The original railway alignment apparently undercut the center of an ancient landslide that had failed into a steep tributary canyon of the Carrizo Gorge. It is unknown whether or not the existing shoofly (bypass) fill is still underlain by remnants of the original landslide debris. Subsequent attempts to clear slide debris at the elevation of the railway grade exposed the landslide failure surface to the left (northwesterly) of the buried west (southwesterly) tunnel portal.

The author was able to access Tunnel No. 7 by entering the east (northeastern) portal at about Milepost 98.9 and then walking about 800 feet back into the 1,234-foot-long bore. The majority of the tunnel is cut through metasedimentary rock, with narrow pegmatite dikes following steeply dipping foliation. All of the original timbering had long since burned, leaving considerable amounts of fallen rock on the tunnel floor.

Tunnel repair and landslide mitigation at the west (southwesterly) portal will require "slot-removals" of landslide debris, as well as buttressing of any slide debris left in place around the portal. Cleanup of loose fire-spalled rock on the floor of the tunnel can be accomplished from the east (northeasterly) portal, as well as from the west (southwesterly) portal — after stability-fills have been placed against landslide remnants. CZRY has received some interest from aggregate companies willing to mine oversize rock in the slide debris to assist in initial excavations.

Care should be taken during excavation of slide debris, by using narrow "slot-excavations" running perpendicular to the slide-mass. (Remnants of the ancient slide, above the tunnel portal, could be undercut and activated by excavations running parallel to slope-contours at the level of the tunnel.) Continuous observation of landslide excavation and buttressing by an engineering geologist will be a necessary requirement for successful repair of Tunnel No. 7 and its west (southwesterly) approach. In addition, the SANDAG report in 1995 recommended supporting Tunnel No. 7 with "full sets and lagging."

Reopening this straight tunnel will help speed train passage and allow the shoofly-bypass to be converted into a valuable siding.

Next to landslides, faults and the transitions between hard pegmatite dikes and relatively weak Triassic metasedimentary rock pose the greatest geotechnical problems for tunnels and the stability of steep cut slopes along the Carrizo Gorge Segment of the San Diego & Arizona Eastern Railway.

Tunnel No. 8, Faults and Water — Milepost 99.1

One of the last tunnels to be repaired and reopened to train traffic was Tunnel No. 8, a short distance beyond the east portal of previously described Tunnel No. 7. The second longest tunnel on the San Diego & Arizona Eastern Railway, at 2,173 feet, it has been a "problem child" since the beginning of construction (Hanft, 1984).

The author was able to visit Tunnel No. 8 after most of the recent repairs by CZRY were completed in late 2003. At that time, the majority of the tunnel had been continuously shored with steel sets and shotcreted in accordance with SANDAG's engineering report of 1995.

Observations by the author indicate that most of Tunnel No. 8 is bored through foliated Triassic metasedimentary rock, similar to that observed in Tunnel No. 7. In the eastern approximately one-fourth of the tunnel, however, strong groundwater seepage was evident along with a gutter and drainpipe to collect water for discharge at the east (northerly) portal. The source of water is thought to be a fault or fracture zone that intersects, and collects water from at least two northwesterly draining canyons that cross the railway at approximate Mileposts 99.6 and 99.7.

The author mapped a N.25°E. striking fault crossing both of the above-noted drainages, hundreds of feet above the tunnel, where natural springs and associated phreatophytes (willows and cottonwoods) occur. Projecting this linear fault, that likely forms an uphill groundwater-barrier along strike, would intersect Tunnel No. 8 near the area of strong seepage. This is also near an area where another fault, projected from the bottom of Carrizo Gorge, over a mile to the northwest, also intersects Tunnel No 8. Through-going fault-lines, strongly developed foliation, intersecting fractures, as well as perched groundwater, could easily explain why there were excavation problems at the beginning of construction, as well as rockfall problems during the railway's early operations.

Recommendations that could mitigate future seepage and shoring-damage problems could include installation of additional drainage systems, via pumps and/or drain-outlets. Horizontal wells could also be drilled into the hillside to capture, and discharge ground water harmlessly away from the railroad alignment. Another option might be to use pressure-grouting, to plug weeping fractures with "grout curtains."

Figure 2. Landslide on the west portal of abandoned Tunnel No. 7 of the San Diego & Arizona Eastern Railroad. This potentially ancient landside occurs within metasedimentary rock of the Triassic Indian Hill-Tule Mountain roof pendant. The largely blocked portal of the tunnel is located just beyond the railroad near the left edge of the image. The ancient landslide was undercut by excavations along the original railroad alignment, causing renewed movement on the slopes above the railroad. The tunnel has been abandoned since the 1930s, but may be reopened by Carrizo Gorge Railway Inc. in the future.

Figure 3. The Goat Canyon Landslide. This huge bedrock landslide destroyed and displaced the old Tunnel 15 in 1932, causing the bypass (or shoofly) around the ridge and the construction of the 600 foot-long and nearly 200 foot-high wooden trestle and Tunnel 15.5 on the far right. Northwest-striking foliation and weak shear zones in metasedimentary rocks are undercut by erosion along the steep canyon walls. Photo by Cathy Copenhaver.

Figure 4. View of Goat Canyon and the unique wooden trestle constructed in 1932. Above the trestle and immediately up the canyon, is the massive bedrock landslide that severed the original Tunnel 15. A northwest-to-southeast striking fault zone paralleling foliation planes is suspected to have contributed to failure-shears. Photo courtesy of Charlie Coval.

More Faulting and a Recent Landslide — Mileposts 100 to 102

Between Tunnel Nos. 9 and 14 (Mileposts 100 to 102) the railroad alignment is cut and bored through Triassic metasedimentary rock that has been folded and faulted, as well as intruded by both felsic (pegmatite) and mafic dikes. In some areas there is evidence of multiple episodes of deformation; for example, pegmatite dikes follow foliation planes, but have also been folded, faulted, and then intruded by mafic dikes. This seems to be most pronounced along a north-northeast-trending zone extending approximately between Mileposts 101 and 102, roughly paralleling the west side of the Railway alignment. This steeply dipping to vertical zone underlies erosional notches in ridges cut by Tunnel Nos. 12, 13 and 14.

Tunnel Nos. 11 through 14, all bored through metasedimentary rock, have exhibited only minor problems in terms of stability and maintenance. Tunnel portals, however, require ongoing support repairs, as well as periodic cleanup efforts following minor rockfalls (SANDAG, 1995).

At approximately Milepost 100.9, between Tunnel Nos. 11 and 12, a relatively "fresh" landslide scar can be seen by rail passengers on the left (westerly) side of the tracks, on the western side of the Carrizo Gorge. This slide does not endanger the alignment, but is typical of major bedrock landslides in the Carrizo Gorge, in that it occurs in metasedimentary rock. A major fault line, below the landslide, trends across the bottom of the Gorge, and extends in a south-southeasterly direction toward the east (northerly) end of Tunnel No. 8.

The influence of faults in metasedimentary rock on the frequency of significant landslides in the Carrizo Gorge is evident from presently known locations of landslides at or near the traces of faults, fault-intersections, and/or intensely fractured and foliated rock.

About halfway between Tunnel Nos. 11 and 12, John D. Spreckels connected the last rails laid with a golden spike in 1919 (at Carriso Station).

Goat Canyon Landslide and Bridge — Mileposts 102.1 to 102.3

Emerging from the east (northerly) portal of 2,597-foot-long, arrow-straight Tunnel No. 14, the long-abandoned west (southerly) portal of original Tunnel No. 15 can be seen on the right (east), at about Milepost No. 102.1. Quickly, the rails enter short Tunnel No. 15 1/2, bored in 1932 as part of the "shoofly" (or bypass) around the huge Goat Canyon bedrock landslide that occurred on March 27, 1932.

Even though the original alignment was built across Goat Canyon on a substantial fill (at least 50 feet thick), the fill was clearly insufficient to buttress the toe of this bedrock slide. And, since 1932, running water and erosion have breached the canyon fill, moving large boulders and slide debris to within a few feet of the bottom supports for the Goat Canyon Bridge.

The Goat Canyon landslide is the largest ever observed in the Carrizo Gorge. Like other landslides in the Gorge, it is located at, or near, intersecting faults/shear zones. The slide-mass itself, appears to be a block, or blocks of metasedimentary rock, estimated to be about 500 feet wide, 600 feet long, and 100 feet thick. Principal elements of the landslide can best be observed from a high vantage point, such as the ridgeline above Tunnel No. 16 (see Figure 3).

Sheared foliation planes dip northeasterly into Goat Canyon. The failure surface, or surfaces, at the base of the slide may transition into a back-scarp that follows a northwesterly striking fault zone observed just west of the west (southerly) portal of bypass Tunnel No. 15 1/2.

After emerging from Tunnel No. 15 1/2, the traveler is almost immediately suspended in space as the train crosses the 636-foot-long, 185-foot-high, curved Goat Canyon trestle. This wooden bridge, perhaps the largest existing curved trestle in the world, may also be the largest wooden trestle on any currently operating railroad. To the right, the train traveler will see the broken and rotated eastern portal of old Tunnel No. 15, severed by the Goat Canyon landslide.

In 1932, the Goat Canyon trestle was intended to be a "temporary" structure, bypassing the landslide until the original alignment and Tunnel No. 15 could be repaired. Although it has been "temporary" for more than 70 years, it has withstood the heaviest train traffic in the railway's history, including World War II and significant amounts of freight traffic up to the 1970s.

SANDAG inspections in 1995 indicated an overall "medium" rating for the condition of the bridge's substructure and superstructure, pointing out highly weathered trestle elements and the need for localized maintenance and repair.

Even using modern heavy-duty excavating equipment, the earthwork necessary to stabilize the Goat Canyon landslide would be daunting.

It may be more economical to maintain and protect the existing wooden trestle against minor sloughing near the toe of the Goat Canyon landslide than to make an attempt to repair the severed tunnel, which would doubtless be at great expense. By installing a large-diameter steel drainage culvert beneath the trestle and extending it upstream beyond the toe of the landslide, exceptional seasonal rainfall (for example, summer thunder storms [cloudbursts]) which might move slide debris down against the bridge could be averted.

Further stabilization might also be accomplished by extending the Tunnel Spur (the remnant of the original main line) in a southwesterly direction as far as the Goat Canyon landslide. Placing filled ground across Goat Canyon, to extend the spur, would act as a partial buttress at the toe of the slide. It would also create a catch-basin for displaced rolling boulders as well as for rockfalls. A canyon-bottom drainage structure, beneath the fill, would convey rainfall runoff safely under both the filled ground and the Goat Canyon Bridge.

There is no geomorphic expression for the presence of a pre-existing ancient landslide at Goat Canyon, such as the geomorphic expression that was observed at the now-abandoned Tunnel No. 7. If an ancient landslide did exist, it may have been in the form of large, highly fractured blocks, with only a slight amount of previous movement. Although as-built construction notes/records were not available to the author, personal observations in original Tunnel No. 15 indicate that it was bored through mostly intact metasedimentary rock, displaying localized zones of intense fracturing and foliation. This conclusion is supported by geologic field work in the vicinity of Goat Canyon, and by comparison of geographic as well as anthropomorphic features visible on pre- and post-1932 railroad alignment images.

Examination of accessible western portions of severed original Tunnel No. 15 also showed evidence for faulting near the east (northerly) end, as well as evidence for compression of the tunnel walls and roof on the inside (right hand) curve, showing large splintered shoring beams.

Tunnel No. 16: The Last Barrier — Mileposts 102.42 to 102.56

Shortly after crossing over the eye-opening, wooden Goat Canyon trestle, the rail traveler will be whisked into the west (southerly) portal of Tunnel No. 16. For the Carrizo Gorge Railway, this tunnel represented the last barrier to trains passing through the Gorge, and finally connecting with the Union Pacific Railroad at Seeley.

When the SANDAG engineering report was released in 1995, both the west (southerly) and east (northerly) portals of this 740-foot-long bore had long since been closed by caving; hence, for all practical purposes, Tunnel No. 16 was inaccessible for inspection. Therefore, the SANDAG report section entitled "Minimum Anticipated Work Necessary to Open" had little more to recommend than, "...muck out, steel sets and grout".

By entering Tunnel No. 16 through a small opening in the collapsed west (southerly) portal, the author was able to accompany Carrizo Gorge Railroad employees during their pre-cleanout tunnel inspection; tunnel clearing and repair would follow in 2003.

Approximately 500 feet of the southern portion of the tunnel showed the effects of a fire that had burned all of the old wooden shoring, resulting in substantial spalling of metasedimentary rock (cut by numerous pegmatite dikes) in the walls and roof. Rails in this part of the tunnel were found to be distorted and twisted by heat from the fire.

Anecdotal and/or secondary information from railroad employees, as well as observations made by the author, indicate that explosives might have been used to collapse the tunnel portals in an attempt to put out the fire. If true, the explosives likely contributed to the amount of debris in need of removal; this scenario would be especially true at the east (northern) portal, where intensely faulted tunnel wall- and roof-rock could have been seriously loosened by blasting.

After the east (northerly) portal of Tunnel No. 16 was cleared of debris, a major through-going northwesterly trending fault was observed in the tunnel walls. (This fault was also noted to be coincident with a sharp crosscutting structure observed during a pre-cleanout inspection of the east [northern] tunnel interior.) A northeasterly striking cross fault is terminated by the above-noted northwestern-striking fault on the right (eastern) side of the east portal-entry. Adversely dipping (out of slope) sheared foliation surfaces at this fault intersection increases the probability that minor slough as well as large debris could fall onto the rails and tunnel floor.

The potential for rockfalls from over-steepened cut slopes near the east (northern) entrance to Tunnel No. 16 could be mitigated by laying back the slope to a flatter angle, creating "catch benches" and/or by employing tie-backs ("soil nails") and shotcrete.

Side-Hill Trestles and the Seven Sisters

Beyond Tunnel No. 16, train tracks, if not in a tunnel, are most usually supported by an almost continuous succession of side-hill trestles. These lopsided wooden structures cling precariously to the very steep, nearly 100% slope (or about 45° below horizontal), walls of the Carrizo Gorge. The *Seven Sisters*, the most striking and picturesque example of these side-hill trestles, consists of a series of seven closely spaced bridges that traverse numerous side-canyons and arroyos between Tunnel Nos. 18 and 19 (approximately between Mileposts 103.7 and 104.7). This series of bridges was named for the first seven Sisters of St. Joseph of Carondelet. It is in honor of their trek through these mountains as Christian missionaries in 1870.

Ancient Gravel Deposit

On the right (easterly) side of the train, just after exiting the east (northerly) portal of Tunnel No. 18, the observant train traveler may be able to spot an exposure of boulder-sized gravel set in a cemented matrix of reddish brown coarse-sand; its origin, perhaps a remnant of an ancient alluvial fan or terrace deposit. Although large sub-rounded boulders protrude from this steep slope, these do not appear to present an erosional problem to the rail alignment.

Petrology and Faulting

After passing the *Seven Sisters* half-trestles, and before reaching Tunnel No. 19, the rails leave Triassic metasedimentary rock, cross over two faults, and enter a bedrock terrane composed almost entirely of tonalite.

These two contact, or boundary faults, are oriented as follows: east-northeast (dipping 60° southerly), and east-west (dipping 40° northerly). The latter fault occurs at about Milepost 104.5, and has an approximately four-foot-wide gouge zone.

These two faults have a significantly different trend (east-west), when compared to other more typically northwest and northeast trending faults and foliation noted in the Triassic metasediments throughout most of the Carrizo Gorge. Is it possible, therefore, that these east-west shears, along with the east-west shears previously noted at Tunnel No. 5, may represent the northern and southern boundaries of a large structural block of metasedimentary rock? Clearly, more research will be required to answer this over-arching question.

Having now traversed the generally south-to-north trending Carrizo Gorge, and having passed safely through fourteen tunnels (Tunnel Nos. 5 through 18 — including the now-abandoned Tunnel No. 7), the rails will soon disappear into Tunnel No. 19. From one end to the other, Tunnel No. 19 passes through 715 feet of solid granitic rock (tonalite).

Bending slightly clockwise to the right (easterly), the tracks in Tunnel No. 19 emerge near Milepost 104.9. From here on, the railway alignment takes the "now seasoned" rail traveler away from the Carrizo Gorge and its many tunnels. But not to despair, for those of you who have become "tunnel junkies" two short tunnels remain, Tunnel Nos. 20 and 21.

Desert Palms

Near Milepost No. 106, an oasis of native palms (Washingtonia) can be easily spotted to the lower left (westerly) of the train, where springs occur in the headwaters of Palm Canyon, a tributary to Carrizo Gorge.

Wild Flowers

In the spring of the year, following favorable winter rains, wild flowers, native to the Colorado Desert Habitat, bathe the landscape in a rainbow of colors.

Bighorn Sheep

Although extremely rare, small groups of bighorn sheep (borregos) may be spotted between Goat Canyon and Tunnel 19 by the sharp-eyed and very vigilant train passenger.

Tunnel No. 21 to Dos Cabezas

After leaving Tunnel No. 21, on its way to the old water tank at Dos Cabezas (Two Heads), the San Diego & Arizona Eastern Railway alignment begins to straighten out, but still crosses repeatedly back and forth over exposures of granitic rock (La Posta pluton) and narrow roof pendants of Triassic metasedimentary rock. Several limestone and marble quarrying operations near Dos Cabezas occur within metasedimentary terrane.

DOS CABEZAS TO SEELEY, THE "DESERT SEGMENT" — MILEPOSTS 109.7 TO 139.0

Dos Cabezas to Sugarloaf Mountain

Past Dos Cabezas (elevation about 1720 feet MSL), the route continues its descent, winding its way steadily through the eastern foothills of the Peninsular Ranges Mountains, crossing boulder-strewn alluvial fans and fields of aeolian dune sand.

Easterly from Dos Cabezas, the rails crosses Palm Canyon Wash three times, by making two large "S" turns to loose altitude. As the rails approach Interstate Freeway 8, on the right (south), Sugarloaf Mountain rises about 200 feet above the tracks on the left (north) side. This isolated hill is underlain by non-marine Miocene conglomerate and arkosic sandstone.

Dune Sand

During reopening work, Carrizo Gorge Railway personnel found substantial deposits of drifting dune sand in this western desert area of the alignment. Laboratory testing has showed that this is a coarse, good quality silica sand, with a sand-equivalent rating over 80.

Dune sand has plagued the railroad ever since construction. Frequent desert wind storms leave large piles (or dunes) of sand on the rails. The wind still blows, and no, this problem has not gone away.

One of the few remaining sources of good quality natural sand in the Southern California region, sand cleared from the tracks is hauled away in gondola cars and sold to the construction industry.

Sugarloaf Mountain to Seeley

About one and one-half miles after passing under Interstate Freeway 8, the community of Ocotillo can be seen across the Freeway, about one mile to the north. Just beyond Ocotillo, the Coyote Mountains (bounded by the Elsinore fault on the near side) rise sharply against the skyline.

About two miles further on, the rails pass the community of Coyote Wells (approximate elevation 280 feet), just before passing back under Interstate Freeway 8.

About a mile after this second passing under Interstate Freeway 8, the railroad grade crosses over Coyote Wash (approximate elevation 240 feet). Here the tracks begin a gradual ascent to an elevation of about 265 feet as they climb up and over low hills underlain by fossiliferous Pliocene/Quaternary sedimentary formations (Morton, 1977). On its descent to Plaster City, the train is traveling in an east-northeasterly direction.

Plaster City, a processing plant for gypsum, is at an approximate elevation of 100 feet msl. Gypsum is mined in the Split Mountain area of the Fish Creek Mountains, located approximately 22 miles northwesterly from Plaster City. A short-line mining railroad is used to transport gypsum ore to Plaster City.

Past Plaster City, the rails descend steadily eastward toward Seeley. About three miles beyond Plaster City, and about two miles before reaching the community of Dixieland, the tracks reach Sea Level (0 feet MSL).

At Seeley, the eastern terminus of the Carrizo Gorge Railway, the rail alignment is at an elevation of about 42 feet <u>BELOW</u> mean sea level (-42 feet MSL).

Underlain by hundreds of feet of marine and non-marine late Tertiary and Quaternary sedimentary strata, Seeley is one of several small agricultural communities in the Imperial Valley. Due south of Seeley, Signal Mountain rises out of the desert sands to an elevation of 2,262 feet.

Easterly from Seeley, the tracks are owned by the Union Pacific Railroad.

REAL TRAINS ON A REAL RAILROAD

More than 20 years had passed, prior to February 19, 2004, with no regular train traffic along the San Diego & Arizona Eastern Railway. Public perception over that time was that the railroad alignment was abandoned. It naturally attracted off-road vehicles, mountain biking and hiking enthusiasts that used the alignment for recreational activities. Because the entire line is now open to train traffic, it is patrolled regularly for public safety. And, NO TRESPASSING signs have been posted along the entire route.

Even though the narrow corridor through Carrizo Gorge cannot now be safely or legally used by hikers or bikers, this historic and scenic route may be transformed to accommodate multiple-uses in the future. A real possibility also exists for the route to establish regular passenger service along with special excursions trains designed for tourism, as well as educational and scientific activities. These are visions for the future, but will come along only after making the necessary improvements to operate a viable railroad business. Carrizo Gorge Railway, Inc. has already made heroic strides toward accomplishing these tasks.

The San Diego Association of Governments (SANDAG) has resolved, by *TEA-21 Demonstration Funds, Work Element #3000900* drafted November 29, 2003, that it has a serious interest in the future of this project. The resolution clearly cites several studies that indicate economic feasibility and marketing potential for reopening the San Diego & Arizona Eastern Railway.

This SANDAG decision was potentially a "sesquicentennial event" in accordance with the earliest of San Diego's resolutions to improve their transportation network some 150 years ago. The old dream of constructing an important, viable and reliable commercial railroad, connecting the interior of the United States with two of the best shipping harbors on the West Coast, San Diego, California, and Ensenada, Baja California, may become a reality within the next few years.

In our "shrinking world," new transportation corridors can only benefit citizens of all nations. Like a healthy tree, San Diego needs an eastern railroad "Main Stem," or it will remain just a branch.

ACKNOWLEDGMENTS

The author is most grateful to Bill Elliott, an old friend, colleague and ubiquitous local consulting geologist, for his meticulous review of the manuscript. His organization and "smoothing" of the initial bizarre mixture of interspersed technical, historical and anecdotal subjects was invaluable.

The principals and employees of Carrizo Gorge Railway, Incorporated (CZRY) also deserve unlimited thanks for their help in allowing access to the Carrizo Gorge segments during their dedicated, even heroic, construction activities to reopen this historic route. Special thanks go to Gary Sweetwood, Geoff Scheurman, John Zentmyer, Byron Wear, Rich Borstadt and many others with CZRY.

Bill Clevenger of Geocon, Incorporated and Charlie Coval of Parsons were initially instrumental in making the author aware of the ongoing activities to reopen the railway in 2003.

Others who were very supportive of the project were the author's family (Gerry, David, Cathy, Melissa, and Amanda), Howard Grebe, Mike Chapin and Rudy Surya of Geocon, Incorporated; Mike Hart, Consultant; Tom Deméré, Hugh Gardner, Steven Walsh and Richard Cerruti of the San Diego Museum of Natural History; Mike Walawender, George Morgan and J.R. Morgan of San Diego State University; Monte and Diane Murbach, Buck McCain, Phil Farquharson and Dave Bloom ... along with practically the entire membership of the San Diego Association of Geologists!

REFERENCES

Abbott, Patrick L., 1999, *The Rise and Fall of San Diego,* Sunbelt Publications, San Diego, California, 231 p.

Ambrose, Stephen E., 2000, *Nothing Like it in the World,* Simon & Schuster, New York, 431 p.

Bain, David H., 2000, *Empire Express,* Penguin Books, New York, 797 p.

Beauchamp, R. Mitchel, 1986, *A Flora Of San Diego County, California,* Sweetwater River Press, National City, California, 241 p.

Biehler, S., 1979, Bouguer Gravity Map of California, San Diego — El Centro Sheet: California Division of Mines and Geology Map Sheet, map scale 1:250,000.

Bowers, Q. David, *The Treasure Ship S.S. Brother Jonathan,* Bowers and Merena Galleries, Inc., Wolfeboro, New Hampshire, 415 p.

Boyle, R.W., *The Geochemistry of Gold and its Deposits,* Canada Geological Survey Bulletin.

Brands, H.W., 2002, *The Age of Gold,* Doubleday, New York, 547 p.

California Geological Survey, formerly Division of Mines and Geology, 1963, *Geology and Mineral Resources of San Diego County, California,* County Report 3, by F. Harold Weber Jr., 309 p.

California Geological Survey, formerly Division of Mines and Geology, 1975, *Geology of the San Diego Metropolitan Area, San Diego County, California,* Bulletin 200, By M.P. Kennedy and G.L. Peterson, Sacramento, California, 56 p.

California Geological Survey, formerly Division of Mines and Geology, 1984, *Preliminary Geologic Map of the California-Baja California Border Region,* DMG Open-File Report 84-59.

California Geological Survey, formerly Division of Mines and Geology, 1988, *Preliminary Map of the Quaternary Faults In Southwestern San Diego County And In Southwestern Imperial County, California,* DMG.

Copenhaver, George C. (Jr.), 1966, *Relationship Between Nickel Mineralization and Overlying Soil Composition in the Cuyamaca Gabbro, San Diego County, California.* Geological Society of America Special Paper 87: 199.

Copenhaver, George C. (Jr.), 1963, *Geologic Map of California, Orange County Sheet.*

Copenhaver, George C. (Jr.), 1970, *Geochemical Prospecting for Nickel in the Julian-Cuyamaca Area, California,* Master of Science Thesis, San Diego State University, San Diego, California, 160 p.

Copenhaver, George C. (Jr.), 1991, *Mining in the San Diego Metropolitan Area, Past and Present,* in *Environmental Perils of the San Diego Region,* joint publication of the San Diego Association of Geologists and Geological Society of America.

Copenhaver, George C. (Jr.), 2005, personal recollections 1948 to present.

Deméré, T., 1985, San Diego Natural History Museum, personal communication, the age of the Otay and Sweetwater formations is Late Oligocene to Early Miocene.

Einaudi, M.T., et al., 1981, *Skarn Deposits*, Economic Geology, 75th Anniversary Volume, pp. 317-391.

Gastil, Gordon, et al., 1971, *Reconnaissance Geologic Map of the State of Baja California*, Geological Society of America Memoir 140, Plate 1-A.

Gray, Andrew B. 1856, *Survey of a Route for the Southern Pacific R.R. on the 32nd Parallel,* Facsimile Editor L.R. Bailey, Westernlore Press, 1963, 240 p.

Hanft, Robert M., 1984, *San Diego & Arizona, The Impossible Railroad*, Trans-Anglo.

Harris, Mary E., 1986, *Geology of Baja California: a Bibliography,* Science Division, San Diego State University Library, San Diego, California, 159 p.

Hart, E.W., and Bryant, W.A., 1997, Fault-Rupture Hazard Zones in California: California Division of Mines and Geology, Special Publication 42, 38 p.

Hoggatt, W.C., 1979, *Geologic Map of Sweeney Pass Quadrangle, San Diego County, California,* U.S. Geological Survey, Open-File Report 79-754.

Hurlbut, C.S., Jr., 1959, Dana's Manual of Mineralogy: John Wiley & Sons, Inc., New York, 609 p.

Kennedy, M.P., 1975, Geology of the San Diego Metropolitan Area, California, Section A — Western San Diego Metropolitan Area: California Division of Mines and Geology Bulletin 200, 39 p., map scale 1:24,000.

Kennedy, M.P., 1977, Geology of National City, Imperial Beach and Otay Mesa Quadrangles, Southern San Diego Metropolitan Area, California: California Division of Mines and Geology, Map Sheet 29, map scale 1:24,000.

Kennedy, M.P., and Welday, E.E., 1980, Recency and Character of Faulting Offshore Metropolitan San Diego, California: California Division of Mines and Geology, Map Sheet 40, map scale 1:50,000.

Kirchner, J.A., 1988, *Baja California Railways,* Dawson's Book Shop, Los Angeles, California, pp. 92-137.

Lindsay, Diana, 2001, *Anza–Borrego A to Z: People, Places and Things,* Sunbelt Publications, San Diego, California, 400 p.

McCain, Fred (Buck), 2006, personal communication.

McGovern, Dan, 1995, *The Campo Indian Landfill War,* University of Oklahoma Press, Norman, Oklahoma, 325 p.

Miller, William J., 1935, *Geomorphology of the Southern Peninsular Range of California,* Bulletin of the Geological Society of America, v. 46, pp. 1535-1562.

Minch, John A., and Abbott, Patrick L., 1973, *Post-Batholithic Geology of the Jacumba Area, Southeastern San Diego County, California,* Transactions of the San Diego Society of Natural History, v. 17, no. 11, pp. 129-135.

Morton, P.K., 1977, Geology and Mineral Resources of Imperial County, California: California Division of Mines and Geology, County Report 7, 104 p., map scale 1:125,000.

Pourade, Richard F., 1966, *The History of San Diego, The Silver Dons,* The Union Tribune Publishing Company, San Diego, California, 286 p.

Pourade, Richard F., 1964, *The History of San Diego, The Glory Years,* The Union Tribune Publishing Company, San Diego, California, 274 p.

Remeika, P., and Sturz, A., eds., 1995, *Paleontology and Geology of the Western Salton Trough Detachment, Anza-Borrego Desert State Park, California,* San Diego Association of Geologists, San Diego, California.

Richardson, Christopher, 2004, Personal Communication re: Great Grandfather Cave J. Couts.

San Diego Association of Governments (SANDAG), 1995, *San Diego & Arizona Eastern Railway, Desert Line, Element No. 1 Engineering Study*.

Strand, R.G., 1962, Geologic Map of California, San Diego — El Centro Sheet: California Division of Mines and Geology Map Sheet, scale 1:250,000.

Todd, Victoria R., 2004, *Preliminary Geologic Map of the El Cajon 30'x60' Quadrangle, Southern California,* U.S. Geological Survey Open-File Report 2004-1361.

United States Geological Survey, 1915, *Guidebook of the Western United States Part C. The Santa Fe Route,* Bulletin 613, ed. By N.H. Darton, Government Printing Office, Washington, D.C., 194 p.

United States House of Representatives, 1853, *Report of The Secretary Of War,* Executive Document No. 1, pp. 17-25.

United States Senate, 1850, *Report Of The Secretary Of War,* Executive Document No. 47, Part I, pp. 118-123.

Walawender, Michael J., 2000, *The Peninsular Ranges,* Kendall/Hunt Publishing Company, Dubuque, Iowa.

Walawender, Michael J., 2003, *Geology of the Elsinore Fault Zone,* San Diego Association of Geologists, San Diego, California, and South Coast Geological Society, Santa Ana, California.

Weber, F.H., Jr., 1963, Geology and Mineral Resources of San Diego County, California: California Division of Mines and Geology, County Report 3, 309 p., map scale 1 inch equals 2 miles.

Wisser, Edward, 1954, *Geology and Ore Deposits of Baja California, Mexico,* Economic Geology, v. 49, pp. 44-76.

Young, Otis E., 1970, *Western Mining,* University of Oklahoma Press, Norman, Oklahoma.

History of the Campo Mill

with Related History of Feldspar and the Local Mining Operations

***Carl Calvert**, Motor Transport Museum,*
31949 Highway 94, Campo, California 91906, 619-478-2492

MILL HISTORY

The 2.5-acre Campo Mill site is found in an area known as Campo Meadows, which is located one mile north of Cameron Corners in eastern San Diego County. It is the current home of the Motor Transport Museum and was once part of a large cattle ranch dating back to about 1865.

When the railroad was completed in 1916, a siding was included at this location called Barrett. This siding was used to off-load construction equipment and materials for the building of Barrett Dam, some 20 miles to the west, which was built from 1919 to 1922. One can still see an original telegraph pole behind the mill identified with the mile post marker of "67," indicating 67 miles to San Diego. The Barrett marker sign also remains standing.

Sometime in the 1920s, Klauber Wangenheim purchased this site and sold it to Manuel and Mary Ortega on September 27, 1926 for $200. Manuel Ortega ran cattle on the ranch adjacent to the mill where his home was located. Five months later on August 28, 1926, Manuel and Mary Ortega sold the land to a New Jersey corporation, Standard Sanitary Manufacturing Company, Pacific Mines Division, for $10.00.

Plans began at that time to construct the mill to process feldspar, a mineral discovered about 6 miles to the west in Hauser Canyon. The mill was constructed in 1928 and 1929 with corrugated sheet metal over a riveted steel structure. It cost $100,000 to build and is an 8,000 square foot building reaching 100 feet tall encompassing nine levels. Forty-nine people were employed during peak production periods.

Feldspar, a quartz-like mineral, was used to make many items. Once ground, it was sold to be used to manufacture sparkplugs, signs, sinks, stoves, bathtubs, polish, abrasive soaps, glass, china, tile, insulators, and as a straining material. A large amount was used by Standard Sanitary Manufacturing Company to manufacture kitchen and bathroom fixtures.

In addition to feldspar, silica was also mined in Hauser Canyon. Once a year, an order for very fine silica was shipped to Hawaii for straining pineapple juice. At one time it was the only mill that could produce the silica to the proper specifications.

The ore was trucked to the mill from the mines in Hauser Canyon. At the mill it was crushed, ground, screened and put in 100-pound bags. These were loaded into railroad cars bound for San Diego and other points of distribution.

In the mill, workmen were allowed to work in the dusty areas for only 20 minutes wearing respirators. They then had to work elsewhere.

The processing of the ore was done in stages. A large jaw-type crusher sat on the west side of the mill and large stones were run through the crusher before entering the mill. The ore was then fed into chutes that passed it into a basement. Elevators carried the mineral to the upper nine stories of the mill. The mineral then passed down through various sized screens. The coarser material was directed into various mills where it was crushed further. It was redirected back to the upper levels to be rescreened and ground again if necessary.

Some of the material was fed into "chaser stones" which were large solid stone wheels from Norway. They measured 6 feet in diameter and 20 inches thick, revolving in pairs around an axle. One of these wheels is still on the site. Additionally, some of the material was placed into 5-foot revolving cylinder ball mills and further ground.

Once the feldspar passed through a 200-mesh screen it was bagged and shipped. At this point it was a fine powder like flour. Some of the material was loaded into box cars in bulk after they were lined with heavy wax paper.

Standard Sanitary Manufacturing Company at one time believed these mines comprised the largest known body of feldspar in the world. In 1934, Phil Jones, Pacific Mine Manager of Standard Sanitary Manufacturing Company, reported the mine had produced 100,000 tons of feldspar valued at $1,500,000. He declined to estimate the amount of feldspar in the claim, however, he indicated the company has 200,000 tons "blocked out." The company was selling all of the feldspar they could mine at that time. They were also concerned about the railroad rates for shipping the milled feldspar. They claimed any increase would be detrimental to continued operations.

The Standard Sanitary Manufacturing Company operated this plant until 1943 when it was closed because of increased labor costs and because of a lack of reserves of easily mined feldspar.

In December 1933, an intermediate crusher caught fire and closed the mill for a short while. The storage room and drying rooms were also damaged.

Records show sometime between 1929 and 1950 the Standard Sanitary Manufacturing Company merged with The American Radiator Company. During the operation of the mill, it was covered with layers of fine white powder. The constant rumble of grinders, mills, shakers and sorters made a deafening noise. In 1938, according to a former employee, J.G. Archibald, one workman's arm sleeve was caught in a large gear, his arm was pulled out of the socket. His name was Enrique Arcala and he subsequently died from the accident.

The big 2-cylinder (4-cylinder?) Fairbanks Morse 180-horsepower vertical diesel engine which provided power to the mill was temporarily shut down. It ran a generator which powered the electric motors throughout the mill. It is believed the engine and generator were removed in the 1950s and sold to the township of Daggett to be used to power electric current.

On January 10, 1950, the mill was sold for $4,100 to Farrar Matthews, a black miner and former general foreman of the mill under the Standard Sanitary Manufacturing Company. Farrar and Gertrude Matthews sold the mill on January 12, 1953 to Associated Mining & Materials Corporation, a Nevada corporation which eventually went bankrupt. On October 29, 1955, the owners, Lafayette Morrison, Richard S. Morrison, and Ralph W. Morrison, lost the mill and numerous mining claims, including Cinder Block and Volcanic Metallite claims at Little Lake in Inyo County; Kaolin Placer Mining claims in San Bernardino County; a pyrophyllite claim and quarry in Rancho Santa Fe; the Arthur Clay Feldspar claims and mines #4, 5, & 6 and Feldspar Lode claims #7, 8, 9, 10, 11, 12, all in the Campo Mining District. Additionally, the dry grinding mill located at the foot of "G" Street in Chula Vista was included in the bankruptcy. It appears these mining claims also served as a source of materials for milling at the mill. On October 26, 1955, William A. Wylie, as trustee, sold the entire estate for $10.00 including a $97,976 mortgage to Block-Sheppard Enterprises. This company was owned by Thomas M. Block, a Long Beach contractor, and John N. Frolich. The mine remained intact at this time.

On December 29, 1959, Block-Sheppard sold their interest in the Campo Mill to Dorothy Benner, a widow, for an undisclosed amount along with the $97,000 mortgage. On July 10, 1967 she added her son, Mason P. Benner, to the title. They also owned Torrance Sand and Gravel Products Company at 25701 Crenshaw Boulevard, Torrance, California. On April 17, 1987, Farrar Matthews again took title to the mill along with his son, Donald Matthews, and Clarice Adams, by deed from Dorothy Benner. He died in Los Angeles on May 3, 1980. On July 17, 1987, the mill was sold to James McGuffie and his mother Flora Harris on a land contract.

Shortly after the McGuffie-Harris acquisition legal problems again arose over the conveyance and release of a prior lien. Title became clouded and although James McGuffie had a vision of using part of the mill for a restaurant, the idea never matured. He was eventually able to clear title after several years of legal arguments.

The mill had been idle since about 1959 and by this time most of the machinery was stripped from the mill. In January, 1988, the mill was acquired for use by the Motor Transport Museum.

MINING OPERATIONS

The feldspar mine in Hauser Canyon is located one-half mile beyond the end of the Hauser Canyon dirt road. Permission is required to pass through a 40-acre private parcel of land owned by Pacific Pipeline Construction Company in Temecula, California. With permission granted, it is an easy hike down one of East County's most beautiful canyon areas. One can follow a dirt road that passes among huge sycamore and oak trees scattered in the wide canyon area. The canyon provides a watershed area for Barrett Dam, built from 1919 to 1922. A well-traveled road once traversed this canyon towards San Diego which avoided the need to pass into Mexico on a more rugged southern route prior to the dam being built.

The first feldspar deposit was discovered by William Stell in 1918. Through the assistance of the R.C. Springer Company, the first claim, the Arthur Clay Spar Number One, was patented. Patent number 944723 was issued by Calvin Coolidge, the president of the United States, and recorded on October 8, 1924. It included an area of 14.481 acres and was issued subject to any rules issued by the California State Legislature about rules for mining and also subject to any local laws or requirement for water rights, easements or road accesses.

From 1921 to 1943, it was reported the deposit yielded approximately 87,000 tons of feldspar and this was the principal source of the mineral in the state of California. The mine was operated during this time by the American Radiator and Standard Sanitary Corporation under the following names:

1921-1923 Pacific Porcelain Ware
1923-1925 Pacific Sanitary Company
1925-1943 Standard Sanitary Manufacturing Company

Peak production was from 1927 to 1929, when 32,000 tons were mined. Sometime after 1949 the underground mines were dynamited, closed by the Forest Service and are therefore no longer open.

Mining efforts began by the Standard Sanitary Manufacturing Company in 1921 and after the mill was constructed on Highway 94 about 4 miles west, material was then trucked to the mill for crushing.

The feldspar mine was located on the northeastern slope of Hauser Canyon, up the steep canyon face about 150 ft. It is situated in the Southwest 1/4 of Section 25, Township 17 South, Range 4 East, San Bernardino Base Line and Meridian.

The original mining efforts were done in an open pit as a quarry. However, when the face of the quarry was about 400 feet deep, the company began to worry about safety so they changed to mining by tunnel and drifting with timbers. In May of 1923 the tunnel was 165 feet deep in solid feldspar.

The hard feldspar mineral was drilled, dynamited and hauled down the canyon side. Some feldspar continued to be dug out on an open pit basis and other material was excavated from underground mines. Six to seven people usually worked at the mine. Ore cars and a hi-line bucket brought the material out of the mine to a tram which was a cable-drawn railway. Two large ore cars were loaded at intervals and as one loaded car descended to the bottom the other car rose to the top to be loaded. The weight of the heavier car brought the unloaded car to the top. To prevent a run-a-way, the tram operator located at the top operated a brake on the cable line. These rail lines were typically found in mining operations and did not require power to operate. They were called ventriculars. A cable-drawn railway operated on an incline in which the weight of one car offset the weight of another. This was a single railway with a divided segment at the halfway point in which the cars could pass each other.

One of the tram operators, Max Senter, still lives in Campo and operated this 2-car rail line in the early 1930s. His wage was about $5.00 per day. He still recalls one incident where a brake was inadvertently released while a workman was ascending to the top in an empty car and tossed him out of the car. The workman, Bob Mesa, suffered some injury but soon returned to work. It was company policy not to ride in the ore car.

In 1937, another employee, J.G. Archibald, worked at the mine as a mechanic. He serviced a diesel engine that ran a conveyor, another one that ran an air compressor, and a third one that ran the water pump in the nearby creek. On May 18, 1940 a large earthquake hit the Imperial Valley area, and he vividly recalls being in the mine at the time one of the aftershocks took place. It took him several hours to get out as dust filled the mine to the point that one could not even see with a light. When he finally got out about two hours later, he looked as white as a ghost.

After the feldspar was transported to the bottom of the hill on the tram, it was put on a conveyor belt known as a picking belt and dumped on a table for sorting and handcobbing to insure delivery of a clean product at the mill after which it was dumped into a bunker. A 10-horsepower vertical diesel Fairbanks Morse engine ran the belt line. From there it was loaded into trucks for transport to the mill.

As the ore was extracted from the mine claims and brought down for sorting, it was segregated into spar and silica. The spar is an opaque chalky white rock, whereas the silica was a white translucent crystalline appearing rock. The spar was used for sparkplugs, sinks, stoves, bathtubs, tile and insulators. The silica was used for glass manufacture and straining material.

The final grinding of the feldspar products were (1) 200 mesh for the pottery industry, and (2) 100-150 mesh for the enameled ware industry.

Quartz products were (1) 200 mesh for the pottery industry, and (2) 65-100 mesh for the glass industry. The grinding capacity of the mill was 50 tons per day.

The remaining lode claims discovered by William Sell were unpatented and adjoined the one patented claim. They were as follows:

- Arthur Clay Spar Mine #4
- Arthur Clay Spar Mine #5
- Pacific Spar Quarry #6
- Feldspar Lode and Claim #8
- Feldspar Lode and Claim #9
- Feldspar Lode and Claim #10
- Feldspar Lode and Claim #11
- Feldspar Lode and Claim #12

ABOUT FELDSPAR

Feldspar is the general name for a group of anhydrous aluminum silicate minerals which contain varying amounts of potassium, sodium, and calcium $[(K,Na,Ca)Al(Al,Si)_3O_8]$. It is the most abundant mineral in the earth's crust.

About 50% of feldspar is used in glass production, 35% in ceramics, 4% in enamel, and 11% in soap, abrasives, alumina and soda production. Feldspar increases the chemical stability on the finished product. It is usually ground to a 20 or 30 mesh for glass use and as fine as 200 mesh for ceramic use.

The use of feldspar in glass was begun in 1925 and became standard use in the glass container industry about 1935 and in the flat glass industry by 1940. A little more than 110 pounds of feldspar is used in an average ton (2000 pounds) of glass containers and about 95 pounds in a ton of flat glass.

Feldspar is used in ceramics because it fuses at lower temperatures than the other ingredients. It acts as a flux by wetting the solid particles and surface tension pulls them together. Feldspar also increases strength, toughness, and durability. It is used in amounts from 10 to 35 percent in glass and from 30 to 50 percent in glaze. Feldspar has been used in ceramics in China since the Tang Dynasty (621-945 A.D.) and in Europe since the 18th century.

Feldspar was first mined in the United States in 1825 in Connecticut and sent to England for use in ceramics. In about 1911 feldspar was mined in Mitchell County, North Carolina. North Carolina became our major producer shortly thereafter. California is also a major producer.

Feldspar mined in the Campo District was by open pit and by underground methods. The broken rock is sorted by hand during loading. The value of this ore in the 1950s and 1960s was about $10.00/ton for crude feldspar; $15.00/ton ground 40 mesh and $20.00/ton for ground 200 mesh. Current 1990 production values are about 3 times higher. Production trends and demand for feldspar has been increasing primarily because of the increased demand for glass products.

ABOUT THE MOTOR TRANSPORT MUSEUM

Now housed at the mill are over 150 antique trucks with such names as Moreland, Federal, Fageol, Mack, White, USA, Linn, Coleman, Brockway, and Packard. Many of these vehicles were used in mining operations and with the Army during World Wars I and II. Most still show the scars of hard use. Many are equipped with hard rubber tires and are chain-drive.

The Motor Transport Museum is open for public viewing from 10 a.m. to 5 p.m. every Saturday.

(Edited reprint of Campo Mill brochure, Calvert.)

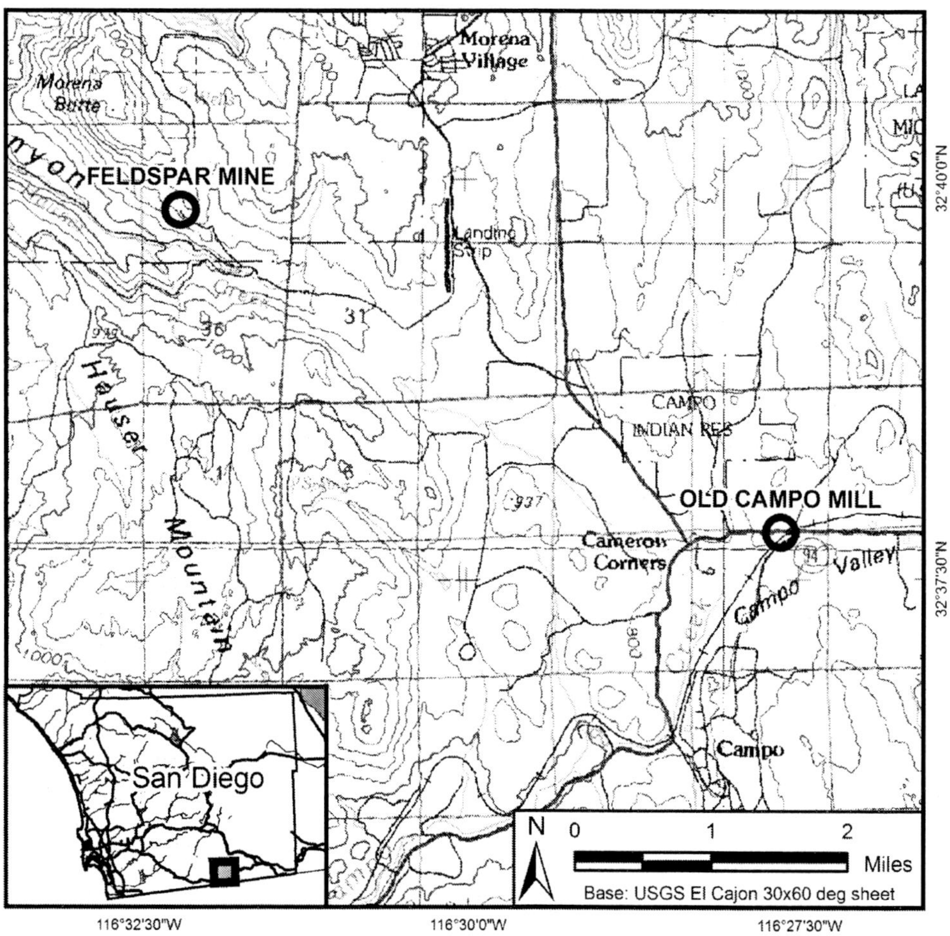

Post-Batholithic Geology of the Jacumba Area, Southeastern San Diego County, California

***John A. Minch**, 1100 Mission Canyon Road, Santa Barbara, California 93105*

***Patrick L. Abbott**, Department of Geological Sciences, San Diego State University, San Diego, California 92182*
professor_pat_abbott@yahoo.com

ABSTRACT

The post-batholithic history of the Peninsular Range is documented by sparse exposures of fluvial and volcanic rocks in widely separated areas. The Jacumba Valley outcrops present one of the most complete stratigraphic records in the range.

The initial erosion surface upon the Southern California Batholith probably formed in the Late Cretaceous. This surface was fairly deeply weathered as was the earliest granitic gravel deposited upon it. Clasts resembling those of the Table Mountain Gravels were transported across the Jacumba area to the Pacific Coastal region, where they appear in the Cabrillo Formation of Late Cretaceous age. Eocene "Poway-type" gravel was transported across an essentially parallel surface just south of Jacumba Valley near La Rumorosa. Erosion partially removed and reworked the gravel until Early Miocene outpourings of basalt and pyroclastic debris filled in much of the Jacumba Valley area. The first basalt flows from eruptive centers within the valley were followed by faulting and the deposition of andesitic pyroclastic and lahar deposits. The source of this andesite may have been the plug-like masses at Round Mountain and Jade Benchmark. Continued faulting offset the volcanic rocks before the eruption of a second series of basalt flows which covered the andesite in the northeast and east portions of the valley. Intermittent erosion within the volcanic sequence is evidenced by fluvial and eolian volcaniclastic deposits. Post-volcanic faulting elevated the Peninsular Range and accelerated erosion to produce the present topography.

INTRODUCTION

The Jacumba Valley area is located at the crest of the Peninsular Range and straddles the Mexico-United States boundary. Interstate 8 and the San Diego and Arizona Eastern Railroad pass through the valley between San Diego and the Imperial Valley. Rainfall is light resulting in sparse vegetation and excellent outcrops. Field work for this report was accomplished in the fall of 1971 and spring of 1972 in conjunction with a field geology class at California State University, San Diego.

The first account of the post-batholithic geology of Jacumba Valley was by Fairbanks (1893), who indicated the presence of the gravels and volcanics at the crest of the Peninsular Range. Miller (1935a) followed with a brief description of the rocks and included speculations on their former widespread extent, exotic origin, and Miocene age.

In the late 1940s and early 1950s field classes from San Diego State College used the area for reconnaissance mapping exercises. These were compiled as map sheet 23 (Brooks and Roberts, 1954) in Bulletin 170 of the California Division of Mines and Geology. Jacumba Valley remains an excellent area for students to map in a terrane exposing a variety of eruptive features resting on sedimentary, plutonic, and metamorphic rocks.

Gastil and Bushee (1961) and Weber (1963) briefly mention the valley. Hawkins (1970) provided a detailed analysis of the chemistry of the volcanic rocks in the valley and tied them into the overall picture of sea-floor spreading in Southern California. Minch (1971) described the Table Mountain Formation and indicated its non-local origin.

BASEMENT ROCKS

The crystalline rocks flooring the valley have been mapped as two separate units (Weber, 1963). The older mass is comprised of metamorphic rocks mixed with granodiorite, diorite, and pegmatites. These metasedimentary rocks are dominated by the quartz- and mica-rich Julian Schist, with minor amounts of gneiss and quartzite and occasional pods of marble. These are cut by abundant pegmatites and plutonic bodies.

The younger plutonics of the Southern California Batholith are here composed mostly of quartz diorite along with granodiorite and minor pods of gabbro.

TABLE MOUNTAIN GRAVELS

The Table Mountain Gravels are light yellow-brown, moderately bedded, fairly well-sorted, very friable, medium to coarse-grained sandstones and conglomeratic sandstones which crop out in and near Jacumba Valley. In addition to local granitic clasts they contain clasts of low-grade green metavolcanic and metasedimentary rocks and quartzites that are not found locally.

Miller (1935a: 138) defined the Table Mountain Formation from exposures at Table Mountain 7 km northeast of Jacumba as: "Moderately consolidated deposits of yellowish to reddish brown gravels and sands. Various kinds of pre-Cretaceous crystalline rock fragments occur in the formation ... These sediments are rather variable in character, crudely stratified, and gently dipping."

Several authors have indicated the exotic nature of these gravels found high in the Peninsular Range (Fairbanks, 1893; Brooks and Roberts, 1954; Weber, 1963). Brooks and Roberts (1954) compare the clasts to the Santiago Peak Volcanics of western San Diego County: "They contain fragments of dacites and other aphanitic rocks that show strong similarities to the Jurassic(?) Santiago Peak Volcanics of western San Diego County. These gravels are partially interbedded with and principally overlain by other volcanic rocks."

Distribution

The Table Mountain Gravels crop out on the erosion surface in a belt about 10 km wide and 25 km long that lies roughly parallel to the axis of the Peninsular Range. They are the remnants of an extensive fluvial deposit. In the Jacumba area they stretch another 5 km down the frontal scarp of the range. The principal outcrops are in and around Jacumba Valley and in the area just west of La Rumorosa in Baja California. The small isolated patches of the gravels which occur at lower elevations on the frontal scarp of the range are the easternmost exposures.

The best exposures of the Table Mountain Gravels are in the area of Jacumba Valley where the Jacumba Volcanics form a resistant cap above the gravels (Figure 1). A typical section measured on a flat-topped hill just north of Jacumba in the northwest corner of Sec. 5, T 18 S, R 8 E consists of 75 m of interbedded light yellow-brown, moderately to thickly bedded, very friable, medium- to coarse-grained sandstone and 0.5 to 1 m thick beds of conglomeratic sandstone. The sandstones within the Table Mountain Formation are sheetwash to fluvially deposited, plutonic lithic arkose. The framework grains are very angular, poorly to very poorly sorted, mineralogically immature, and are cemented by poikilotopic, very coarsely crystalline calcite where unleached. Common grains include both fresh and heavily sericitized plagioclase and orthoclase, polycrystalline quartz, schist and plutonic rock fragments along with hornblende, biotite, muscovite, and other accessory minerals.

The gravel-sized clasts are subangular to subrounded and average 2.5 to 5 cm in diameter with clasts commonly to 10 cm and rarely to 30 cm. Fifty percent of the clasts are extra-regional, low-grade, green metavolcanic and metasedimentary rocks. Other significant components of these gravels are quartzites and resistant sandstones (25%), and local granitic and gneissic basement rocks (12%). The thickness of the gravels is quite variable, ranging from a thin mantle on the surface to greater than 75 m. Most exposures are less than 30 m in thickness.

Other good exposures of these gravels occur on the northeast side of Jacumba Peak (N. Center Sec. 7, T 18 S, R 8 E), on Table Mountain (T 17 S, R 8 E), and in Myer Valley (NE 1/4, Sec. 26, T 17 S, R 19 E) about halfway down the frontal scarp.

In the Jacumba area the Table Mountain Gravels seem to have been highly eroded before the deposition of the Jacumba Volcanics. This is evident because the basalts and breccias have an irregular basal contact which, in one case, rests on 75 m of gravel at one end of a hill but sits on the granitic erosion surface 120 m lower in altitude at the other end of the hill.

Age

The Table Mountain Gravels are certainly older than the 18.5 m.y. Jacumba Volcanics which overlie them, and they are younger than the rocks of the Peninsular Range Batholith upon which they rest (90-105 m.y., Bushee et al., 1963; cooling age of 65-80 m.y., R.G. Gastil, pers. comm.). They were deeply eroded before the deposition of the Jacumba Volcanics, suggesting that they may be significantly older than the Miocene volcanics.

In the San Diego coastal area the conglomerate of the Upper Cretaceous Cabrillo Formation (Kennedy and Moore, 1971) contains a small percentage of black to blue-gray quartzite, light-colored quartzite, and chert pebbles, and several types of metavolcanic clasts not associated with the local basement. These clasts are not found in the locally derived gravel of the underlying Lusardi Formation, indicating that they are extra-regionally derived clasts. These exotic clast types are also found in the Table Mountain Formation.

The presence of the extra-regional clasts in the Cabrillo Formation, which has been dated as Maestrichtian (Kennedy and Moore, 1971), suggests a relation to the Table Mountain Gravels, which in turn suggests that the two formations might be the same age. If so, this would place the initial deposition of the gravels in the Late Cretaceous.

[Note added to text in 2006 by Abbott: "With regards to the age of the Table Mountain Gravels, shortly after this paper was submitted, I undertook a rather exhaustive search for clasts of the Miocene Jacumba Volcanics within the Table Mountain Gravels. They are not common, but

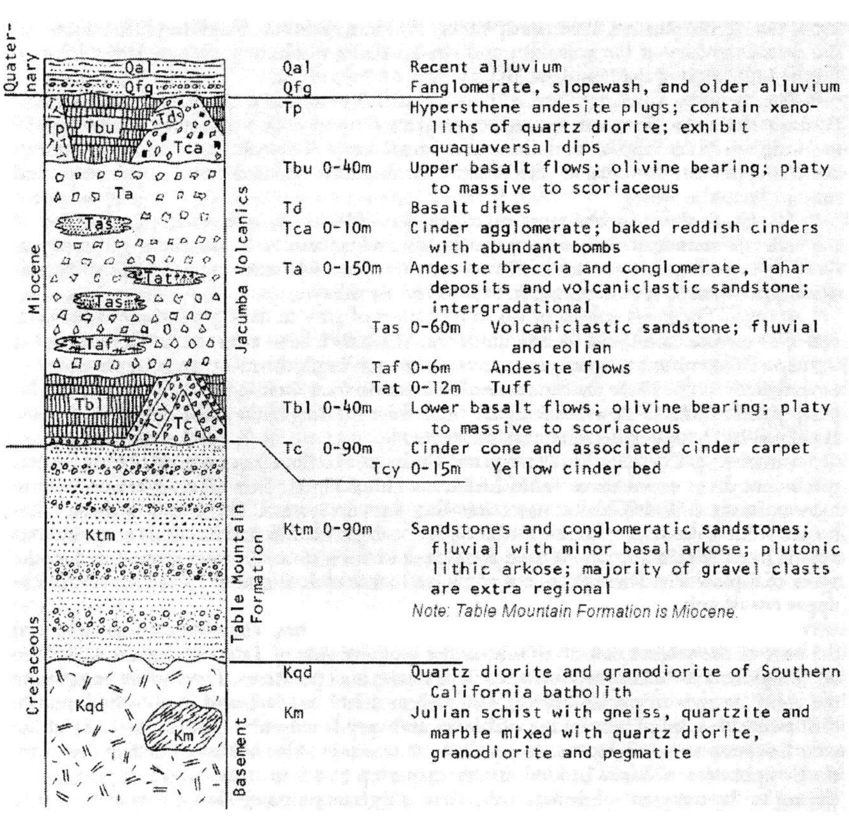

Figure 1. Generalized Columnar Section.

they can be found including clasts low in the stratigraphic section. Bottom line: The Table Mountain Gravels are Miocene in age."]

JACUMBA VOLCANICS

The Jacumba Volcanics were defined by Miller (1935a: 138-139) for "...the extensive rocks which are excellently exposed in the several areas north to east of Jacumba ... Some volcanic breccias or pyroclastics here occur toward the bottom of the lava beds." In the present report the Jacumba Volcanics are subdivided into three basic units. They are: (1) basalt flows and cinder-cone deposits, (2) basaltic-andesite plugs, and (3) andesite breccia, lahar deposits, conglomerate, and volcaniclastic sandstone.

The same authors who discussed the Table Mountain Gravels also generally discussed the Jacumba Volcanics (Fairbanks, 1893; Miller, 1935a, b; Brooks and Roberts, 1954; Gastil and Bushee, 1961; and Weber, 1963). In addition, Hawkins (1970) discussed the petrochemistry of the volcanics and cited a K/Ar whole rock date of 18.7±1.3 m.y. for the lower part of the basalt sequence.

The Jacumba Volcanics form a 20 by 55 km belt of outcrops parallel to the axis of the Peninsular Range. However, the principal areas of outcrop are in a narrow belt 5 to 10 km wide by 55 km long on and along the frontal scarp of the range, with the majority of the outcrops near the base of the scarp. The Jacumba Volcanics also crop out in and around Jacumba Valley.

The basalt flows are the most extensive part of this unit, with outcrops over most of the valley. The andesite breccia, lahar deposits, and volcaniclastic sandstone are as extensive but more limited in outcrop. Remnants of at least five cinder cones and two prominent hypersthene andesite plugs are exposed in the valley.

Flows

The lavas consist of up to 40 meters of gray to dark gray, platy to massive, vesicular olivine basalt and basaltic andesite. The basalt flows crop out at the base and at the top of the volcanic sequence. As many as five or six flow units may be represented in any given outcrop. These flow units tend to be massive in their upper portions but exhibit platy jointing due to flow shearing at their bases or where they abut other rocks (see Hawkins, 1970, for detailed chemical description).

Dikes and Cinder Cones

Squaw Tit is an excellent example of one of several prominent dikes exposed on Table Mountain (Map: Figure 2). This dike system appears to have partially followed a northwest-trending fracture system. The basalt in these dikes ranges from a dense gray-green to dark gray, well jointed olivine basalt to a scoriaceous olivine basalt which appears to have been close to the surface of a vent represented by the spine of Squaw Tit. These dikes cut the lower basalt unit and may be the source for the upper basalt unit.

Associated with these dikes are at least two cinder cones. The oldest cinder cone is at the base of the section east of a fault on the southern side of Table Mountain. It is 75 to 90 m high and is well exposed as a result of quarrying operations. The cinders comprising the cone are red- to purple-brown, well sorted, thinly bedded, and tend to be lapilli to dust size with a small percentage of blocks and very few bombs. The upper 15 m of the cone has been altered by gasses to a yellow-brown color. Also deposited during this pyroclastic episode is a thinly bedded cinder

A
Qfg
Tbl
Ktm
Km
Kqm
Tp
Ta
Qal
Kqd
Prospect
JEEP TRAIL
Dubber Spur
Spring
Springs
Gray Mountain
TAB
ANZA-BORREGO DESERT STATE PARK
EASTERN
Titus
Jacumba Valley
Gravel Pit
Jacumba Peak
ARIZONA
Creek
Jacumba
AIRPORT
CALIFORNIA
BAJA CALIFORNIA
BM 2993
BM 2798
BM 2829
BM 2919
BM 2967
Res
CONTOUR INTERVAL 40
DOTTED LINES REPRESENT 20-FOOT
DATUM IS MEAN SEA LEV
CALIF
QUADRANGLE LOCATION
TRUE NORTH
MAGNETIC NORTH
14°
APPROXIMATE MEAN DECLINATION, 1959

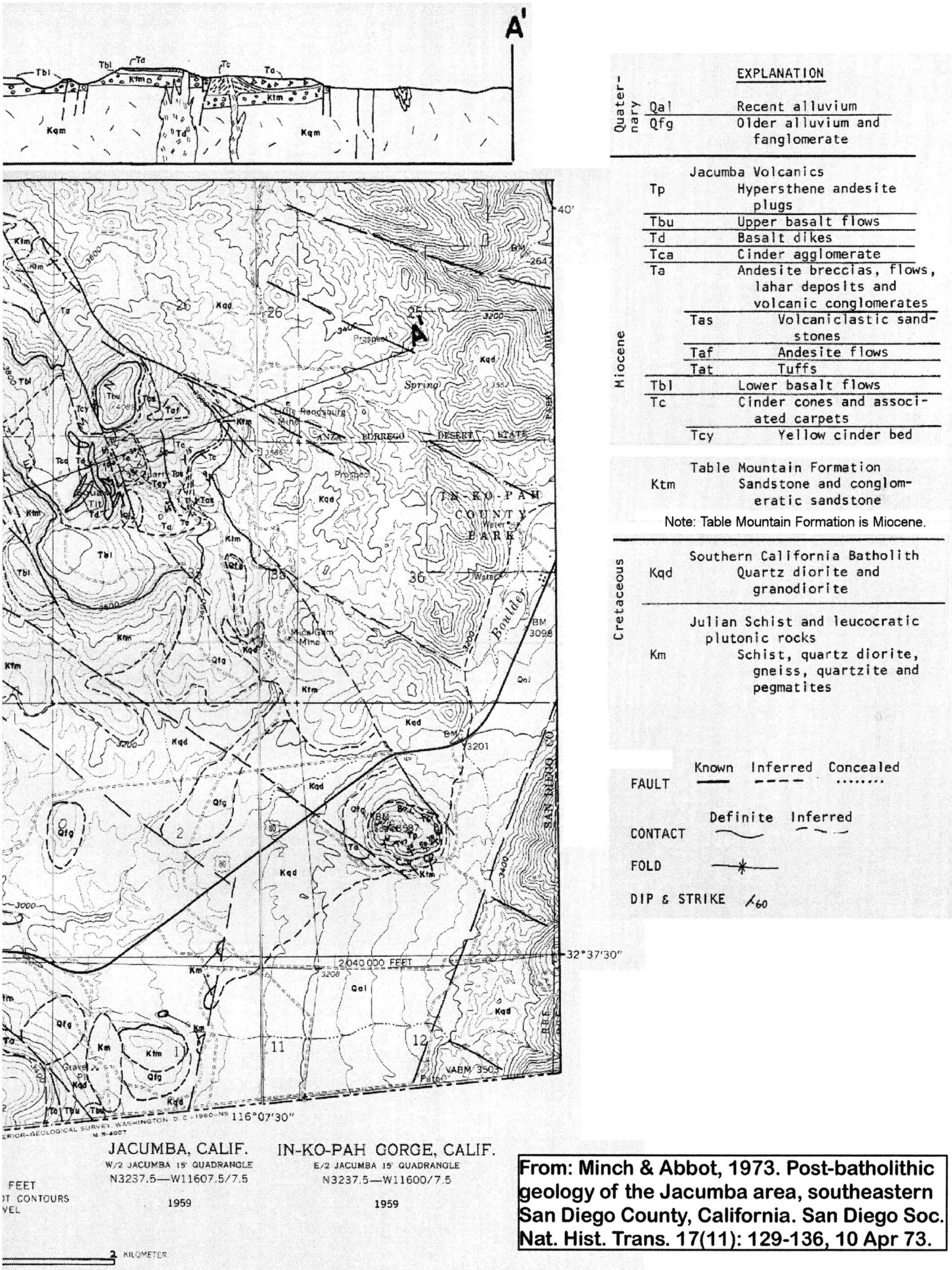

From: Minch & Abbot, 1973. Post-batholithic geology of the Jacumba area, southeastern San Diego County, California. San Diego Soc. Nat. Hist. Trans. 17(11): 129-136, 10 Apr 73.

Figure 2. GEOLOGIC MAP OF JACUMBA VALLEY

carpet up to 4.5 m thick, which appears to be thicker to the east and southeast of the cone, suggesting a paleowind direction similar to the present prevailing wind pattern. This cone may have been the source for the earlier basalt flows. A fault truncates its west side and moves the western portion right-laterally about 90-150 m.

Transitional lithologies occur where the first pyroclastic eruptives are mingled with the granitic wash mantling the Table Mountain Formation. Commonly found here are volcanic glass-cemented, spherical concretions of slightly granular, bimodal, very coarse- and very fine-grained, volcanic lithic arkose. The coarser mode contains numerous plutonic rock fragments along with microcline, orthoclase, and perthite from the surrounding highlands. The finer mode comprises idiomorphic volcanic plagioclase, relict shards, lamprobolite, zircon, biotite, and apatite. The non-concretionary arenite in places contains montmorillonite formed from altered pyroclastics.

West of the major fault and interlayered with but partially overlying the earlier flows is a series of agglomerate and cinder deposits. These pyroclastics are distinguished from the earlier cone materials because they overlie the lower basalt, are coarser, lack sorting, and contain a large percentage of bombs and agglutinate material. Some of the bombs and blocks in this younger cone are up to 1.3 m in diameter, although most average 3 to 5 cm. Several basalt flow units appear to have issued from the base of the younger cone along the west and northwest side of Table Mountain. The cone morphology is not obvious as it has largely been destroyed or obscured by later deposits. Cinder and scoria deposits suggestive of smaller eruptive centers are found on the east side of the border hill (VABM 3572), in a pit along Carrizo Gorge road just north of old Highway 80, in a pit east of Round Mountain and south of Interstate 8, and in the low hills north of the freeway and west of the Jacumba turnoff. All have an abundance of calcite cement, and in the eruptive center just south of the freeway the calcite percentage is so high that the cinders initially appear to lack a supporting framework.

Andesite Plugs

Round Mountain and the hill at Jade Benchmark (Map: Figure 2) are plugs whose chemistry differs significantly from the other volcanic rocks in the valley (Hawkins, 1970). They contain significant percentages of hypersthene in place of the hornblende typical of the basalt flows. These plugs are largely intact and have typical onion-skin jointing and a bulbous dome-like structure. Both plugs contain xenoliths of granodiorite, quartz diorite, and schist from the basement rock.

The plugs are younger than at least that part of the andesite breccia sequence which they rest upon. No clear-cut evidence can be presented for their minimum age and they could be much younger than the lavas and andesites. Similar appearing plugs of Pliocene age (R.G. Gastil, pers. comm.) occur in other parts of the Peninsular Range near the international boundary.

Andesite Breccia, Lahar Deposits, Conglomerate, and Sandstone

A heterogeneous sequence of andesitic lahar deposits, breccia, conglomerate and sandstone form a significant portion of the Jacumba Volcanics. The andesites are up to 150 m thick in some localities and generally average 90 m thick over much of the area. The breccia appears to be the dominant form of the andesite, but other forms are more common locally. The intergradation of the component rock types and the presence of a heavy lag gravel over the andesite prevented the mapping of the individual andesitic rock units.

The best exposure of the andesite breccia is just east of the Table Mountain quarry in NE 1/4, Sec. 35, T 17 S, R 8 E (Map: Figure 2). There the breccia consists of red-brown to brown to gray massive hornblende andesite which has been intensely sheared, broken, and comminuted to form a flow breccia. No bedding or stratification can be discerned at the outcrop.

The andesite breccias grade into lahars (mudflows of volcanic detritus). In the vicinity of the Table Mountain quarry an approximately 9 m thick lahar deposit consists of a red-brown to brown, massive andesitic breccia with andesite and cinder particles ranging from dust to small boulder size. The complete lack of bedding and sorting coupled with monolithologic andesite fragments are characteristic of a lahar deposit. Intermittent fluvial action has reworked the breccia and lahar deposits producing volcaniclastic conglomerate and sandstone which occur throughout the section. These fluvial units resemble the breccia in clast composition, and they are distinguished by the presence of bedding and of increased sorting of the clasts. A typical sandstone exposed above the quarry on Hill 4089 (SW 1/4, Sec. 26, T 17 S, R 8 E) is a very poorly sorted, very angular to subangular, pebbly medium sandstone. This and similar sandstone beds are lithic arkoses and are loaded with volcanic rock fragments and extrusive euhedral minerals, while plutonic-derived sediment is usually absent.

Mixed plutonic-volcanic lithic arkoses are found between tuffs (below) and volcanic mudflows (above) on the south side of Round Mountain. This mixed-provenance lithology illustrates the intermittent nature of volcanic sedimentation that allowed inworking of granitic-derived debris.

Intercalated within the volcanic sequence west of Gray Mountain is a volcanic lithic arkose mass exhibiting 0.6 to 1.3 m thick planar-wedge sets of cross-laminae. These fine sandstone grains are mostly plagioclase, volcanic-rock fragments, hypersthene, hornblende, and biotite that are moderately sorted, skewed toward the fines, and subrounded to subangular. The sedimentary structures, presence of abrasion, fine grain size, and best sorting in the valley all indicate an episode of reworking of volcanic sediment into eolian dunes.

The various units of the Jacumba Volcanics are considered to be of Early Miocene age. The K/Ar whole rock age date of 18.7 ±1.3 m.y. (Hawkins, 1970) for a basalt on Jacumba Peak corresponds with concordant hornblende and plagioclase dates of 18.5 ±0.9 m.y. and 18.6 ±0.8 m.y., respectively, obtained from a clast in the andesite breccia on Table Mountain (K/Ar laboratory CSUSD). Thus, the bulk of the Jacumba Volcanics were erupted in the Early Miocene.

STRUCTURE

The structure of the Jacumba area is dominated by a series of northwest-trending normal and reverse faults. Two east-trending faults bound the area on the north and south. Only a few northwest-trending faults could be traced beyond the east-trending faults.

The northwest-trending set of faults produces a horst and graben effect within the valley. Throws of up to 300 m are necessary to produce some of the observed features. The fault which bisects Table Mountain has approximately 90 to 150 m of right-lateral separation and 30 to 60 m of vertical separation. In many cases a fault which can be traced for some distance with certainty in the volcanic rocks is lost within a few feet in the granitic rocks. In other cases a single fault in the volcanic rocks splits into several segments as it enters the granitic rocks.

An interesting feature noted along several large faults is the inclusion of a thin sliver of Table Mountain Formation in the fault zone. In a number of cases the presence of a fault was recognized by the thin strip of conglomerate between the granitic and the volcanic rocks.

The two east-trending faults are primarily in granitic rocks and have helped erosion produce long linear valleys. The northern fault exhibits a 0.6 to 1.5 m shear zone along the north side of Table Mountain.

The structure in the Jacumba Valley closely parallels the regional structure of the northwest side of the Peninsular Range. The valley itself is directly along the projected trend of the Elsinore fault zone as it passes out of the main mountain mass near Banner Grade. The right-lateral separation on at least one fault in the Jacumba area may suggest that some of the displacement on the Elsinore is taken up by movement along its projected trend in the Jacumba Valley area.

ACKNOWLEDGMENT

We would like to thank George W. Moore for his helpful review of this manuscript.

LITERATURE CITED

Brooks, B., and E. Roberts, 1954. Geology of the Jacumba area, San Diego and Imperial Counties. *In* Jahns, R.H., ed., Geology of Southern California. Calif. Div. Mines and Geology Bull. 170, map sheet 23.

Bushee, J., J. Holden, B. Geyer, and G. Gastil, 1963. Lead-alpha dates for some basement rocks of southwestern California. Geol. Soc. Amer. Bull. 74: 803-806.

Fairbanks, H.W., 1893. Geology of San Diego County. Calif. State Mining Bur. Rept. 11: 76-120.

Gastil, G., and J. Bushee, 1961. Geology and geomorphology of eastern San Diego County, pp. 8-22. *In* Field trip guidebook, San Diego County. Geol. Soc. Amer. (Cordilleran Section) 57th Ann. Mtg.

Hawkins, J.W., 1970. Petrology and possible tectonic significance of late Cenozoic volcanic rocks, southern California and Baja California. Geol. Soc. Amer. Bull. 81: 3323-3338.

Kennedy, M.P., and G.W. Moore, 1971. Stratigraphic relations of Upper Cretaceous and Eocene formations, San Diego coastal area, California. Amer. Assoc. Petrol. Geol. Bull. 55: 709-722.

Miller, W.J., 1935a. Geologic section across southern Peninsular Range of California. Calif. Jour. Mines and Geology 31: 115-142.

Miller, W.J., 1935b. Geomorphology of the southern Peninsular Range of California. Geol. Soc. Amer. Bull. 46: 15351561.

Minch, J.A., 1970. Early Tertiary paleogeography of a portion of the northern Peninsular Range, pp. 83-87. *In* Pacific slope geology of northern Baja California and adjacent Alta California. Amer. Assoc. Petrol. Geol. (Pacific Section) Fall Field Trip Guidebook.

Weber, F.H., Jr., 1963. Mines and mineral resources of San Diego County, California. Calif. Div. Mines and Geology County Rept. 3, 1-309.

The Carrizo Fault: Geomorphic and Petrologic Implications of a Well-Hidden Geologic Structure

Dr. Michael J. Walawender, *Department of Geological Sciences*
San Diego State University, San Diego, California 92182

ABSTRACT

Carrizo Gorge hides a major unrecognized geologic structure that separates the metasedimentary rocks of the Indian Hill-Tule Mountain roof pendant into a western flysche-type sequence and an eastern shelf-type assemblage. K-Ar biotite cooling ages jump abruptly from an average of 89 Ma in the western portion of the enveloping La Posta Pluton to an average of 76 Ma on the eastern side. Likewise, amphibole geobarometry indicates that the western half of the La Posta Pluton is a graben that was down dropped 8 kms relative to its eastern half. The movement on the proposed Carrizo fault may be due to either rapid evacuation of the La Posta magma chamber or to shallowing of the subduction plane at the onset of the Laramide tectonics.

INTRODUCTION

An excursion along the route of Carrizo Gorge Railway will take us past the old but improved SD&AE track system and through the wildness that is Carrizo Gorge. This remarkable topographic feature begins calmly enough in an alluvium-filled Miocene valley and heads northward through some of the most remote and rugged country in southern California. Its northerly course is interrupted twice by resistant masses of igneous rock that force offsetting right-angle jogs before the gorge returns to its original direction. From its head in Jacumba Valley to where the canyon flattens out into Carrizo Wash, a distance of about 16 miles, the canyon bottom drops nearly 2000 feet. It cuts across the bottom of Walker Canyon which has a much steeper gradient (600 feet in 3.4 miles). The fault in that canyon, or at least a similar structure on the same trend, comes in from the west and cuts across the gorge just below Dubber Spur. Tule Canyon, another east-west feeder into the gorge, lies less than one mile north of Walker Canyon. It has the steepest gradient as the canyon bottom drops more than 1100 feet in three miles. These two canyons are the only major drainage channels that feed runoff into the canyon from the west.

All of this leads to some simple geomorphologic questions. Why is there a canyon there at all? Why does it run north-south? Why does it capture secondary channels with steeper gradients? Are there more deep-seated (pun intended) controls on these surface features?

REGIONAL GEOLOGIC SETTING

Carrizo Gorge lies almost entirely within the Indian Hill-Tule Mountain roof pendant, an elongate body of metamorphic and associated igneous rocks that forms the eastern fringes of the In-Ko-Pah and the core of the Jacumba Mountains in eastern San Diego County. This lithologic package extends southward at least 50 kilometers into Baja California. North of the international border, the roof pendant occupies the medial portion of the La Posta Pluton (Figure 1) in the eastern zone of the Peninsular Ranges Batholith (Gastil, 1975; Walawender and others, 1990). The roof pendant is intruded by abundant pegmatitic dikes that have been dated at between 95 and 100 Ma elsewhere in the batholith (Foord and others, 1991). That radiometric age range is consistent with the relative age of the dikes which are terminated at their contact with the enveloping 94 Ma La Posta Pluton (Henninger, 1996, Walawender, 2003). The metamorphic package is also intruded by a series of small garnetiferous, two-mica granitoids that include the Indian Hill Pluton and the Boulevard Pluton (Walawender and others, 1991; Walawender, 2003). Locally, the pegmatitic dikes are cut off at their contacts with these granitic bodies, indicating that the dikes are the oldest of the plutonic rocks that intrude the roof pendant. The Boulevard Pluton has been dated at 93 ±1 Ma, an age consistent with the field evidence indicating it is younger than the La Posta Pluton (Clinkenbeard, 1990). The Indian Hill Pluton, which is entirely enclosed within the roof pendant, has a zircon U-Pb age of 84.4 ± 6 Ma and a Rb/Sr isochron age of 89.6 ± 2.6 Ma (Parrish, 1990). The zircon age, with its unusually large error estimate, led Walawender and others (1990) to suggest a "preferred" age of 89 Ma for that pluton. Similar plutons that are spatially associated with the roof pendant occur further south in Baja California and have ages that are both slightly older and slightly younger than the associated La Posta-type plutons (Gastil and others, 1990).

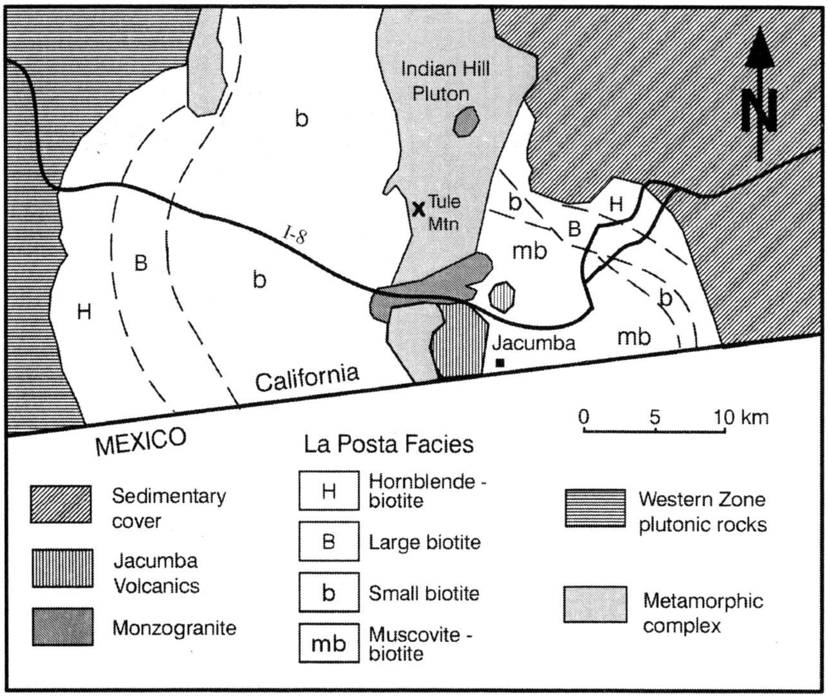

Figure 1. Generalized geologic map of the La Posta Pluton.

Parrish (1990) defined two granitic facies in the Indian Hill Pluton. The main facies is an undeformed garnet-bearing, two-mica granite with SiO_2 content between 73 and 76%. Modal analyses, however, indicate that on the granite classification system (Streckeisen, 1976), the rocks fall across the granodiorite-monzogranite boundary. Plagioclase and alkali feldspar comprise about 50 to 60% of the rock in a roughly 2:1 ratio. Quartz is the most abundant mineral and averages about 35%. Muscovite forms subhedral to euhedral books, which, along with its uniform size, suggest a primary origin. Sericite is present in small amounts and is typically intergrown with biotite. The second facies is also a two-mica granitoid with similar grain size and undeformed character. It is characterized in the field by a more deeply weathered appearance and the lack of accessory garnet. Alkali feldspar and muscovite are less abundant than in the main facies of the pluton.

In the southern portion of the roof pendant near Dubber Spur, Peterson (1985) described a kilometer-wide contact zone between garnet-bearing, two-mica monzogranite and the roof pendant as a *mélange* of undeformed aplitic dikes and monzogranite. The aplitic dikes are similar in composition to the monzogranite but are mineralogically and texturally distinct from the pegmatitic dikes found within the roof pendant. As reported for the Indian Hill Pluton, field relationships indicate that the aplitic dikes and monzogranite are younger than their pegmatitic counterparts.

The pattern of emplacement ages suggests a genetic as well as spatial relationship between the metamorphic assemblage and the igneous units described above. The geochemistry of the two facies of the Indian Hill Pluton and that of the spatially associated pegmatitic dike system forms a continuous smooth trend with each rock unit forming a distinct group (Parrish, 1990). This pattern and the spatial relationships suggested that the garnet-bearing monzogranite was a separate magmatic unit generated by partial melting of the surrounding metasedimentary package. It has been argued (Walawender and others, 1991; Walawender, 2004) that the large 94 Ma La Posta Pluton was the heat source for the development of the granitic pegmatite dikes and the somewhat younger garnetiferous two-mica granitoids associated with the roof pendant. As the rising La Posta melt engulfed the supracrustal metasedimentary package, muscovite-controlled partial melting within pelitic assemblages led to the development of the water-rich pegmatitic fluids. These rose from the now partially depleted (migmatitic) prebatholithic assemblage and into the overlying thermally softened, high-grade metamorphic rocks. As temperatures increased further in the now muscovite-depleted metasedimentary rocks, biotite-controlled melting in the same source region gave rise to larger batches of garnetiferous two-mica granitic magmas such as the Indian Hill and Boulevard plutons which then rose into the existing dike complex.

Henninger (1996), using the ER Mapper program and a SPOT panchromatic image, plotted the distribution of leucocratic dikes in the southern end of the Indian Hill-Tule Mountain roof pendant. The image, with its 30-meter resolution, showed a concentric distribution of dikes about the south-central portion of the roof pendant. The dikes generally appeared to dip outward at moderate to steep angles based on limited field observations by this author. This antiformal structure suggests an underlying intrusion of either the garnetiferous two-mica granite or, less likely, the La Posta Pluton.

The roof pendant occupies a unique position within both the La Posta Pluton and the Peninsular Ranges Batholith and appears to contain a structural feature of unknown origin and length. Krummenacher and others (1975) described the regional K/Ar cooling pattern and noted that the biotite and hornblende cooling ages decreased more or less regularly from west to east across the batholith.

However, the cooling ages within the La Posta Pluton, rather than decreasing regularly from west to east, fall into two groups. Four biotite ages west of the roof pendant average 89(±2) Ma whereas five samples in the eastern section average 76(±2) Ma. This abrupt shift can be interpreted in two different ways. Either the La Posta Pluton consists of two different plutons separated by the roof pendant with the easternmost one having been emplaced at greater depth, or the two halves of one pluton are separated by a near-vertical fault within the roof pendant with the eastern half having been uplifted relative to the western half. Walawender and others (1990) reported essentially identical 94 Ma zircon U-Pb (emplacement) ages for the eastern and western portions of the pluton. The larger differential between emplacement (U/Pb) and cooling (K/Ar) ages in the eastern half suggests that the La Posta Pluton is cut by a near vertical medial fault that has raised the eastern half relative to the western half. Not only does the Indian Hill-Tule Mountain roof pendant fall in a convenient position to house this structure but it is interesting to note that the pattern of K/Ar cooling ages continues southward on either side of the roof pendant for at least another fifty kilometers.

The aluminum-in-hornblende geobarometer and amphibole-plagioclase geothermometer were used to estimate pressures and temperatures of equilibration for the La Posta Pluton and adjacent western tonalites. Initial calculations reported by Clinkenbeard and Walawender (1989) were recalibrated to a modified geobarometer and reported in Symons and others (2003). Final equilibration pressures for four samples in the western-zone tonalites adjacent to the western contact of the La Posta Pluton ranged from 3.4 to 3.9 kb (12.2 to 14.0 kms) and averaged 3.6 kb (13 kms). Four samples from the western Hornblende-Biotite Facies and one sample from the western Large Biotite Facies of the La Posta Pluton ranged from 2.3 to 2.9 kb (8.3 to 10.4 kms) and averaged 2.5 kb (9.0 kms). Four samples from the eastern Hornblende-Biotite Facies and one from the eastern Large Biotite Facies ranged from 4.1 to 5.2 kb and averaged 4.8 kb (17 kms). This disparity in emplacement depths indicates that the western zone of the La Posta Pluton represents a down-dropped block bounded by the Cuyamaca-Laguna Mountain shear zone on the west and a "mid La Posta" fault on the east. Symons and others (2003) arbitrarily placed this hypothetical fault on the eastern margin of the Indian Hill-Tule Mountain roof pendant. *However, the medial portion of the roof pendant, approximated by the topographic feature known as Carrizo Gorge, may be a more likely position for this proposed structure.*

METAMORPHIC PETROGRAPHY

West of Carrizo Gorge

Metamorphic rocks within the Indian Hill-Tule Mountain roof pendant have been the subject of several Senior and Masters theses at San Diego State University (Figure 2). Eaton (1982) examined the portion of the roof pendant west of the gorge near Sacatone Springs and delineated a series of lithologies that, broadly characterized, fall into three protolith categories. Quartz-rich pelitic rocks, which include muscovite, sillimanite, and biotite as the main metamorphic minerals, represent clay-bearing sedimentary materials such as shale, mudstone, and clay-rich sandstone. These appear to be the most abundant lithologies in this portion of the roof pendant. Quartzites, with modal abundances of quartz greater than 90%, occur interlayered within the pelitic section. These contain minor amounts of feldspar, muscovite, biotite and sillimanite, and likely represent relatively clean quartz-rich sedimentary rocks. Meta-igneous rocks comprise the third protolith category and includes both amphibolite and meta-andesite. The amphibolites described by Eaton contain up to 80% mafic minerals (amphibole, diopside, biotite, and sphene) and were included in a geochemical study by Brent Foster (1994, this volume). This study concluded that the amphibolites intercalated within the metamorphic section in the Peninsular Ranges Batholith are back-arc basalts. The variation in mafic mineral content observed by Eaton may be the result of crystal fractionation processes in the original basaltic protoliths.

In addition to the amphibolites, two samples studied by Eaton have mineralogical and textural characteristics that suggest an igneous parentage other than basalt. One sample contains plagioclase megacrysts with both normal igneous zoning and well-developed polysynthetic twinning. The second is coarse grained with sparse amphibole megacrysts and matrix plagioclase exhibiting well developed polysynthetic twinning and normal zoning. Plagioclase megacrysts (phenocrysts?) and moderate amounts of modal quartz in both samples suggests an igneous parent with andesitic rather than basaltic affinities.

Odum (1982) and Diercks (1978) also examined the metamorphic assemblages along the western side of the Indian Hill-Tule Mountain roof pendant. Odum, working just north of I-8, observed and described a series of pelitic schists similar to those near the Sacatone Springs area. Thinly bedded, sillimanite- bearing quartzite is interlayered with the pelitic assemblages and suggests a deep-water, flysche-like sequence as suggested by Todd and others (1987) for the prebatholithic rocks in the medial zone of

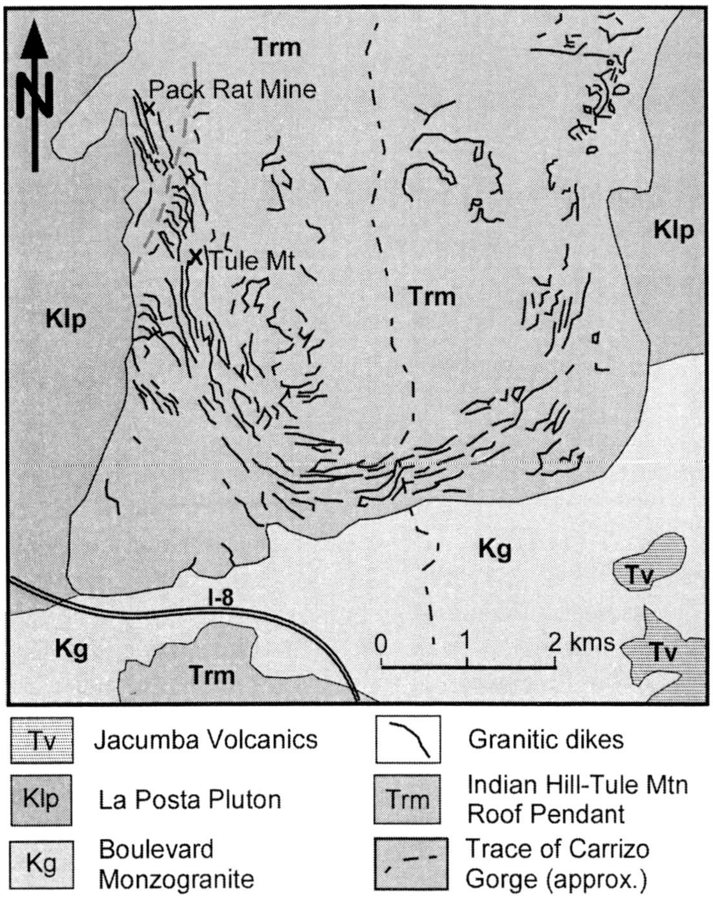

Figure 2. Generalized geologic map of the Carrizo Gorge area.

the Peninsular Ranges Batholith. Of particular note is the occurrence of up to 15% graphite in the pelitic rocks in this area. Dierks (1978), working just south of I-8 and east of Old Highway 80, also described thinly bedded quartzites interlayered with sillimanite-bearing pelitic schists, as well as a number of intricately folded amphibolite units similar to those described by Eaton. Just west of Jacumba Peak in the southern portion of his study area, Dierks reported small amounts of garnet-rich calc-silicate rocks whose mineral association included wollastonite and calcite. This marks the westernmost occurrence of carbonate-bearing rock yet documented in the roof pendant.

East of Carrizo Gorge

The metamorphic rocks in the northern portion of the roof pendant but east of Carrizo gorge were described by Parrish (1990) as part of his study of the Indian Hill Pluton, by several field classes from SDSU, and by Hoggatt (1979). These rocks provide a marked contrast to the metasedimentary package described west of the Gorge. In this area, coarse-grained gneissic rocks form the core of the roof pendant and enclose variable amounts of amphibolite, quartzite, and marble. The gneiss typically contains 50-60% quartz, 25-30% micaceous minerals, and 10-15% feldspar and trace amounts of fibrolitic sillimanite (Parrish, 1990). Plagioclase is the main feldspar and occurs as larger (3-5 mm) grains with irregular outlines whereas alkali feldspar is minor with myrmekitic intergrowths. The predominance of quartz and relative paucity of feldspar suggests that the protolith to this unit was a relatively shallow-water feldspathic wacke with a granitic source region.

The other metamorphic rocks occur as small to moderate-sized isolated blocks within the gneiss. The quartzite is massive to coarsely layered and devoid of all but minor amounts of accessory minerals. The amphibolite is also massive but may contain small, irregular to lensoidal bodies of calc-silicate material (Foster, 1990; this volume). The marbles form larger masses that tend to stand in relief in the current arid environment. They have been mapped by Weber (1960), Hoggatt (1979), and Parrish (1990) as isolated units surrounded either by granitic rocks or by the gneiss or both. Layering within the marbles is defined by changes in grain size or color, and is commonly discordant to the foliation in the enclosing metamorphic rocks. *This suggests the marbles and the pod-like bodies of quartzite and amphibolite have been tectonically rotated within the gneiss.*

Several exposures of the marbles in and around the Dos Cabezas siding were exploited between 1950 and 1960 for roofing granules, chicken grit and agricultural limestone (Weber, 1960). Although the bulk of the mined marble is relatively pure, averaging less than 6% total SiO_2 $+Al_2O_3$ and less than 1% MgO (Weber, 1960), smaller bodies and discontinuous lenses within the larger exposures contain a variety of accessory minerals such as wollastonite, diopside, garnet, monticellite, and calcic plagioclase. These minerals testify both to the conditions of metamorphism as well as to the centimeter-scale migration of selected elements between the Ca-rich marbles and the enclosing igneous rocks.

Protolith Ages

The ages of the two metasedimentary packages on either side of the gorge are not well constrained. Hoggatt (1979) assigned both the siliciclastic rocks and the carbonate units to a tentative Triassic age based on their similarity to mapped units in the western zone of the batholith. Miller and Dockum (1983) reported Ordovician conodonts in dolomitic marbles of the Coyote Mountains. Gastil and others reported similar fossil ages in marbles in northern Baja California. The two protoliths, therefore, may have drastically different ages so that their apparent juxtaposition must then be structurally controlled.

Conditions of Metamorphism

The grade of metamorphism in the eastern and western portions of the Indian Hill-Tule Mountain roof pendant appears to be uniform. On both sides of Carrizo Gorge, mineral parageneses indicate middle-to-upper amphibolite grade conditions. All three previously cited studies in the western section estimate peak P-T conditions at between 600-650° C and 2.5 to 4.5 kb (6.5 to 11.7 km). The eastern assemblage is more problematic due to a paucity of pelitic assemblages. However, the calcite-wollastonite-diopside-garnet-calcic plagioclase association found in the Dos Cabezas marbles, the diopside-hornblende assemblage in the associated amphibolite, and the trace amounts of sillimanite reported by Parrish (1990) in the enclosing gneiss (meta-arkose) suggest similar conditions of metamorphism. The range of pressure (depth) estimates is large enough to include the depths reported by Symons and others (2003) for the eastern as well as the western portions of the La Posta Pluton.

Structure

The Indian Hill-Tule Mountain roof pendant records at least two periods of deformation, the earliest of which left a dominant north-northeast foliation. The limited measurements available to Eaton (1982) in the narrow fragment of metamorphic rock near Sacatone Springs revealed a series of near vertical isoclinal folds whose axes were oriented N28E and plunged approximately 52 degrees to the northeast. Limited observations of a younger set appeared to be more gently folded but plunging approximately 60 degrees to the southeast. Diercks (1985) recorded similar north-northeast fold axes south of I-8 with hinge lines that plunged steeply to the southeast. Peterson (1985) also observed isoclinal folds at the southern end of the gorge whose axial planes are oriented approximately N10E and dip steeply to the northeast.

DISCUSSION AND CONCLUSIONS

The medial zone of the Indian Hill-Tule Mountain roof pendant hides the transition from a deeper-water, flysche-type sequence west of Carrizo gorge to a shallower-water, shelf-type environment on its eastern flank. The nature of that boundary, as well as whether or not this transition continues southward into Baja, is unknown. Although the mineralogy of both sets of rocks indicates that they were both metamorphosed to approximately the same metamorphic grade, the wide range of conditions implied does not indicate if that boundary was in place during the metamorphic event or whether the two environments were juxtaposed as a result of the emplacement of the eastern zone of the batholith.

The discordant relationship of the isolated blocks of marble, amphibolite, and quartzite found in the gneissic unit on the eastern side of the gorge suggests that these units were displaced from their original stratigraphic position in the quartzo-feldspathic gneissic unit during metamorphism. In contrast to the preserved stratigraphy of the western side of the roof pendant, the eastern metamorphic package appears to have behaved in a more ductile manner. As the rising La Posta magma released its latent heat to the surrounding country rock, the less refractory micaceous gneiss underwent an increase in ductility and eventually partial melting (Walawender, 2004) to produce the early granitic pegmatite dikes. The more refractory rocks (marble, quartzite, and amphibolite) remained rheologically intact but were displaced from their original stratigraphic position as this portion of the roof pendant underwent ductile transformation. In effect, they acted like chocolate chips as the cookie dough was being mixed! This behavioral distinction between the two sides of the roof pendant implies different thermal conditions despite the lack of resolution using mineralogical indicators.

Grove and others (2003) supplemented the original K-Ar study of Krummenacher and others (1975) and confirmed that the two halves of the La Posta Pluton have experienced different cooling rates. The apparent structural displacement of 9 km between the two halves as indicated by the amphibole geobarometry (Walawender and others, 1990; Symons and others, 2003) and the difference in cooling histories can be accommodated by a near-vertical, east-side-up fault. Since mapping in the pluton failed to reveal any large-scale structures, the only real constraint on the position of this long-hypothesized fault is that it cut the La Posta Pluton in the vicinity of the Indian Hill-Tule Mountain roof pendant. Symons and others (2003) arbitrarily placed this structural feature along the eastern side of the roof pendant but detailed mapping by Hoggatt (1979) and Parrish (1990) failed to reveal any major structural break in the Cretaceous or older rock units along the eastern side of that structure. It seems more likely, therefore, that this feature lies within the metamorphic assemblage of the roof pendant and approximates the position of Carrizo Gorge (Figure 3). Such a position conveniently hides

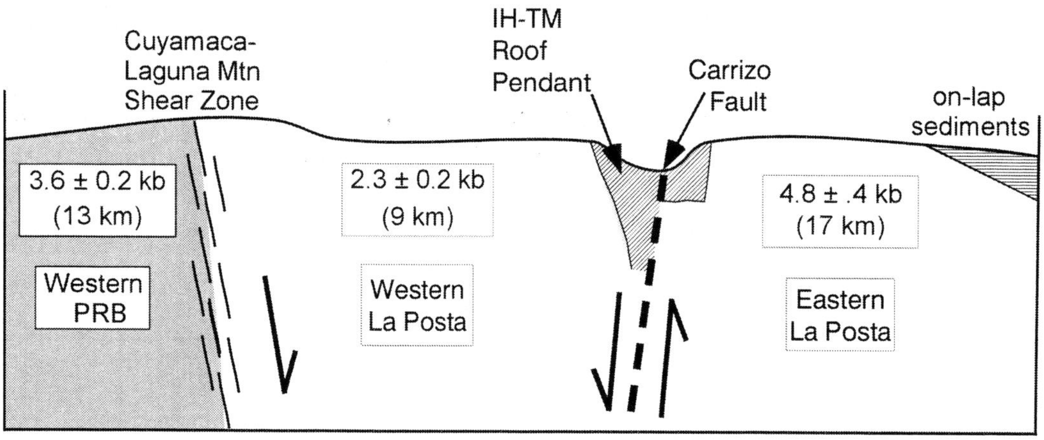

Figure 3. Generalized cross section showing emplacement depths and major structural features in the La Posta Pluton.

the southern end of this structure in the 18 Ma extensional terrane now occupied by the Jacumba Volcanics, whereas the northern end disappears, just as conveniently, into the alluvial fill of Carrizo Valley.

The fault, if it indeed exists, may raise the entire eastern half of the pluton relative to the western half or it may simply rotate the southern portion of the uplifted block relative to the northern portion. In doing so, the two different metamorphic protoliths could be raised to the same erosional level. Either scenario is compatible with the data array which is limited to the southern portion of the La Posta Pluton where it is traversed by I-8. The predominantly north-south foliation pattern within the roof pendant, and the relative susceptibility of the micaceous metamorphic rocks to erosion may be one of the dominant controls on the position and orientation of the gorge. The mysterious and elusive boundary between the two metamorphic protoliths within the gorge and the inferred structural control on their contact suggests that it also played a major role in the development of the topographic feature. *Thus, the gorge is where it is because that's where the roof pendant is! It runs north-south because that's the orientation of the foliation and the well-hidden Carrizo fault!*

The timing of movement on the fault is as elusive as the fault itself. North of the international border, it must be younger than 93 Ma, the approximate age of the La Posta Pluton and is probably younger than the 89 Ma "preferred" age of the Indian Hill Pluton. The roof pendant that appears to hide the fault stretches another 50 kms southward into Baja California. If this proposed structure is contained within the metamorphic package, it must be a major player in the tectonic and cooling history of the batholith. Does it reflect shallowing of the subduction plane and the onset of the Laramide orogeny that began approximately 80 million years ago? Is it limited to the La Posta Pluton and, if so, does it hide a record of massive eruptions that depleted the central magma chamber and down-dropped its western zone?

Carrizo Gorge may hold the answers but it will require (*shudder*) field work (*groan*) by younger geologists. Aerial photographs appear to show that the roof pendant has different segments with variations in dike abundance and orientation. The fault, therefore, may be recognized not only by superposition of the two metamorphic protoliths but by intrusive patterns within the metamorphic complex itself. Geotypes in this brave new world may have to leave their laptops behind and venture into the unknown and fearsome wilderness to actually look at metamorphic rocks.

The final question to be answered is why the gorge begins so gently and yet appears to capture streams with much steeper gradients. Jacumba Valley, at the head of Carrizo Gorge, is an extensional terrane whose topographic expression extends southward for another four or five kms into Baja California. It is bounded by a series of northwest-trending normal faults with modest throws of up to 300 meters (Minch and Abbott, 1973). The Table Mountain Formation consists of fluvial sandstones and lesser conglomeratic lenses that have been shown to be at least 150 meters thick in the central portion of the valley (Swenson, 1981). It underlies the 18 Ma Jacumba Volcanics and lies directly on igneous rocks of the batholith. The base of the Table Mountain Formation is in depositional contact with basement rocks at elevations in excess of 3200 feet along the west side of the valley near the head of Jacumba Gorge and at over 3400 feet near Table Mountain along the eastern side of the volcanic province (Minch and Abbott, 1973). Thus, prior to Miocene extension, the area now occupied by Jacumba Valley was at least at that elevation and may have been a basement highland that fed a pre-18 Ma equivalent to the modern gorge. Since that time, volcanism and fluvial sedimentation filled the down-dropped valley so that it now provides only a gentle reminder of the ancient drainage basin whose waters created what is now Carrizo Gorge.

EPILOGUE

To add one final bit of blatant conjecture to an already fuzzy idea of a fault that hasn't been found, what does it mean to have the eastern La Posta Pluton represent a deeper segment of that body of igneous rock? The lithologic zones mapped east of the roof pendant are much narrower in map view (Figure 2). If that is a picture of the deeper portion of the western La Posta, then in the third dimension of the palinspastically (I love that word!) reconstructed pluton, the zones taper downward into a funnel-shaped body. This geometry implies that it was intruded as an upward expanding diapir being fed from a large central conduit. That idea is contrary to the currently fashionable concept of pluton growth by dike emplacement from multiple intrusive centers and should irritate almost everyone in the petrologic community. I am quietly proud!

ACKNOWLEDGMENTS

This paper relies heavily on the work of the following former SDSU students, a grand bunch all! I am grateful to them for their efforts, for their past and continued friendship, and for their companionship in the field. Chris Diercks (1977), Sherri Eaton (1982), Sharon Isaac (1982), Heidi Odum (1982), John F. Peterson (1985), Kent Parrish (1990), Eric Augustine (1996), and Carol Henninger (1996).

REFERENCES

Augustine, E., 1998. Structure and petrology of a portion of the Sacatone Springs roof pendant at Tule Mountain [M.S. Thesis]: San Diego State University, 21 p.

Clinkenbeard, J.P., 1987. The mineralogy, geochemistry, and geochronology of the La Posta Pluton, San Diego and Imperial Counties, California [M.S. Thesis]: San Diego State University, 215 p.

Clinkenbeard, J.P., and Walawender, M.J., 1989. Mineralogy of the La Posta Pluton: Implications for the origin of zoned plutons in the eastern Peninsular Ranges Batholith, southern and Baja California: American Mineralogist, v. 74, pp. 1258-1269.

Diercks, C., 1985. The Structure and Petrology of the Southern Portion of the Jacumba Roof Pendant: [B.S. Thesis]: San Diego State University, 23 p.

Eaton, S.L., 1982. Petrology of a Portion of the Sacatone Springs Roof Pendant, In-Ko-Pah Mountains, California: [B.S. Thesis]: San Diego State University, 23 p.

Foord, E.E., London, D., Kampf, A.R., Shigley, J.E., and Snee, L.W., 1991, Gem pegmatites of San Diego County, California, *in* Walawender, M.J., and Hanan, B.B., eds., Geological Excursions in southern California and Mexico: Department of Geological Sciences, San Diego State University, pp. 128-146.

Gastil, R.G., 1975, Plutonic zones in the Peninsular Ranges of southern California and northern Baja California: Geology, v. 3, pp. 361-363.

Gastil, R.G., Phillips, R.P., and Allison, E.C., 1975, Reconnaissance geology of the state of Baja California: Geological Society of America Memoir 140, 170 p.

Gastil, R.G., Diamond, J., Knaack, C., Walawender, M., Marshall, M., Boyles, C., and Chadwick, B., 1990, The problem of the magnetite-ilmenite boundary in southern and Baja California, *in* Anderson, J.L., ed., The nature and origin of Cordilleran magmatism: Geological Society of America Memoir 174, pp. 19-32.

Germanario, M.P., 1982, The depositional and tectonic environments of the Julian Schist, Julian, California [M.S. Thesis]: San Diego State University, 95 p.

Girty, G.H., Thomson, C.N., Carmichael, D.L., Netto, S.L., and Wagemakers, R., 1998, Normal-sense mylonites of the Scove Canyon segment (SCs), Cuyamaca-Laguna Mountains shear zone (CLMSZ), California: *in* Snoke, A., Tullis, J.A., and Todd, V.R., eds., Atlas of Mylonitic and Fault-Related Rocks, Princeton Press.

Gromet, L.P., and Silver, L.T., 1987, REE variations across the Peninsular Ranges Batholith: Implications for batholithic petrogenesis and crustal growth in magmatic arcs: Journal of Petrology, v. 28, pp. 75-125.

Henninger, C., 1996, Structure of the Mt. Tule dike swarm, Jacumba, California [B.S. Thesis]: San Diego State University, 13 p.

Hoggatt, W.C., 1979, Geologic map of the Sweeney Pass Quadrangle, San Diego County, California: Pamphlet to accompany Open-File Report 79-754, 28 p.

Isaac, S., 1982, Peraluminous Granitoid Rocks in the Tule Canyon Area of the In-Ko-Pah Mountains, Southern California: [B.S. Thesis]: San Diego State University, 26 p.

Krummenacher, D., Gastil, R.G., Bushee, J., and Doupont, J., 1975, K-Ar Apparent Ages, Peninsular Ranges Batholith, Southern California and Baja California: Geological Society of America Bulletin, v. 86, pp. 760-768.

Miller, R.H., and Dockum, M.S., 1983, Ordovician conodonts from metamorphosed carbonates of the Salton Trough, California: Geology, v. 11, pp. 410-412.

Minch, J.A., and Abbott, P.L., 1973, Post-batholithic geology of the Jacumba area, southeastern San Diego County, California: Transactions of the San Diego Society of Natural History, v. 7, No. 11, pp. 129-136.

Odum, H.A., 1982, Metamorphism and Depositional Environment of the Tule Canyon Roof Pendant: [B.S. Thesis]: San Diego State University, 16 p.

Parrish, K.E., Walawender, M.J., Clinkenbeard, J.P., and Wardlaw, M.S., 1986, The Indian Hills granodiorite, a peraluminous, garnet-bearing granitoid, Eastern Peninsular Ranges: Cordilleran Section, Geological Society of America Abstracts with Program, v. 18, no. 2, p. 168.

Parrish, K.E., 1990, Geology, petrology, geochemistry, and isotopic character of the Indian Hill Pluton, Jacumba Mountains, California [M.S. Thesis]: San Diego State University, 136 p.

Peterson, J.F., 1985, Geology of the southern portion of the Carrizo Gorge, Jacumba Mountains, the Peninsular Ranges Batholith, Southern California: [Senior Thesis]: San Diego State University, 13 p.

Streckeisen, A., 1976, To each plutonic rock its proper name: Earth Science Review, v. 12, pp. 1-33.

Swenson, G., 1981, Ground water hydrology of Jacumba Valley, California and Baja California: [Masters Thesis], San Diego State University, 260 p.

Symons, D.T.A., Walawender, M.J., Smith, T.E., Molnar, S.E., Harris, M.J., and Blackburn, W.H., 2003, Paleomagnetism and geobarometry of the La Posta Pluton, California: *in* Johnson, S.E., Paterson, S.R., Fletcher, J.M., Girty, G.H., Kimbrough, D.K., and Martin-Barajas, A., eds., Tectonic evolution of northwestern Mexico and southwestern U.S.A., Geological Society of America Special Paper 374, pp. 135-155.

Todd, V.R., Erskine, B.G., and Morton, D.M., 1987, Metamorphic and tectonic evolution of the northern Peninsular Ranges Batholith, southern California, *in* Ernst, W.G., ed., Metamorphism and crustal evolution in the western United States, Rubey Volume VII; Prentice Hall, pp. 894-937.

Thomson, C.N., and Girty, G.H., 1994, Early Cretaceous intra-arc ductile strain in Triassic-Jurassic and Cretaceous continental margin arc rocks, Peninsular Ranges, California, Tectonics, v. 13, pp. 1108-1119.

Walawender, M.J., 2000, The Peninsular Ranges; A geological guide to San Diego's back country: Kendall/Hunt Publishing Company, Dubuque, Iowa, 114 p.

Walawender, M.J., 2003, Garnet-bearing granitoids and pegmatite dikes in the Indian Hill-Tule Mountain roof pendant: Implications for the origin of gem-bearing pegmatites in southern California, *in* Murbach, M.L., and Hart, M.W., eds., Geology of the Elsinore fault zone, San Diego Region, SDAG/SCGS Field Trip Volume 31, San Diego Association of Geologists, San Diego, California, and South Coast Geological Society, Santa Ana, California, pp. 27-38.

Walawender, M.J., Gastil, R.G., Clinkenbeard, J.P., McCormick, W.V., Eastman, B.G., Wernicke, R.S., Wardlaw, M.S., Gunn, S.H., and Smith, B.M., 1990, Origin and evolution of the zoned La Posta-type plutons, eastern Peninsular Ranges Batholith, southern and Baja California, *in* Anderson, J.L., ed., The nature and origin of Cordilleran magmatism: Geological society of America Memoir 174, pp. 1-18.

Walawender, M.J., Girty, G.H., Lombardi, M.R., Kimbrough, D., Girty, M.S., and Anderson, C., 1991, A synthesis of recent work in the Peninsular Ranges Batholith, *in* Walawender, M.J., and Hanan, B.B., eds., Geological Excursions in southern California and Mexico, Department of Geological Sciences, San Diego State University, pp. 297-318.

Weber, F.H., 1960, Mines and mineral resources of San Diego County, California; County Report No. 3, California Division of Mines and Geology, San Francisco, California, 309 p.

The Desert View Tower

***Mike Retz**, Editor,* Mountain Heritage, *Mountain Empire Historical Society, 31130 Highway 94, Campo, California 91906*

From Mountain Heritage — The Back Country's Historical Digest, Voice of the Mountain Empire Historical Society, *Campo, California, Vol. 18, No. 2, Winter 2004*

Photo by Phil Farquharson.

How many of you have driven east on I-8 and passed the tower that overlooks the desert at In-Ko-Pah and wondered what it was? Undoubtedly, some of you have stopped and found out. Ben Schultz was a young boy in the early 1960s when his folks used to take him and his four siblings out to the desert from San Diego along Old Highway 80. They usually stopped at the Desert View Tower for a break in the long trip. It was those stops at the place with animals carved out of the boulders that enthralled young Ben. He never forgot them and last year Ben bought the tower and the "animals" around it.

To start the story of the tower, we need to go back to the year 1917. Bert Vaughn, who was mayor of San Diego, thought that the federal government was going to build a border crossing at Jacumba in eastern San Diego County. So, he bought the "town." In reality, there wasn't much there, but he thought the community had potential. At the time, Vaughn was also involved in the concrete paving of Highway 80.

In 1922 or 23, Bert Vaughn conceived the idea of building a tower to commemorate the many heroic pioneers, teamsters and railroad builders who helped open the area. The site he chose was only a mile from the old Mountain Springs Station that had provided oxen to help the teams over the very steep grade coming up out of the desert. The San Diego & Arizona Railway, that was completed just a few years before, and the new concrete highway were revolutionizing travel in this part of the country. Vaughn worked on the 70-foot tower more than ten years but never completed it. He used local stone that was found in abundance to build the four-foot thick walls. An article in the December 22, 1937 *San Diego Union* stated that the tower's total cost of construction was $500 since Vaughn did much of the work himself. He also did some "bartering" with the highway construction crews for some of the needed materials. For example, the window frames were constructed of wood from the Old Plank Road that crossed the sand dunes between El Centro and Yuma.

In 1927, Bert Vaughn gave the tower to his son, Dick, to be used as a hunting lodge. Dick attended Ohio State

Photo by Phil Farquharson.

Aerial photo from USGS, via TerraServer, through QuakeMap. Desert View Tower at center.

Also in 1928, Vaughn's fortunes suffered a setback when the government decided not to build the border crossing at Jacumba. By then, he had built much of what is present-day Jacumba.

The resort town was constructed around the existing hot springs and provided a relatively cool respite for the people in the Imperial Valley in the days before air-conditioning. The railroad provided easy access for those coming "up the hill" from El Centro.

During the depression of the 1930s, an out-of-work civil engineer, Merle Ratcliff, visited the tower and thought that the boulders in the area resembled actual and mythological animals. He spent the next several years with a mallet and chisel, creating the mysterious creatures that now inhabit Boulder Park, a few hundred feet from the tower. In exchange for his labors, Bert Vaughn gave him free rent in a house in Jacumba.

University where he became friends with Dennis Newman. They were fraternity brothers at Beta Theta Pi. In 1928, they drove back to San Diego in a 1928 Reo automobile and Dennis was shown the uncompleted Desert View Tower. He fell in love with the tower and its surroundings.

In 1938, Dennis Newman, who was born in England, watched the events in Europe spiral towards war. Since the United States was not yet involved, he decided to join the Royal Canadian Air Force. After flight training he immediately traveled to England and joined the Royal Air Force. In 1943, while on a mission against Rommel's African Corps, he was shot down but managed to crash land the Wellington bomber without any loss of life. He and his crew were captured by the Italians and imprisoned first at Bari and then at a place called Chitie. Hearing that the Germans were coming to take them to a stalag in Germany, they escaped and evaded the German and Italian forces for a month before making it to a British occupied area.

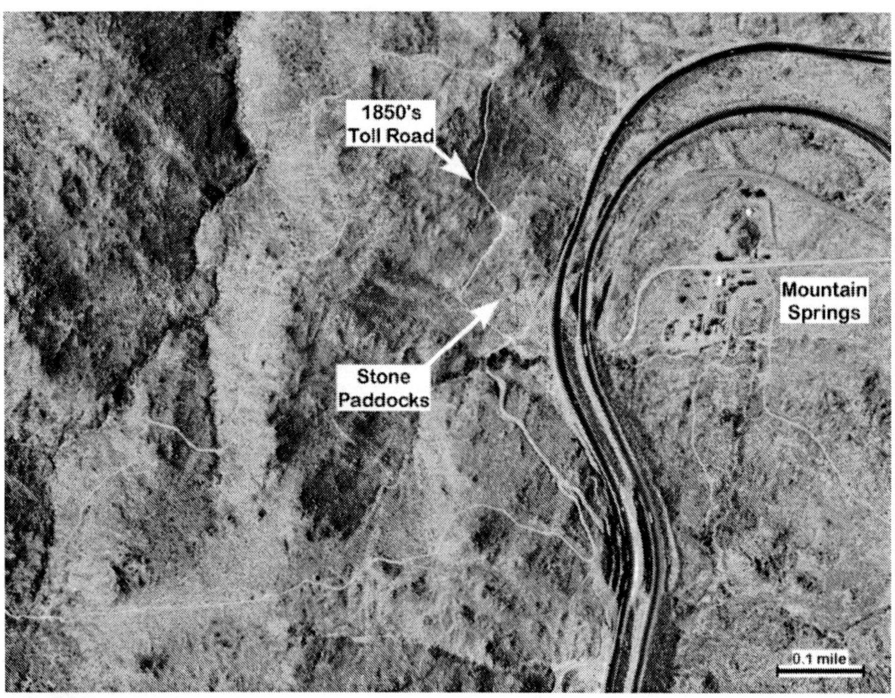

Aerial photo from USGS, via TerraServer, through QuakeMap. Stone Paddocks right of center; Mountain Springs at right (east) part of photo.

Newman was reinstated into the RAF and was flying the 4-engine Lancaster bomber when the war ended. For his part in the war effort, he was given the Distinguished Flying Cross and many other awards. He returned to the United States, and with his mother, moved from Texas to San Diego where he prospered as a general contractor.

In 1947, Newman purchased the Desert View Tower from his friend, Dick Vaughn. He used his contracting skills to enclose the top level, install the stairs and extend the ground floor to its present size. These things were done to make the tower tourist-friendly. Water and electricity were added; the cafe and several cabins were spruced up. After three years of construction, the tower was opened to the public in 1950.

Construction of Interstate 8 began in 1962 and took 10 acres of Newman's land, eliminating the cafe, cabins and water well. He was forced to dig another well and was surprised to hit water at 75 feet. To be sure, they went down to 200 feet and had a well that produced 80 gallons per minute. The number of tourists that stopped at the tower dropped after the Interstate opened.

Postcard from the collection of Phil Farquharson.

Marv and Gilda Overman were managing the store in 1976, still welcoming tourists from across the nation and around the world. Another long time employee at the tower was Ralph (Hap) Home who was known as the rock man. Home and his wife managed the cafe and cabins before the Interstate eliminated them. He then turned his talents to scouring the desert for geodes and other minerals which were sold in the tower's gift shop.

During the Dennis Newman tenure (1950 to 1979), the tower became an eclectic collection of historic and personal items. Among the many things found on the first floor were longhorn cattle horns, an 18th-century bed warmer, and a locomotive bell from the 19th century. The second floor contained American Indian artifacts. The third was dedicated to wars past, primarily the Vietnam War and World War II. On the fourth floor was Newman's lamp collection, many over 100 years old. The top or fifth floor was dedicated to viewing the desert. After all, the

Postcard from the collection of Phil Farquharson.

incredible view on a clear day was the reason for building the tower.

By 1975, Newman and his wife were thinking of retiring and traveling, and so put the tower up for sale. He first attempted to sell it to the state for a park, at a reduced price, but couldn't interest anyone in the state's Park and Recreation Department. In late December 1979, Jane Knap, and other principals who wished to remain anonymous, purchased the tower. The cost of Newman's artifacts, however, was deemed too high and they were not included in the sale. Most were subsequently trucked down to the Scottish Rite Temple in San Diego's Mission Valley in June 1980 and auctioned off to the highest bidders.

Unlike Dennis Newman, Jane Knap was a private person. In an interview shortly after purchasing the tower, she said she enjoyed the solitude and beauty of the place, but would rather not talk about her personal background. "Talk about the tower, not me," she said. Knap was a widow in her early 60s at the time of the purchase and nearing completion of studies for a law degree.

Ben Schultz was born in San Diego in 1958, one of five children. His family was big on outdoor experiences

Postcard from the collection of Phil Farquharson.

Desert View Tower on the mountain skyline just left of center. Postcard from the collection of Phil Farquharson.

and often camped at Dos Cabezas Springs and Mortero Palms. Enroute to these camp sites, the family always stopped at the Desert View Tower to "stretch their legs." He said the kids always rode in the back of an old truck, which would be illegal today, but they loved it.

One of Ben's first childhood memories was of the tower, especially the little stuffed kit fox on display at the door for the last fifty years. He's getting a little tattered now but still strikes one as a fierce predator. As a young boy, Ben was always scared by that little fox.

After living in northern California for many years, Ben moved back to San Diego and occasionally stopped at the tower. He noticed that it was always "for sale." In fact, it was for sale for nine years! By 1998, Jane Knap was living at the tower full time. She was still selling antiques on the ground floor; the other floors were a museum. But her health was failing rapidly and she was anxious to sell. On one of his visits, in early 2002, Ben asked Jane how much she wanted for the tower. The price was right.

To quote Ben, "Now I am astonished to wake up in the morning and look out my door at the grimly beautiful tower and know that it is my lot to steward this place for the next generation of little 4-year-olds who I hope will have the same powerful memories of our cultural and ecological heritage."

Ben Schultz
The Desert View Tower
In-Ko-Pah Road
Boulder Park, California
619-766-4612

Postcard from the collection of Phil Farquharson.

Postcard from the collection of Phil Farquharson.

Geology of the La Posta Pluton in the In-Ko-Pah Gorge and Mountain Springs Area, San Diego and Imperial Counties, California

***Len Sinfield**, P.G., 8909 Scorpius Way, San Diego, California 92126*
geologylen@Earthlink.net

INTRODUCTION

The late Mesozoic (94 Ma) La Posta Pluton lies within the eastern portion of the Peninsular Ranges Batholith (Figure 1). This portion of the batholith is characterized by large concentrically zoned plutons of intermediate composition (Walawender et al., 1989) with deep levels of emplacement and associated high-grade metamorphic rocks. The La Posta Pluton is the largest of these plutons, consisting of leucocratic hornblende-biotite tonalite in the outer zones to granodiorite inward, and has experienced little deformation or alteration (Clinkenbeard and Walawender, 1989). The La Posta Pluton and others like it represent the deepest level of the exposed batholith. The plutons of the western portion of the batholith are older (120 to 104 Ma), deformed, smaller, range in composition from gabbro to monzogranite, have lower depths of emplacement, and have associated lower grade metamorphic rocks (Gastil, 1975).

Along Interstate 8 (I-8), the La Posta Pluton occurs from Cameron Station eastward to the metamorphic screen in the Jacumba Mountains, emerging again north of Jacumba and west of the Desert View Tower, extending through Mountain Springs and In-Ko-Pah Gorge, and disappearing beneath the desert floor west of Ocotillo (Figure 2). The rugged and steep boulder-mantled slopes of the inhospitable In-Ko-Pah Gorge and the surrounding area offer excellent exposures of the eastern facies of the La Posta Pluton. The study area occurs within the mountainous desert visible from the Desert View Tower, along and between the eastbound and westbound split of I-8 (Figure 3).

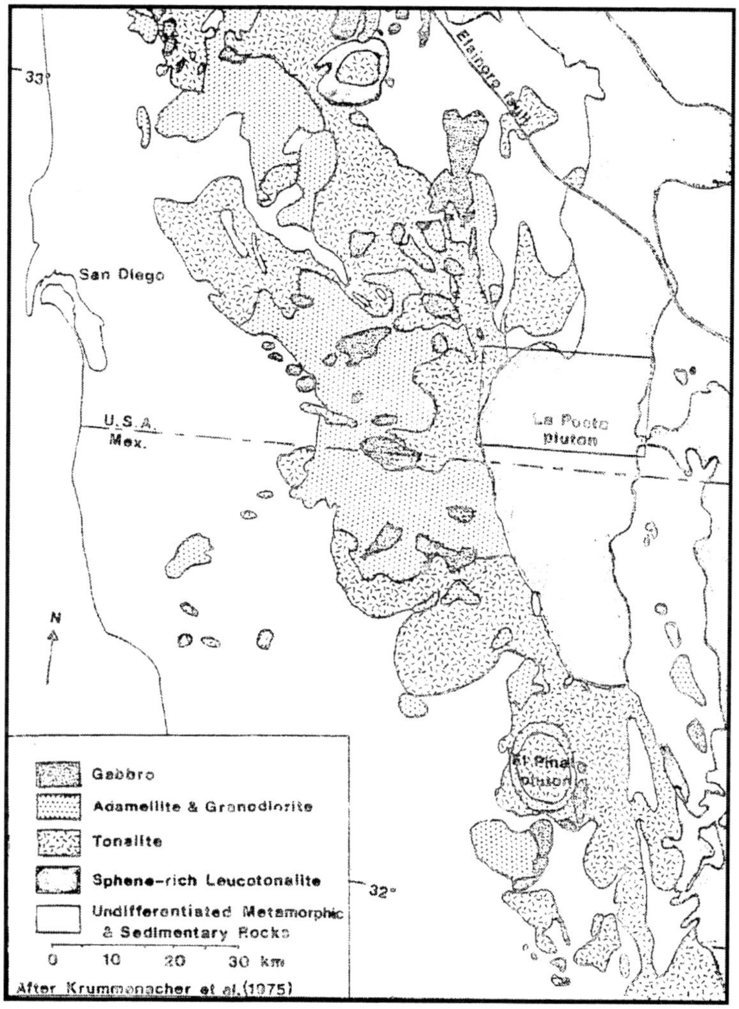

Figure 1. Geology of the Peninsular Ranges Batholith (after Krummenacher et al., 1975).

As an interesting side note, three generations of roadway are visible in the study area and road cuts from all three were used to map the La Posta Pluton facies. In-Ko-Pah Gorge has been used for a route between the desert and San Diego since the Yuma and In-Ko-Pah Indians used it to get to their winter fishing areas near San Diego. During mapping of the area, a few small and scattered Indian pottery fragments were found along the eastbound portion of I-8 in the gorge.

Small portions of an early 1900s narrow paved road can be seen in the hillside above Mountain Springs along I-8 and in In-Ko-Pah Gorge itself. Portions of this road, along with old U.S. Highway 80, are shown on a California Division of Mines and Geology map of the Jacumba area by Brooks and Roberts (1954) and are included on Plate 1. In 1940, the modern U.S. Highway 80 was completed in the area and follows the path of the current eastbound lanes of I-8 through the gorge. Portions of concrete pavement from old Highway 80 can be seen in and around Mountain Springs and the Desert View Tower. The

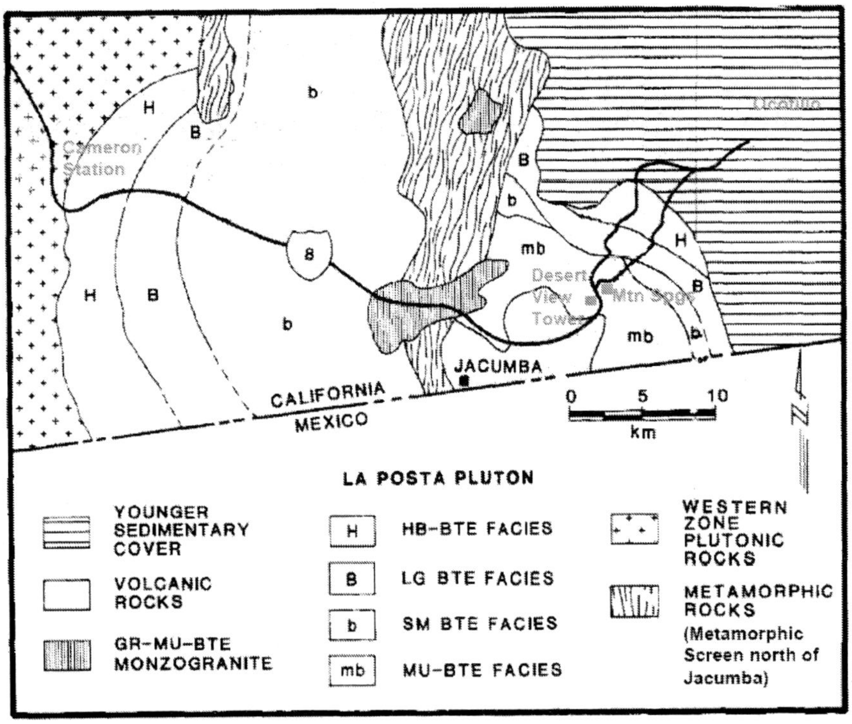

Figure 2. Geologic map of the La Posta Pluton (modified from Clinkenbeard and Walawender, 1989).

westbound lanes of I-8, which ascend Devil's Canyon, were added in 1964. At that time, U.S. Highway 80 was renamed I-8. Much of the old highway was obliterated in favor of the Interstate lanes, which were completed in 1966.

PREBATHOLITHIC METASEDIMENTARY ROCKS

Three metamorphic rock types were recognized in the study area; marbles (and associated tactites), schistose quartzite, and minor amounts of mica schist. The metamorphic rocks lie in a northwest-trending narrow belt from the east side of Myer Valley to Interstate 8.

Marble

This unit is easily recognized by the brilliant milky-white color of the variable-sized calcite crystals (1 to 20 mm). Saccharoidal marble also occurs with numerous small (1 to 10 mm) quartzite streaks, with occurrences every few centimeters. The marble outcrops are 1 meter to several 10s of meters wide and are elongated along several hundreds of meters.

Tactite has formed along contacts with the younger La Posta Plutonic units and is characterized by large (5 to 10 mm) euhedral to subhedral almandine garnet, small (1 to 3 mm) columnar to acicular epidote, and large (5 to 10 mm) tabular euhedral scheelite crystals. Tactite occurs in a few isolated outcrops throughout the Jacumba region (Brooks and Roberts, 1954; Morton, 1977).

At the metamorphic-plutonic contact, plutonic rocks grade into an epidote-bearing milky-white quartzite with inclusions of marble. This quartzite averages 50 cm thick and grades into a layer of scheelite-garnet calc-silicate rock, 5 to 100 cm thick. Almandine garnet occurs as large aggregates or as large isolated crystals. The scheelite occurs as tabular masses or as large isolated crystals within the garnet aggregates. This layer grades into marble over a few centimeters.

Quartzite

This is a dark-brown to black, massive to thinly bedded schistose-quartzite. Schistosity and foliation are defined by small (1 to 2 mm) biotite layers every 2 to 15 cm. In outcrop, the weathered quartzite displays shiny schistose surfaces that give the appearance of a schist. Fractures form preferentially along foliation planes and cause the exposure of the schistose surfaces. The quartzite is fairly massive in some outcrops and is always found in close association with the marble.

Mica Schist

This is a fissile light-brown to dark brown, thinly bedded, fine-grained biotite-schist with abundant boudinage. Foliation is defined by biotite. A few small (1 to 2 mm) layers of small (1 to 3 mm) subhedral quartz and feldspar occur. Numerous quartz dikes parallel both the foliation and the folds in the foliation. The dikes are also strung-out in boudins. Folds and boudinage indicates that the schist has undergone multiple episodes of deformation. The mica schist usually occurs with the marble and quartzite, but it can occur as isolated outcrops throughout the study area.

LA POSTA PLUTON

The La Posta Pluton is divided into two parts by a large metamorphic screen extending from the vicinity of Jacumba northward (Figure 2). The screen is composed of sillimanite-bearing pelitic schists and migmatitic gneisses along with lesser amounts of marble, quartzite, and amphibolite (Clinkenbeard and Walawender, 1989). The western facies of the La Posta Pluton, from Cameron Station to

Figure 3. View of study area with steep boulder-mantled slopes and an arid climate.

the Jacumba Mountains, were mapped as part of a SDSU Master's thesis by JoAnn Kimzey (Kimzey, 1982). The study found that the area is subdivided into five facies, with the interior facies named and defined by biotite, a ubiquitous mafic mineral. Kimzey (1982) found, from west to east, the following facies: a border zone, hornblende-biotite, large-biotite, small-biotite, and a muscovite-biotite core. The border facies forms the western contact of the pluton and is restricted to a narrow band up to 100 meters wide along the contact. All internal contacts between the facies are gradational, steep to vertical, and are subparallel to the trend of the western contact.

EASTERN FACIES OF THE LA POSTA PLUTON

The eastern facies of the La Posta Pluton within the In-Ko-Pah Gorge region were mapped as part of an undergraduate senior thesis project at SDSU (Sinfield, 1983) under the steady guidance of Dr. Walawender. The western facies were found to be repeated, in reverse order. Both the host rock and the border facies were not exposed in the eastern facies in the In-Koh-Pah Gorge study area.

Hornblende-Biotite Facies

Driving west from the dusty town of Ocotillo along Interstate 8, impressive steep boulder-mantled slopes emerge from beneath the sedimentary cover of the Imperial Valley (Figure 2). These slopes are composed of the oldest eastern facies found in the La Posta Pluton, the hornblende-biotite facies, a massive coarse-grained leucocratic sphene-hornblende-biotite tonalite.

Euhedral hornblende, biotite, and sphene are ubiquitous within this facies with large distinctive crystals up to 8 mm in length. Plagioclase forms large subhedral to euhedral crystals up to 10 mm, exhibit weak oscillatory or patchy zoning, and commonly have cores composed of two or more crystals joined in a synneusis relationship and are mantled by less calcic rims. In thin sections from the study area, cores have average anorthite contents of An_{37} and rims, An_{27} (Sinfield, 1983). Alkali feldspar is interstitial exhibiting "gridiron" twinning. Locally, myrmekitic intergrowths occur at the contact between the two feldspars. Quartz exists as discrete or composite grains with weak undulatory extinction and moderately sutured boundaries (Clinkenbeard and Walawender, 1989). Large (10 to 150 cm) hornblende and biotite bearing xenoliths occur throughout the hornblende-biotite facies rocks.

Large-Biotite Facies

The hornblende-biotite facies grades to the southwest into the large-biotite facies. The facies outcrops west of Devil's Canyon, and south and west of Myer Valley (Figure 2 and Plate 1). Both the hornblende-biotite and large-biotite facies form large (3 to 7 m) medium-brown boulders with well developed joint sets. The contact between these two facies trends to the northwest and is straight along most of its length (Figure 2). Moving across the gradational contact, hornblende gradually decreases and the composition shifts from a tonalite to a coarse-grained (5 to 9 mm) leucocratic sphene-biotite granodiorite with large (5 to 9 mm) ubiquitous euhedral biotite books. These books give the rock a distinctive "salt and pepper" appearance. Also, pseudomorphs of biotite after hornblende can be seen in hand sample.

Plagioclase forms subhedral to euhedral crystals that average 2 to 7 mm and commonly exhibit oscillatory zoning. In thin sections from the study area, in comparison with the hornblende-biotite facies, plagioclase cores have lower average anorthite contents of An_{35} and rims, An_{25} (Sinfield, 1983). Quartz appears to be slightly coarser in this facies. Oikocrysts of alkali feldspar enclose earlier

Figure 4. Dikes of small-biotite facies intruding the large-biotite facies. Photo of an outcrop within In-Ko-Pah Gorge.

formed minerals: plagioclase, biotite, and quartz (Sinfield, 1983 and Clinkenbeard and Walawender, 1989). As in the hornblende-biotite facies, sphene is the main accessory mineral with some zircon.

Small-Biotite Facies

In contrast with the western La Posta Pluton facies where the large-biotite grades into the small-biotite facies, both the hornblende-biotite and large-biotite facies are mostly intruded by the small-biotite facies in the In-Ko-Pah Gorge study area (Figure 2). The small-biotite facies is characterized by a medium-grained (3 to 7 mm) biotite granodiorite with small (1 to 3 mm) ubiquitous subhedral to euhedral biotite.

Excellent outcrops of this facies occur on the eastern wall of Devil's Canyon to In-Ko-Pah Gorge. The contact between the small-biotite facies and the earlier formed hornblende-biotite and large-biotite facies can be readily seen from far distances due to the smaller size of the small-biotite facies boulders (1 to 4 m) and their lighter color (Figure 4). Next to contacts and in small-biotite dikes, the small-biotite facies defines a foliation that follows the trend of these contacts (Figure 5). Inclusions of pre-batholithic plagioclase crystals are smaller (1- to 3-mm) than the first two facies with similar metamorphic rocks, gabbroic country rocks, hornblende-biotite facies, and large-biotite facies occur within the small-biotite facies.

Close to contacts with earlier formed La Posta units, a few large (3 to 8 mm) scattered aggregates of subhedral biotite occur. These are small-biotite pseudomorphs after hornblende and large-biotite synneusis relationships. Crystals are subhedral. Weak oscillatory zoning is present, but the patchy zoning is missing. Quartz is texturally similar, but exhibits stronger undulatory extinction. As in other facies, alkali feldspar typically forms oikocrysts with well-developed "gridiron" twinning (Sinfield, 1983; Clinkenbeard and Walawender, 1989).

A thin northeast-trending band of high-grade metamorphic rocks occurs along the intrusive contact with the large-biotite faces between In-Ko-Pah Gorge and the east side of Myer Valley. These pre-batholithic marbles (with associated tactites), schistose-quartzite, and minor amounts of mica schist occur in pods and steeply dipping highly folded bands 10 to 300 m wide.

Muscovite-Biotite Facies

The small-biotite facies grades to the southwest into the muscovite-biotite (two-mica) facies, a medium-grained (1 to 6 mm) muscovite-biotite granodiorite to adamellite. This facies is similar to the small-biotite facies except for the presence of small (1 to 3 mm) subhedral muscovite crystals and the absence of sphene. In thin sections from the study area, in comparison with the hornblende-biotite and large-biotite facies, plagioclase cores have lower average anorthite contents of An_{24} and rims, An_{11} (Sinfield, 1983). The muscovite-biotite facies outcrops from Devil's Canyon southward to the vicinity of the Desert View Tower and just north of Jacumba. It has the same outcrop characteristics as the small-biotite facies and intrudes the large-biotite facies.

Dikes

In the northern section of the study area, several large northeast trending fine-grained (1 to 2 mm) to medium-grained sphene-biotite granodiorite dikes intrude the hornblende-biotite and large-biotite facies. The dikes were first mapped by Brooks and Roberts (1954). The dikes do not intrude other La Posta facies. Typically, the dikes are easily recognized in the field by a dark-brown color, 5 to 10 m thickness, strike N80°E, and high dips to the northwest and southeast. Foliation within the dikes parallels the orientation of the dikes and is defined by biotite. At the

Figure 5. Foliation within the small-biotite facies forms parallel to contacts and wraps around inclusions. Photo from outcrop on Devil's Canyon.

margins, the dikes lack sphene and are fine grained. The cores have abundant sphene and are medium grained. The dikes are similar to the small-biotite facies in composition and texture and may originate from the small-biotite intrusion. Typically, dikes from the small-biotite facies only intrude a few tens of meters into the surrounding rocks. If these are from small-biotite intrusive events, then these sphene-rich dikes represent a major fracturing and deep penetration of the hornblende-biotite and large-biotite facies rocks.

Small (20 to 100 cm) pegmatite dikes cut all of the La Posta units. The pegmatites in the interior facies tend to be larger than the ones in the outer facies. The dikes usually are restricted in mineralogy, have large euhedral alkali feldspars with interstitial quartz, and have large (5 to 50 mm) biotite and muscovite books that are commonly elongate. Large (5 to 7 mm) almandine garnets also occur in a few places.

ALVERSON ANDESITE (MIOCENE)

The Alverson Andesite outcrops on the northern portion of the study area near Dos Cabezas and in the southeastern portion of the site, just south of Myer Valley. This unit was named by Dibblee (1954) for volcanic flows in Alverson Canyon. The flows are characteristically dark-mauve-gray plagioclase-hornblende andesite. They are interbedded with gray andesite tuff and were extruded onto the eroded La Posta Pluton surface. South of Myer Valley, the Alverson Andesite occurs up to 120 m thick, and in the northern portion of the study area, it occurs up to 85 m thick. Olivine basalt from the northern Jacumba Mountains yielded a K-Ar whole-rock age of 16.9 ± 0.5 Ma (Hoggatt, 1979). This age is in accord with isotopic ages of volcanic rocks elsewhere in the region that range from about 14 to 22 Ma (Kerr and Kidwell, 1991).

DISCUSSION

Based upon the crosscutting relationships, crystallization sequences, anorthite contents of the plagioclase, and consistent and linear geochemical trends, the pluton crystallized from the margins inward. On the outer margin, the hornblende-biotite facies has euhedral high temperature minerals (hornblende and biotite) with plagioclase cores having average anorthite content of An_{37}. Inward from the margin, within the large-biotite facies, hornblende content decreases and the plagioclase cores have average anorthite content dropping to An_{35}. Lower temperature alkali feldspar grains also increase in size.

These two facies are cut by the later formed small-biotite facies which contains smaller-grained high temperature minerals plagioclase and biotite. The biotite no longer forms euhedral crystals. Further inward, in the muscovite-biotite facies, the anorthite content decreases to An_{24}. Geochemical trends are linear and are indicative of a comagmatic source for all the facies studied.

The eastern facies has leucocratic lithologies, crystallization histories, compositions, structure, and petrographic trends similar to the western facies of the La Posta Pluton. A notable field difference is that all the western facies contacts are gradational whereas the eastern small-biotite facies, in most part, has an intrusive contact with the earlier formed outer facies. The eastern small-biotite facies is gradational with the later formed muscovite-biotite facies. The eastern and western facies of the La Posta Pluton have undergone similar differentiation processes and are comagmatic (Sinfield, 1983).

Chemically, the I-type and S-type granites described by Chappell and White (1974) reflect their different origins. I-type granites typically have A/(C+N+K)[1] values less

---

[1] A/(C+N+K). A = weight% Al_2O_3/atomic weight Al_2O_3; C = wt.% CaO/atm. wt. CaO; N = wt.% Na_2O/atm. wt. Na_2O; and K = wt.% K_2O/atm. wt. K_2O.

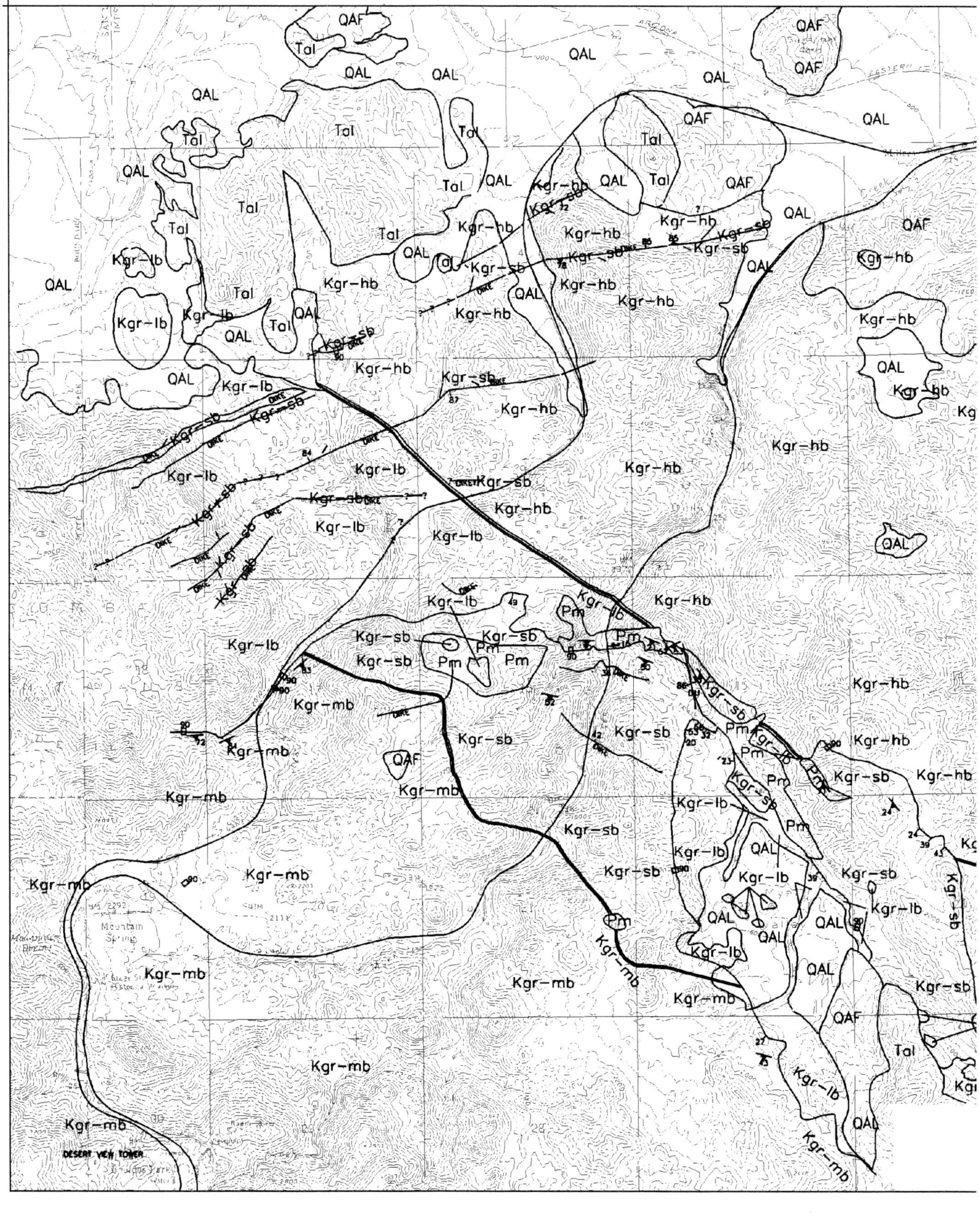
QAF
Tal
QAL
Kgr-hb
Kgr-sb
Kgr-lb
Kgr-mb
Pm
DIKE
Mountain Spring
DESERT VIEW TOWER

GEOLOGY OF THE EASTERN FACIES OF THE LA POSTA PLUTON IN-KO-PAH GORGE AREA, SAN DIEGO AND IMPERIAL COUNTIES

by LEONARD D. SINFIELD, 1983

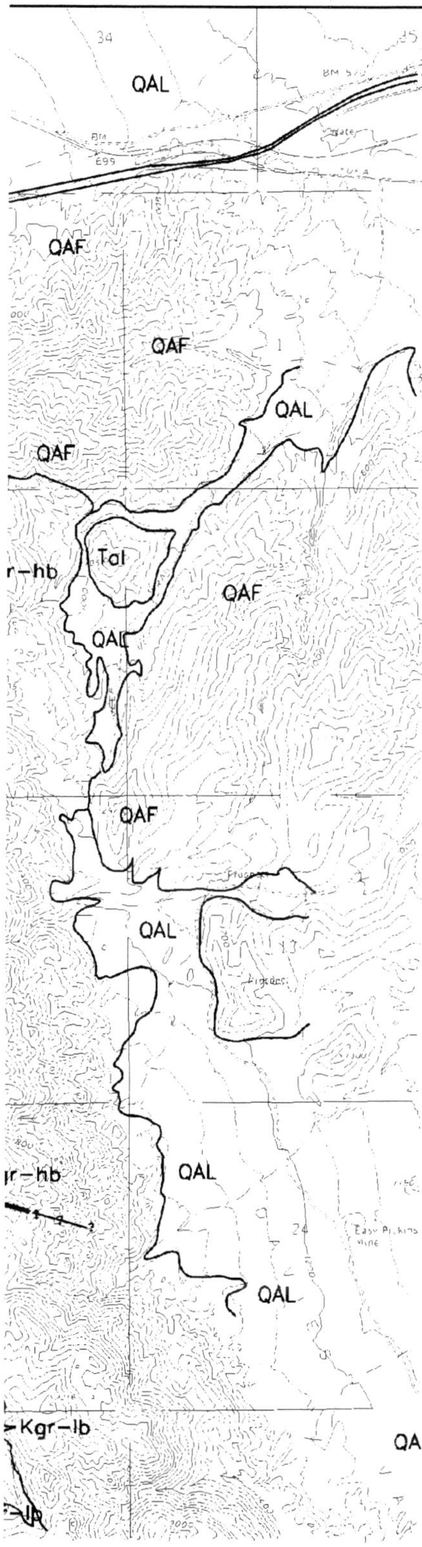

DESCRIPTION OF MAP UNITS

QUATERNARY

QAL Quaternary Alluvium – Sand and gravel in narrow washes and alluvial fans.

QAF Quaternary Fanglomerate – Older sand, gravel, and boulders alluvium on slopes.

TERTIARY

Tal Alverson Andesite (Miocene) Dark-mauve-gray plagioclase-hornblende andesite, interbedded with gray andesite tuff. Consists of massive flows, breccia, intercolated volconiclastic deposits, and minor air-fall cinder deposits.

CRETACEAOUS

Kgr-hb La Posta Pluton: Hornblende-biotite Facies – Massive coarse-grained leucocratic sphene-hornblende- biotite tonalite.

Kgr-lb La Posta Pluton: Large-biotite Facies – Coarse-grained (5 to 9 mm) leucocratic sphene-biotite granodiorite with large (5 to 9 mm) ubiquitous euhedral biotite books. In outcrop, biotite gives the rock a distinctive "salt and pepper" appearance.

Kgr-sb La Posta Pluton: Small-biotite Facies – Medium-grained (3 to 7 mm) biotite granodiorite with small (1 to 3 mm) ubiquitous subhedral to euhedral biotite. Forms smaller boulders with lighter color than large-biotite facies.

Kgr-mb La Posta Pluton: Muscovite -biotite Facies – Medium-grained (1 to 6 mm) muscovite-biotite granodiorite to adamellite. Similar to small-biotite facies except for the presence of small (1- to 3-mm) subhedral muscovite crystals and the absence of sphene.

PALEOZOIC (?)

Pm Prebatholihic Metamorphic Rocks – High-grade (amphibolite facies) metamorphic rocks. Consists of marbles with associated tactites, schistose-quartzite, and minor amounts of mica schist. Marble grades into epidote-bearing milky- white quartzite and scheelite-garnet calc-silicate rock. Almandine garnet occurs as aggregates of large isolated crystals.

GEOLOGIC SYMBOLS USED

Gradational contact between La Posta Pluton facies.

Strike and dip of foliation.

Contact with dip direction and attitude.

Fault, relative displacement indicated (U – Upside, D – Down)

Vertical contact

4,000FT 2,000FT 0 1 MILE

SCALE

SOURCE MAP: U.S.G.S 1959, IN-KO-PAH GORGE 7.5 MINUTE TOPOGRAPHIC QUADRANGLE MAP, PHOTOREVISED 1975.

PLATE 1

than 1.0 (metaluminous) with high sodium to potassium ratios (Na/K). When rocks are subjected to weathering, sodium is preferentially removed leaving the rock relatively enriched in aluminum and potassium, hence, S-type granites typically have A/(C+N+K) values greater than 1.2 (peraluminous) and have low Na/K ratios (Walawender, M.J., per. comm., 1983). Weakly peraluminous granites have A/(C+N+K) values between 1.0 and 1.2. Samples from the eastern facies of the La Posta Pluton in the study area are metaluminous to weakly peraluminous indicating I-type and transitional S-type sources. This range in source types may indicate some limited mixing of different magma sources to produce the La Posta Pluton.

According to Gastil (1975), the parental La Posta melt rose between the older western zone of the batholith and the leading edge of the North American craton. Walawender et al. (1989) described anomalous isotopic patterns in the muscovite-bearing cores of several La Posta-type plutons in southern California, USA, and Baja California, Mexico. These patterns represent mixing of a parental melt (derived primarily from subducted oceanic crust) and a continental component. As in the case for the La Posta Pluton, it was proposed that as the diapir rose and expanded into the overlying metasedimentary apron, a portion of it, adjacent to the deeper segments of the craton, reacted with crustal materials. Inward crystallization within the outer portion of the diapir produced the border facies, hornblende-biotite facies, and large-biotite facies, but prior to the solidification of the small-biotite facies, the tail of the diapir entered the magma chamber and mixed with the derivative liquid. The resulting hybrid melt could only undergo limited mixing and diffusion with the already partly solidified outer zones of the pluton and cooled *in situ* to form the muscovite-bearing core with its gradational contacts.

REFERENCES

Brooks, B., and Roberts, E., 1954, Geology of the Jacumba Area, San Diego and Imperial Counties: California Division of Mines and Geology, Bulletin 170, Map sheet 23.

Chappell, B.W., and White, A.J.R., 1974, Two contrasting granite Types: Pacific Geology, v. 8, pp. 173-174.

Dibblee, T.W., Jr., 1954, Geology of the Imperial Valley region, California: *in* Jahns, R.H., ed., Geology of California, Chapter 2: California Division of Mines Bulletin 170, pp. 21-28.

Gastil, R.G., 1975, Plutonic zones in the Peninsular Ranges of southern California and northern Baja California. Geology, 3, pp. 361-363.

Hoggatt, W.C., 1979, Geologic map of the Sweeney Pass quadrangle, San Diego County, California: U.S. Geological Survey Open-File report 79-754, 28 p., scale 1:24,000.

Kerr, D.R., and Kidwell, S.M., 1991, Late Cenozoic sedimentation and tectonics, western Salton Trough, California: *in* Walawender, M.J., and Hanan, B.B., eds., Geological excursions in southern California and Mexico, Guidebook for the 1991 Annual Meeting of the Geological Society of America, San Diego, California, pp. 397-416.

Kimzey, J.A., 1982, Petrology and geochemistry of the La Posta granodiorite, M.S. Thesis, San Diego State University, San Diego, California, 81 p.

Krummenacher, D., R.G. Gastil, J. Bushee, and J. Doupont, 1975, K-Ar Apparent Ages, Peninsular Ranges Batholith, Southern California and Baja California: Geological Society of America Bulletin, v. 86, pp. 760-768.

Morgan, P.K., 1977, Geology and Mineral Resources of Imperial County, California: California Division of Mines and Geology, County Report 7, pp. 96-97.

Sinfield, L., 1983, The geology of the eastern facies of the La Posta Pluton, 44 p. Undergraduate Research Report. San Diego State University, San Diego, California.

Streckeisen, A.L., 1973, Plutonic rocks — classification and nomenclature. Geotimes, v.18, no. 10, pp. 26-30.

Walawender, M.J., and Clinkenbeard, J.P., 1989, Mineralogy of the La Posta Pluton: Implications for the origin of zoned plutons in the eastern Peninsular Ranges Batholith, southern and Baja California. American Mineralogist, v. 74, pp. 1258-1269.

Walawender, M.J., Gastil, R.G., Clinkenbeard, J.P., McCormick, W.V., Eastman, RG., Wernicke, R.S., Wardlaw, M.S., Gunn, S.H., and Smith, R.M., 1989, Origin and evolution of the zoned La Posta-type plutons, eastern Peninsular Ranges Batholith, southern and Baja California. Geological Society of America Memoir 174.

Dos Cabezas Amphibolites

Brent Foster, *California Geological Survey,*
888 South Figueroa Street, Suite 475, Los Angeles, California 90017

INTRODUCTION

Amphibolites are a notable component of the prebatholithic rocks exposed in screens and roof pendants throughout the Peninsular Ranges Batholith. They are found in the Julian Schist and associated rocks of the western zone of the batholith as well as the higher grade pelitic schists, quartzites, and carbonates of the eastern zone. In 1994 a San Diego State University master's thesis investigated the Peninsular Ranges amphibolites including the outcrops at Dos Cabezas (Foster, 1994).

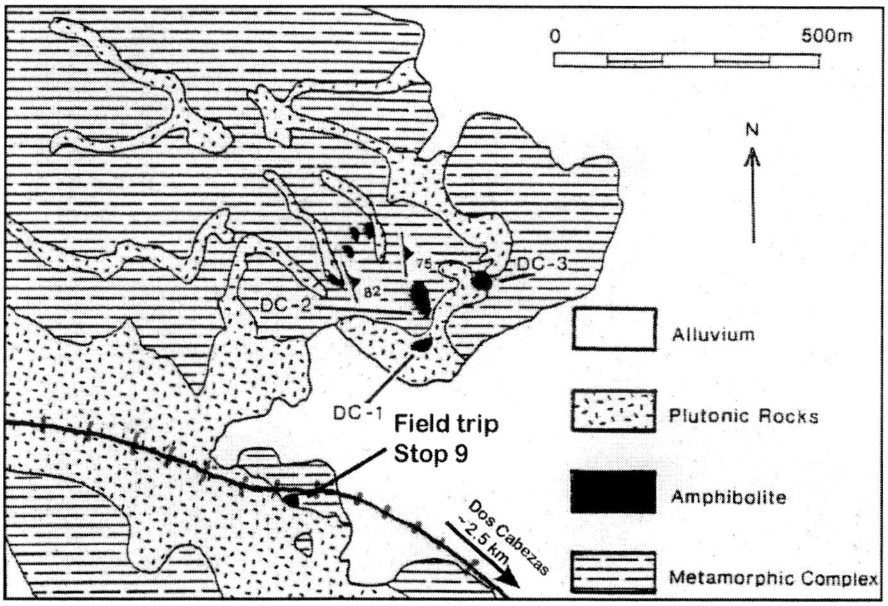

Figure 1. Location map.

Amphibolites may be of igneous or sedimentary origin. Those of igneous origin are called ortho-amphibolites, and those of sedimentary origin are called para-amphibolites. Amphibolites are often assumed to be of igneous origin because their major element chemistry is similar to that of basalt, a very common igneous rock. A sedimentary precursor would have to have similar a chemical composition, and the best candidates are calcareous or dolomitic shales (Leake, 1964), a less common alternative. Thin layers of amphibolite may also develop through chemical exchange between adjacent carbonate and clay layers (Orville, 1969).

In the Peninsular Ranges, amphibolites have generally been considered to be of igneous origin, based primarily on rock associations and lack of evidence to the contrary. Some however, particularly in the eastern Peninsular Ranges, have been interpreted as para-amphibolites based on their association with mio-geoclinal rocks (Todd and Shaw, 1979). These higher grade eastern zone amphibolites typically occur as lensoid outcrops several meters to tens of meters thick with no clear intrusive or depositional relationship to adjacent rocks.

THE DOS CABEZAS AMPHIBOLITES

The area of the Dos Cabezas amphibolites (named for the Dos Cabezas Mine) was mapped by Parrish (1990) in a study of the Indian Hill Pluton (Figure 1). Here, the pluton intrudes a metamorphic complex that consists of highly deformed migmatitic gneiss, with sporadic lenses of marble, quartzite, and amphibolite. Amphibolite is exposed in discontinuous lensoid outcrops which are elongated parallel with foliation in surrounding rocks and with the local structural trend.

The amphibolite is fine-grained and generally equigranular, although grain size variations are encountered between outcrops. Average composition is about 68% hornblende, 24% plagioclase, with lesser amounts of quartz, sphene, ilmenite, and trace zircon. Relict plagioclase phenocrysts are a common feature (Figure 2). They are typically 0.3 mm to 1 mm and are lath-shaped. Internal recrystallization is very extensive. The phenocrysts are only recognizable by their outlines and by faintly preserved albite or Carlsbad twin.

PROTOLITH

The distinction between ortho- and para-amphibolite is made possible by the preservation of relict igneous or sedimentary features. These features may be physical or

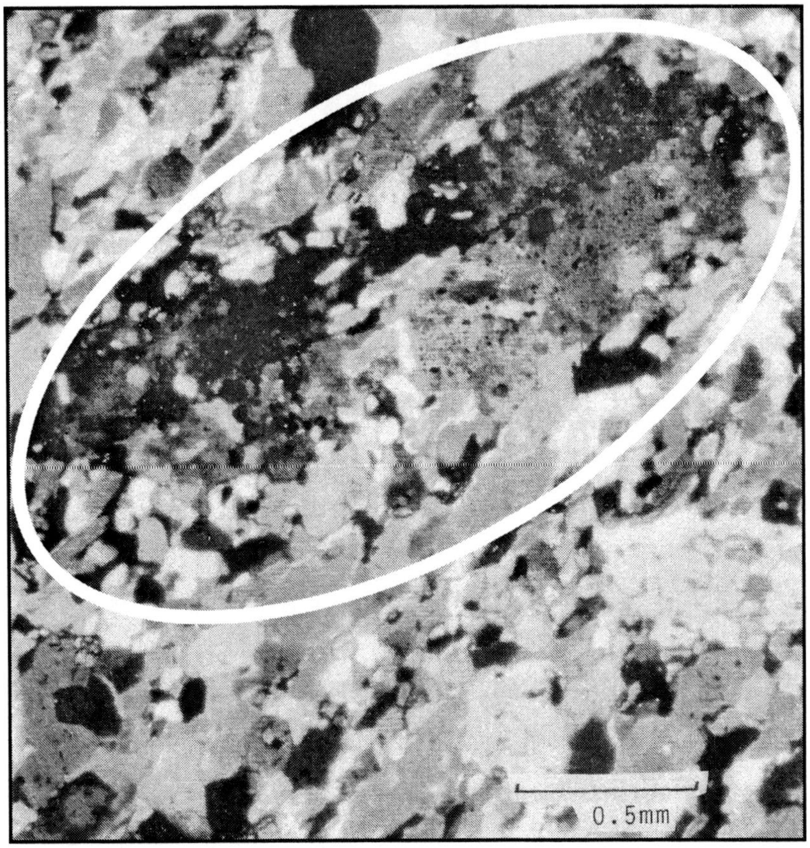

Figure 2. Plagioclase relict phenocryst from sample DC-3.

chemical in nature. In either case relict features become more difficult to recognize with increasing metamorphic grade. Major-element chemistry is likely to be similar for ortho- and para-amphibolites because of similar mineralogy. Also, metamorphism is likely to influence major-element chemistry. Immobile trace-element trends, inherited from igneous or sedimentary precursors, may be the only features that survive intense regional metamorphism.

Many Peninsular Ranges amphibolites, and specifically the Dos Cabezas amphibolites, clearly demonstrate their igneous origin simply by their petrography. Relict phenocrysts are observed in the Dos Cabezas amphibolites and in most of the amphibolites studied by Foster (1994). Plagioclase is the most common mineral that occurs as relict phenocrysts; however pyroxene also occurs in this form at other Peninsular Ranges amphibolite localities.

Geochemical evidence is also consistent with an igneous origin. Geochemical methods of distinguishing between ortho- and para-amphibolites are based on the recognition of various trends that result from magmatic differentiation, and those that reflect sedimentary processes such as mixing of carbonate and clay end members. Such methods point to igneous derivation for the Peninsular Ranges amphibolites, and particularly the Dos Cabezas amphibolites (Foster, 1994).

TECTONIC SETTING OF THE PARENT ROCKS

Having established a mafic igneous origin for the Peninsular Ranges amphibolites, the relationship of these rocks to the metasedimentary rocks that host them can be better understood if more is known of the plate tectonic setting in which they formed. Geochemical methods can be useful in this regard. In particular, study of the high field-strength elements (HFSE) can be especially useful.

The HFSE are a group of elements that have a relatively high ratio of charge to ionic radius. Because of this distinction the, most of the HFSE are highly incompatible, that is they are generally excluded from the structures of the most common rock forming minerals, and as a result are generally concentrated in residual liquid as a magma crystallizes. There are however notable exceptions, for example Titanium. Ti may be preferentially removed from a magma if Ti-rich phases such as ilmenite or sphene are crystallizing. Because of the incompatible and occasionally compatible behavior of these elements, study of their variations can provide much information regarding magma genesis and tectonic setting. Also because of their peculiar HFS characteristics, these elements tend to be relatively insoluble and immune to redistribution by secondary processes, and are therefore especially valuable in the study of meta-igneous rocks.

In contrast to the HFSE are the low field-strength elements (LFSE), many of which are also referred to as large ion lithophile (LIL) elements. These elements are also highly incompatible, but unlike the HFSE, their large size and low valence state makes them vulnerable to redistribution by secondary processes.

A variety of data-analytical methods have made use of HFSE abundances and variations to shed light on tectonic setting for mafic rocks (Pearce and Cann, 1973; Pearce, 1975; Pearce and Norrey, 1979). Most of these are binary or ternary graphical plots of various trace element abundances. Analysis of modern basaltic rocks from known tectonic environments has shown a tendency for compositional grouping when their trace element abundances are plotted on these diagrams. This allows fields to be drawn on these plots around groups for different tectonic settings.

These fields represent the expected composition of basaltic rocks from that tectonic setting.

Figures 3 and 4 are from Foster (1994). Figure 3 is a Ti/Zr/Y diagram (Pearce and Cann, 1973) used to discriminate tectonic setting, with the compositions for Peninsular Ranges amphibolites plotted. The Dos Cabezas amphibolites are in the indistinguishable mass of points in the middle of the ocean floor basalt field. Figure 4 is a Ti/Cr diagram (Pearce, 1975) used to distinguish between ocean floor basalts and island arc basalts (low potassium tholeiites). The Dos Cabezas amphibolites plot with the rest of the Peninsular Ranges amphibolites in the ocean floor basalt (OFB) field.

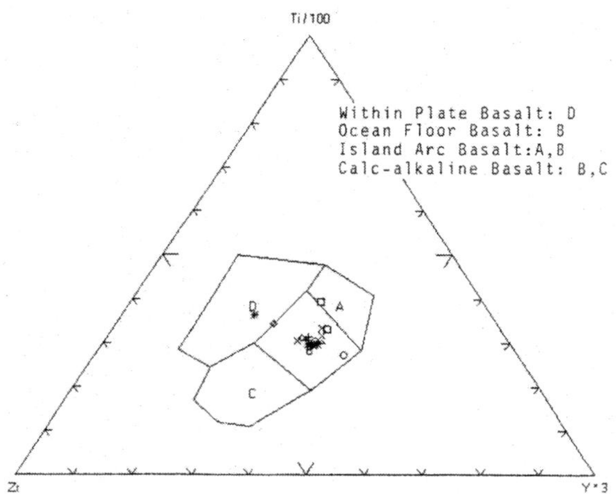

Figure 3. Ti/Zr/Y Diagram.

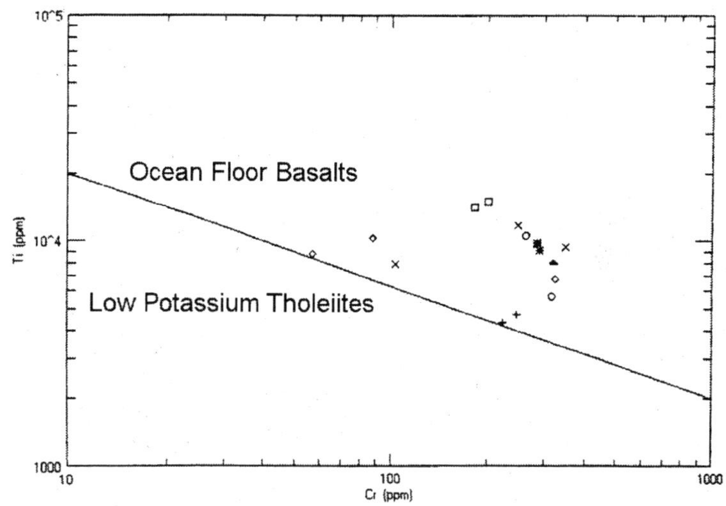

Figure 4. Ti/Cr Diagram.

Although the data appear to indicate an ocean floor origin for the Peninsular Ranges amphibolites, additional geochemical evidence is presented in Foster (1994) that indicates a component of enrichment that is characteristic of subduction zones. Chondrite-normalized rare earth element (REE) patterns for most of the amphibolites display a light-REE (LREE) depleted pattern that is typical of normal mid-ocean ridge basalt (N-MORB). However some of the eastern amphibolites, including those from Dos Cabezas, display slight LREE enrichment. Also when trace element abundances are plotted normalized to N-MORB, the eastern amphibolites display a characteristic pattern of LFSE enrichment and HFSE depletion that is often interpreted as a subducted-slab component (Pearce, 1981; Saunders and Tarney, 1984; 1991). The source of arc magmas is thought to be enriched with fluids derived from a dehydrating subducted slab. These fluids contain abundant LFSE which are very mobile. HFSE depletion reflects the depleted nature of the source.

Two modern tectonic environments are known to produce OFB with a component of enrichment in certain trace elements. These are back-arc basins and elevated or anomalous mid-ocean ridge segments. Back-arc basins are chemically transitional between N-MORB and volcanic arc basalt and, like arc basalts are thought to contain a component derived from a subducted slab (Saunders and Tarney, 1984; 1991). Enriched or E-MORB, from anomalous ridge segments, is believed to derive its enrichment from mantle plumes, and may be transitional with oceanic-island basalt (Pearce, 1981; Floyd, 1991). The trace element enrichment of E-MORB is characterized by HFSE enrichment, reflecting a more fertile mantle source.

TECTONIC IMPLICATIONS

The Peninsular Ranges amphibolites represent a unique magmatic sequence unrelated to any other igneous rocks of the region. Although bearing a slab signature, they are ocean floor basalts, and not products of arc magmatism. They are chemically distinct from the other mafic rocks of the Peninsular Ranges, such as the Santiago Peak Volcanics and the western zone gabbros, which display volcanic arc affinity (Gorzolla, 1988; Walawender and Smith, 1980).

The occurrence of ocean floor basalt (OFB) within the pre-batholithic metasedimentary succession of the Peninsular Ranges places an important constraint on tectonic reconstructions in the region. Barring the possibility that the amphibolites are slivers of ocean floor tectonically introduced into the overlying sediments, a source for OFB, either a mid-ocean ridge or a back-arc ridge, must have been nearby at the time of formation. It implies that the Julian Schist and associated rocks were deposited in an extensional basin.

Deposition of the western pre-batholithic metasedimentary rocks such as the Julian Schist probably occurred in the Triassic to Early Jurassic. It would be difficult to construct a geologic history that places a mid-ocean ridge near the continental margin during that time interval. The rifting which formed the continental margin probably occurred in the Late Precambrian or early Paleozoic and it is unlikely that the ensuing mid-ocean ridge would remain near the continent throughout the Paleozoic and into the Triassic. A back-arc ridge is a more probable source for the amphibolites.

REFERENCES

Foster, B.D., 1994, Origin and Tectonic Significance of Peninsular Ranges Amphibolites [M.S. Thesis]: San Diego State University, San Diego, California, 112 p.

Floyd, P.A., 1991, Oceanic Islands and Seamounts, *in* Floyd, P.A., ed., Oceanic Basalts: Blackie, Glasgow, pp. 174-218.

Gorzolla, Y.R., 1988, Geochemistry and Petrology of the Santiago Peak Volcanics, Santa Margarita and Santa Ana Mountains, Southern California [M.S. Thesis]: San Diego State University, San Diego, California, 145 p.

Leake, B.E., 1964, The Chemical Distinction between Ortho- and Para-amphibolites: Journal of Petrology, v. 5, pp. 238-254.

Orville, P.M., 1969, A Model for Metamorphic Differentiation of Thin Layered Amphibolites: American Journal of Science, v. 267, pp. 64-86.

Parrish, K.E., 1990, Geology, Petrology, Geochemistry and Isotopic Character of the Indian Hill Pluton, Jacumba Mountains, California [M.S. Thesis]: San Diego State University, San Diego, California, 105 p.

Pearce, J.A., 1975, Basalt Geochemistry Used to investigate Past Tectonic Environments in Cyprus: Tectonophysics, v. 25, pp. 41-67.

Pearce, J.A., 1982, Trace Element Characteristics of Lavas From Destructive Plate Boundaries, *in* Thorpe, R.S., ed., Orogenic Andesites: New York, John Wiley, pp. 528-548.

Pearce, J.A., and Cann, J.P., 1973, Tectonic Setting of Basic Volcanic Rocks Determined Using Trace Element Analysis: Earth and Planetary Science Letters, v. 19, pp. 290-300.

Pearce, J.A., and Norry, M.J., 1979, Petrogenetic Implications of Ti, Zr, Y, and Nb Variations in Volcanic Rocks: Contributions to Mineralogy and Petrology, v. 69, pp. 33-47.

Saunders, A.D., and Tarney, J., 1984, Geochemical Characteristics of Basaltic Volcanism within Back-arc basins, *in* Kokelaar, B.P., and Howells, M.F., eds., Marginal Basin Geology: Blackwell, Oxford, pp. 59-76.

Saunders, A.D., and Tarney, J., 1991, Back-arc Basins, *in* Floyd, P.A., ed., Oceanic Basalts: Blackie, Glasgow, pp. 219-263.

Todd, V.R., and Shaw, S.E., 1979, Structural, Metamorphic and Intrusive Framework of the Peninsular Ranges Batholith in San Diego County, California, *in* Abbott, P.L., and Todd, V.R., eds., Mesozoic Crystalline Rocks: San Diego State University, Department of Geological Sciences, Guidebook for the Geological Society of America Annual Meeting, San Diego, California, pp. 177-231.

Walawender, M.J., and Smith, T.E., 1980, Geochemical and Petrologic Evolution of the Basic Plutons of the Peninsular Ranges Batholith, Southern California: Journal of Geology, v. 88, pp. 233-242.

The Fossil Treasures Project

George T. Jefferson, *Colorado Desert District Stout Research Center, California Department of Parks and Recreation, 200 Palm Canyon Drive, Borrego Springs, California 92004*
gjefferson@parks.ca.gov

Lowell Lindsay, *Sunbelt Publications, 1256 Fayette Street, El Cajon, California 92020*
LLindsay@SunbeltPub.com

INTRODUCTION

The following paper describes the motivation, intent and development of the book *Fossil Treasures of the Anza-Borrego Desert: The Last Seven Million Years* (Jefferson and Lindsay, 2006a). It is based on the article *A Project in Science Education and Publishing in the Public Interest* (Lindsay, 2006), published by the authors in *Desert Update* (Lindsay and Jefferson, 2006), Anza-Borrego Foundation and Institute (ABF&I).

> *Scientists are surrounded by a protective jargon and write in inaccessible journals. There is an abyss between things that happen in the natural world and where scientists publish their work. Some people say that we need more scientists who can work on the boundary...Bringing science to common citizens, not only government officials, is something that is desperately needed.*
>
> —Exequiel Ezcurra, Director,
> Biodiversity Research Center,
> San Diego Museum of Natural History
> *San Diego Union-Tribune* (Ezcurra, 2006)

Dr. Ezcurra expressed this concern in conjunction with his recent acceptance of a prestigious Pew Fellowship in marine conservation. Such a point of view reprises much of the motivation, even passion, which drove our project team over the last five years to produce *Fossil Treasures of the Anza-Borrego Desert: The Last Seven Million Years* and associated publications.

The visionary seed for the book came from the ABF&I, key California State Park (CSP) staff and Colorado Desert District (CDD) paleontology program volunteers. The Fossil Treasures Project (FTP) was conceived in 2000 as an educational and interpretive component of the paleontology program that is headquartered at the CDD Stout Research Center of the CSP. It blossomed with the mobilization of twenty-three leading scientists and specialists who shared their knowledge of paleontology and geology in the Anza-Borrego region. It came to fruition with a similar number of editors, photographers, artists, and other contributors plus endorsements from leading paleontologists and institutions throughout the U.S. The result is a 416-page full-color publication about which the San Diego Union-Tribune *Books* notes (McCaffery, 2006), "they have not only done a superb job in gathering together a world-class, state-of-the-art collection of essays, but they have done so in a manner that combines scientific accuracy with readability."

Now begins the next phase, which is fulfilling the vision of the FTP as an example of bridging that "boundary" noted above, that border which lies between the daunting complexities of technical research and understanding the implications of such research for education, interpretation, and public policy. Good science, not communicated to the general public, fails in its mission to guide sound decision-making. A public policy that is ignorant or uninformed by its underlying science ranges from being wasteful of resources to being dangerous to communal health and welfare.

Transitioning from these general issues to the specific project, the FTP presents the paleontology of the Anza-Borrego Desert region, part of the larger geomorphic area of peninsular California. Based on the science and using such multi-media tools as publications, field trips, museum displays, audio-visual presentations and lectures (see ABF&I, 2005), the project is an adventurous exploration of the last seven million years in the region. The intent is to interpret to readers and audiences the *educational themes* and *public policy implications* that have emerged during the course of project.

EDUCATIONAL THEME I

The Anza-Borrego Desert occupies a unique and complex geologic setting on the western edge of the Salton Trough rift valley where a thick depositional sedimentary sequence yields a rich marine and terrestrial paleontologic record.

Policy Implication: Establishment and protection of large natural preserves in their undisturbed state is essential to scientific research as well as other values of the wilderness experience.

1. The Salton Trough, a northwest extension of the Gulf of California, is a major rift basin that has developed since late Miocene time. Major sub-basins (e.g., now the Borrego, Vallecito Creek, and Fish Creek Badlands of Anza-Borrego Desert State Park [ABDSP]) formed along the western margin of the Trough (Dorsey, 2006).

2. Erosion of the Colorado Plateau and the Grand Canyon brought sediment into the region and terminated a marine intrusion about 4.2 Ma, resulting in the mostly below-sea-level Salton Trough covering an area of over 5000 km^2.

3. Initiation of major strike-slip faulting, primarily along the San Jacinto Fault Zone and relative uplift of the Peninsular Ranges increased local erosion and sedimentation rates, exposing the basin deposits to reveal their treasures (e.g., the Fonts Point area of the Borrego Badlands was a former lake bed and then an alluvial apron ca. 1 Ma). A late Pleistocene rain shadow contributed to desert development with sparse vegetation cover facilitating earth science fieldwork.

4. Geologic and paleontologic evidence mutually reinforces a picture of changing landscapes, climates, environments, and flora and fauna over the past 7 million years. This record spans the beginning of the Ice Ages, and reveals the development of inter-continental and inter-oceanic connections, (e.g., the Panama Isthmus, the Central American Seaway, and the Bering Straits and Land Bridge), affording two-way passages for flora and fauna (e.g., the Great American Biological Interchange [McDonald, 2006a]).

5. ABDSP protects over 2,590 km^2 of mountain, upland valley, and desert environments, two thirds of which is designated State Wilderness with another adjacent 1,300 km^2 in U.S. Bureau of Land Management (BLM) Wilderness Areas.

EDUCATIONAL THEME II

Paleontology opens figurative "windows into the past," enabling the reconstruction of paleoclimates, environments, and ecology. Patterns and agents of change can be teased from the stratigraphic and fossil record. The Anza-Borrego sequence reveals distinct paleoenvironments that include open and near shore marine, estuarine, major river delta, lacustrine and playa, savanna, riparian woodland, and desert landscapes.

Policy Implication: Analysis of the prehistoric record may have explanatory and predictive value as a baseline for present and future conditions (e.g., human factors driving climate change).

1. The Imperial Sea, 6.5 to 4.5 Ma (Figure 1), was a warm and tropical environment with affinities to the modern Caribbean Sea via the Central American Seaway. (Envision today's southern Gulf of California around La Paz and mainland Mexico.) The extinct corals, clams, and snails have living Caribbean relatives (Rugh, 2005; Deméré and Rugh, 2006). Walrus fossils from the desert contrast dramatically with the image of their modern ice-loving arctic cousins. A baleen whale rib, from high in the BLM Coyote Mountains Wilderness Area, suggests an evolutionary lineage to a modern baleen relative, the California gray whale (Deméré, 2006).

2. The Colorado River Delta, 4.5 to 3 Ma (Figure 2), rapidly spread west across the ancestral gulf as a result of increased rainfall and a rising Colorado Plateau. The delta floodplain hosts a temperate woodland forest with taxa that suggest a four-season climate (Remeika, 2005 and 2006). (Think of today's San Francisco Bay area delta lands and California central coast woodlands.) The forest browser *Megalonyx* is the first of three waves of bear-sized giant ground sloths to arrive from South America (McDonald, 2006b). Fossil fish, all from the Colorado River watershed, indicate no connection to other river systems in the Californias (Roeder, 2005). A sequence of freshwater lakes and playa lakes form in the Salton Trough north of the delta (Figure 3). In recent time, these lakes hosted Kumeyaay and Cahuilla Indian populations.

3. Savanna plains and riparian woodlands, 3 to 0.5 Ma (Figures 4 and 5), emerge in the rain shadow of the rising Peninsular Ranges. (Think of the famed Serengeti Plain of east Africa and its herds of megafauna.). *Megalonyx harlani* (Harlan's ground sloth), a grassland grazer and mixed feeder, represents the second wave to arrive via the now-closed Isthmus of Panama (McDonald, 2006b). Numerous extinct taxa of camels and horses thrive in open grasslands while llamas emigrate to the south (Webb et al., 2006).

The presence of ostracodes, tiny shrimp-like creatures, is evidence for significantly higher levels of rainfall than today (Cosma, 2002). Water birds, including flamingo, that comprise over half of the fossil avian record, are further evidence of the lakes and ponds that preceded today's arid environment, while the Anza turkey and a fossil woodpecker suggest the continuing presence of woodlands.

Aiolornis incredibilis, known primarily from Anza-Borrego and noted for its 5.3 m wingspan, is the largest bird ever to soar the skies of the Northern Hemisphere (Jefferson, 2006).

Nothrotheriops shastensis (Shasta ground sloth) appears as the third and final immigrant of this remarkable group of great beasts from the south. Its eclectic diet and ability to survive in a more arid or even desert environment is a herald of major environmental changes that persist into recent time (McDonald, 2006b).

The fossil record of the last 500,000 years, while largely removed by erosion from Anza-Borrego, is well preserved at other sites in southern and Baja California including the famed La Brea Tar Pits in Los Angeles, and the Valley of the Mastodons near Hemet, California. These "Rancholabrean" age sediments complete the paleontological package through the end of the Pleistocene Ice Ages some 10,000 years ago, coincident with the arrival of humans in the Americas.

As to the controversial question of the primary cause of the disappearance of the much of the American megafauna, paleoecologist Anthony Barnosky (2006) from the University of California, Berkeley now believes that "humans and climate were the one-two punch that drove the late Pleistocene extinctions." San Diego State University biologist David Archibald (1991) similarly notes that, "the extinction of big animals occurs about the same time humans show up in North America and Australia. One argument is, what we're doing to our environment now will make the Cretaceous extinction look like a picnic."

EDUCATIONAL THEME III

The Fossil Treasures Project is intended to be a communication link between earth science researchers working in the Anza-Borrego region and the public via educational and interpretive tools using a variety of media and experiential methods. It endeavors to explain, "How do we know what we know?"

Policy Implication: Good science must inform and guide good policy via an informed and enthusiastic citizenry, both youth and adults.

1. FTP publications have included not only the Fossil Treasures book but also an artwork packet, suitable for classroom or interpretive display (Jefferson and Lindsay, 2006b), a poster that summarizes "An Unparalleled Continuous Plio-Pleistocene Record of Ecologically Diverse Fossil Assemblages" (Jefferson and Marrs, 2005), and the Paleontology Symposium program (ABF&I, 2005) which contains field trip descriptions and lecture abstracts, supplementary to the book.

Additional publications include the ABDSP annual magazine (Sussman, 2006), and the ABF&I Desert Update (Lindsay and Jefferson, 2006) which also featured paleontology and geology in the California Parks Online Resource for Teachers and Students (PORTS) program.

2. Presentations to date include the 65th annual convention of the Society of Vertebrate Paleontology (Jefferson and Marrs, 2005), San Diego Natural History Museum, the 20th annual California State University Desert Symposium (Lindsay, 2006), the Sierra Club, the Pacific Division of the American Association for the Advancement of Science (Jefferson, 2006), and the 2006 Annual Meeting of the Geological Society of America (Lindsay, in press).

3. Educators from San Diego-area universities and community colleges are using the book artwork and poster on field trips and in the classroom.

4. Dioramas and other displays in the new ABDSP Visitor Center exhibits employ paintings, text, and photographs from the book.

5. Reviews have included: *Quest* San Diego Union-Tribune (Brown, 2005), *Books* San Diego Union-Tribune (McCaffery, 2006), *Town Focus* San Diego Union-Tribune (Lepper, 2006), KPBS radio (7 Feb. 2006), KFMB television (9 Nov. 2005), *Cranium* Dutch Society for Pleistocene Mammals (Mol, 2006), *Midwest Book Review* (Cox, 2006), *Foreword Magazine* (Hallberg, W., 2006), *Desert Magazine* (Pinkston, 2006), *Taxon* (Schmid, 2006), and the *Journal of Vertebrate Paleontology* (Lundelius, in press). Additional on-line reviews appear in the Discovery Channel Book Club Review and Amazon.com Editorial Reviews.

REMARKS

From the book's Epilogue, "A paramount task of paleontology is to laboriously and exactingly unearth the fossil record, one tooth and one animal track at a time. Such evidence of prehistory is essential to research and is the basis of conclusions that will help us, as stewards of our environment, to make the best decisions for the future health of the planet. *Fossil Treasures of the Anza-Borrego Desert* is the story of this paleontologic work, performed by hundreds of volunteers and dozens of professional researchers over the last 150 years. It presents the results of this initial phase of discovery and analysis, paves the way for future paleontologic studies, and reveals past patterns of biotic and environmental change crucial to understanding our own impacts on the global ecosystem."

We have opened a window on the natural prehistory of the Anza-Borrego Desert. Now we are fulfilling the educational vision that has emerged during the course of the project. The Colorado Desert District paleontology program continues to recover significant and exciting new fossils (Figure 6). Study of these materials clearly adds to the scientific knowledge of the natural prehistory of southern California. But in addition, the fossils serve as excellent tools for the interpretation of earth sciences to a general audience. Related developments may include a children's version of "Fossil Treasures" and a Park paleontology Web site with live webcams in the laboratory and on field trips. Such tools become a bridge for students of today to learn about the creatures and landscapes of yesterday which may enable better environmental stewardship of tomorrow.

REFERENCES CITED

Anza-Borrego Institute 2005. Fossil treasures of the Anza-Borrego Desert, a Symposium Exploring North America's Richest Continuous Fossil Record for the Last Seven Million Years. Anza-Borrego Foundation and Institute, Borrego Springs, California, Program Abstracts, 28 p.

Archibald, D. 1991. Dinosaur dilemma. *In* Daily Californian, San Diego, California, 3 Nov., p. D1.

Barnosky, A. 2006. Mammoth Mystery. *In* OnEarth Magazine, winter 2006. Natural Resources Defense Council, New York, pp. 14-19.

Brown, S. 2005. Badlands, good finds: Once an ancient sea, the Anza-Borrego Desert is still awash in fossils. *In* Quest, San Diego Union-Tribune, 16 Nov., pp. F1, F4.

Cosma, T.N. 2002. Late tertiary climate variations inferred from ostracode data, Anza-Borrego Desert southern California, USA. Master Thesis, Department of Geology, Kent State University, Kent, Ohio, 109 p.

Cox, J. 2006. *In* Science Shelf. Midwest Book Review, Apr.

Deméré, T.A. 2006. The Imperial Sea: marine geology and paleontology. *In* The Fossil Treasures of the Anza-Borrego Desert, edited by G.T. Jefferson and L. Lindsay, Sunbelt Publications, San Diego, California, pp. 29-41.

Deméré, T.A., and N.S. Rugh 2006. Invertebrates of the Imperial Sea. *In* The Fossil Treasures of the Anza-Borrego Desert, edited by G.T. Jefferson and L. Lindsay, Sunbelt Publications, San Diego, California, pp. 43-69.

Dorsey, R. 2006. Stratigraphy, tectonics and basin evolution in the Anza-Borrego Desert Region. *In* The Fossil Treasures of the Anza-Borrego Desert, edited by G.T. Jefferson and L. Lindsay, Sunbelt Publications, San Diego, California, pp. 89-104.

Ezcurra, E. 2006. Recipients of Pew grants. *In* San Diego Union-Tribune, 15 Feb., p. B-1.

Hallberg, W. 2006. *In* ForeWord This Week, 12 Apr., p. 10.

Jefferson, G.T. 2006a. The fossil birds of Anza-Borrego. *In* The Fossil Treasures of the Anza-Borrego Desert, edited by G.T. Jefferson and L. Lindsay, Sunbelt Publications, San Diego, California, pp. 151-158.

Jefferson, G.T. 2006b. Paleontology of the Anza-Borrego Desert, an Uninterrupted Marine Late Miocene and Terrestrial Pliocene through Pleistocene Record from the Western Salton Trough. *In* American Association for the Advancement of Science, Pacific Division, 87th Annual Meeting, Program with Abstracts, 2006, University of San Diego, p. 72.

Jefferson, G.T., and L. Lindsay (editors) 2006a. The Fossil Treasures of the Anza-Borrego Desert, Sunbelt Publications, San Diego, California, p. 394.

—— 2006b. Selected artwork from Fossil Treasures of the Anza-Borrego Desert the Last Seven Million Years. Sunbelt Publications, San Diego, California, unpaginated.

Jefferson, G.T., and B. Marrs 2005. A continuous Plio-Pleistocene vertebrate record from the Anza-Borrego Desert State Park region, southern California, a historic perspective. *In* Journal of Vertebrate Paleontology 25(3):74A.

Lepper, R. 2006. Book covers 7 million years of desert history. *In* Town Focus (North County), San Diego Union-Tribune, 26 Apr., p. B-4.

Lindsay, L. 2006. Communicating natural science to the general public: a case study. *In* Making Tracks Across the Southwest, edited by R.E. Reynolds, Abstracts from the 2006 Desert Symposium California State University, Desert Studies Consortium and LSA Associates, Inc., p. 76.

Lindsay, L., and G.T. Jefferson 2006. Project in science education and publishing in the public interest. *In* Anza-Borrego Foundation and Institute, Desert Update, 50:3-4.

McCaffery, L. 2006. 'Fossil' energy. *In* Books, San Diego Union-Tribune, 29 Jan., pp. 2, 4.

McDonald, H.G. 2006a. Anza-Borrego and the Great American Biotic Interchange. *In* The Fossil Treasures of the Anza-Borrego Desert, edited by G.T. Jefferson and L. Lindsay, Sunbelt Publications, San Diego, California, pp. 329-337.

—— 2006b. The ground sloths: invaders from South America. *In* The Fossil Treasures of the Anza-Borrego Desert, edited by G.T. Jefferson and L. Lindsay, Sunbelt Publications, San Diego, California, pp. 161-174.

Mol, R. 2006. *In* Cranium, Mar. 2006. Workgroep Pleistocene Zoogdieren (Dutch Society for Pleistocene Mammals), Culemborg, Netherlands.

Pinkston, M. 2006. Desert Treasures—how time and man have changed the landscape. *In* Desert Magazine, Palm Springs, California, Apr., p. 75.

Remeika, P. 2005. Ancestral woodlands of the Colorado River delta plain. *In* Fossil treasures of the Anza-Borrego Desert, a symposium Exploring North America's Richest Continuous Fossil Record for the Last Seven Million Years, Anza-Borrego Foundation, Borrego Springs, California, Program Abstracts pp. 13-14l, and 2006 *In* The Fossil Treasures of the Anza-Borrego Desert, edited by G.T. Jefferson and L. Lindsay, Sunbelt Publications, San Diego, California, pp. 75-87.

Roeder, M.A. 2005. Fossil fishes of the Anza-Borrego region. *In* Fossil treasures of the Anza-Borrego Desert, a symposium Exploring North America's Richest Continuous Fossil Record for the Last Seven Million Years, Anza-Borrego Foundation, Borrego Springs, California, Program Abstracts, pp. 16-17.

Rugh, N.S. 2005. The Imperial Group invertebrate fossils, their modern relatives and some thoughts on conservation. *In* Fossil treasures of the Anza-Borrego Desert, a symposium Exploring North America's Richest Continuous Fossil Record for the Last Seven Million Years, Anza-Borrego Foundation, Borrego Springs, California, Program Abstracts, pp. 10-11.

Schmid, R. (editor) 2006. Reviews and Notices. *In* Taxon 55(3) August 2006. University of California, Berkeley.

Sussman, S. 2006. Fossil treasures of Anza-Borrego Desert. *In* Anza-Borrego Desert State Park, Annual Magazine, pp. 16-17.

Webb, S.D., K. Randall, and G.T. Jefferson 2006. Extinct camels and llamas of Anza-Borrego. *In* The Fossil Treasures of the Anza-Borrego Desert, edited by G.T. Jefferson and L. Lindsay, Sunbelt Publications, San Diego, California, pp. 292-310.

Figure 1. The Imperial Sea 6.5 to 4.5 Ma, Valenictus imperialensis *(Imperial walrus), the view today is to the northeast from the Fish Creek Wash campground. Painting by J. Francis, from Jefferson and Lindsay 2006, © CSP.*

Figure 2. The Colorado River Delta, 4.5 to 3 Ma, Gomphotherium *(gomphothere) and* Dinohippus *(Pliocene horse), the view today is to the southeast from the west end of Loop Wash. Painting by J. Francis, from Jefferson and Lindsay 2006, © CSP.*

Figure 3. The shore of Lake Borrego about 1 Ma, Tapirus merriami *(Merriam's tapir) and* Felis rufus *(bobcat), view today is east from the Truckhaven Rocks turnout on Highway S22. Painting by J. Francis, from Jefferson and Lindsay 2006, © CSP.*

Figure 4. Borrego Valley riparian forest about 1 Ma, view today is from downtown Borrego Springs to the east. Painting by J. Francis, from Jefferson and Lindsay 2006, © CSP.

Figure 5. Borrego Valley about 1 Ma, Equus scotti *(Scott's horse) and* Camelops *(extinct camel) view today is from the ABDSP Visitor Center. Painting by J. Francis, from Jefferson and Lindsay 2006, © CSP.*

Figure 6. Equus *(extinct horse) vertebral column and pelvis, approximately 800 ka, recovered from the Ocotillo Conglomerate in the Borrego Badlands. Photo by J. Johnstone.*

STRATIGRAPHY OF A MARINE RIFT BASIN: NEOGENE OF THE WESTERN SALTON TROUGH, CALIFORNIA

Charles D. Winker
Shell E & P Technology Co.
P. O. Box 481
Houston TX 77001 U.S.A.

Susan M. Kidwell
Department of Geophysical Sciences
University of Chicago
5734 S. Ellis Ave.
Chicago IL 60637 U.S.A.

INTRODUCTION

The Salton Trough was formed by the late Cenozoic interaction of three major geologic systems: the Gulf of California, Colorado River, and San Andreas fault (Fig. 1). On this field trip we will examine the stratigraphic record of these three systems as exposed in the western Salton Trough. Each day will be spent in one of the three localities -- Split Mountain, Fish Creek Basin, and Coyote Mountains -- that provide the best leverage on chronology, facies, and paleogeography of the early Salton Trough. Although the Salton Trough is not a significantly petroliferous basin, we will also examine sedimentary geometries, lateral variability, and reservoir-seal relationships as they might pertain to exploration and production in analogous basin settings.

The Salton Trough long been a popular area for field trips emphasizing its strike-slip tectonic setting (e.g. Robinson and Threet, 1974, Crowell and Sylvester, 1979, Kerr and Kidwell, 1991). In contrast to previous field trips this one will emphasize:
(1) origins of the Salton Trough as part of the Miocene proto-Gulf of California rift basin (pre-dating the San Andreas fault system); (2) recognition of clastic sediment derived from the Colorado River as a key to paleogeographic reconstruction; and (3) differentiation of Neogene basement topography from that of the present.

Gulf of California

The Gulf of California was initiated by middle Miocene extension ~12-14 Ma, based on several recent studies (Stock and Hodges, 1989; Sawlan and Smith, 1984; Sawlan, 1991; Lyle and Ness, 1991). In contrast to the widespread extension that characterized much of southern California during the Miocene, Gulf of California extension was focused on a single long and relatively straight rift basin parallel to the pre-existing continental margin. Baja California began to detach from intact North America essentially as a single rigid block (Sawlan, 1991).

Marine inundation occurred as early as 13 Ma, based on molluscan biostratigraphy and radiometric dates at Isla Tiburón (Smith, 1991) (Fig. 2). By around 6.5 Ma, marine waters had extended northwest to San Gorgonio Pass (Fig. 1), approaching within 50 km of the marine Los Angeles Basin but never quite reaching it. Prior to 5 Ma the Gulf of California apparently also extended through the Yuma Basin into the Bouse Embayment (now the lower Colorado River valley), based on estuarine macrofauna (Smith, 1960,

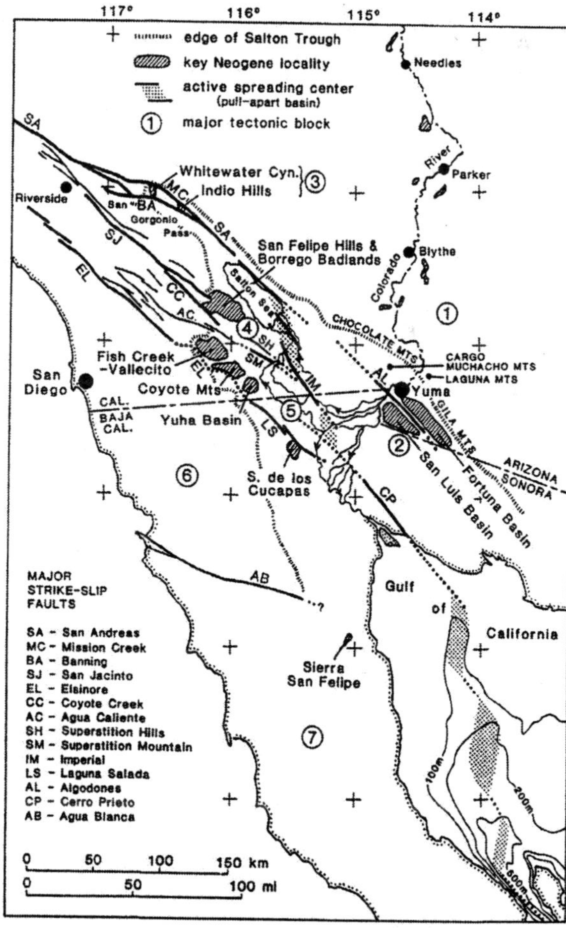

Figure 1. Index map to key Neogene localities and strike-slip faults of the Salton Trough and vicinity. This field trip covers the Fish Creek-Vallecito and Coyote Mountains localities.

in Patrick L. Abbott and John D. Cooper., eds., 1996,
Field Conference Guide 1996. Pacific Section A.A.P.G, GB 73,
Pacific Section S.E.P.M, Book 80, p.295-336.

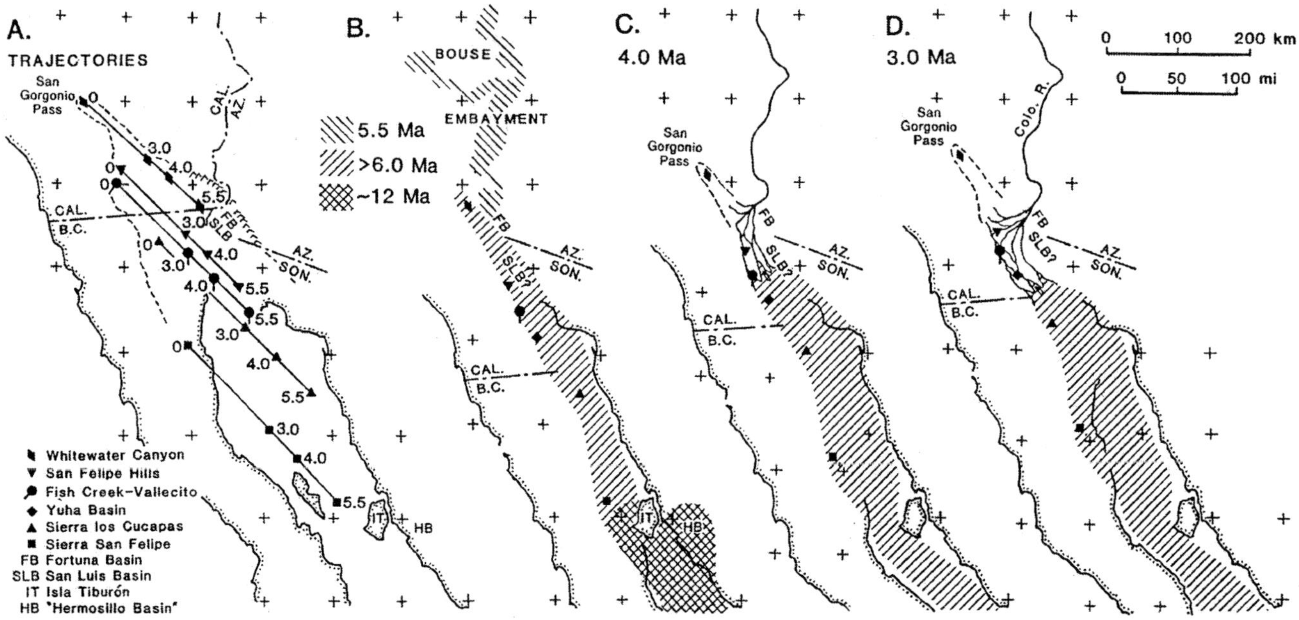

Figure 2. Neogene palinspastic paleogeographic reconstructions of Salton Trough and northern Gulf of California, from Winker and Kidwell (1986). Note narrow width of initial Gulf of California, and tectonic transport of field trip area (Fish Creek-Vallecito) past apex of delta plain.

1970; Metzger, 1968; Winterer, 1975; Buising, 1989). More tenuous microfossil evidence suggests that marine waters may have spilled over into several basins of the Mojave Desert including Cadiz, Bristol, Danby, and Panamint (Smith, 1970).

Colorado River

In early Pliocene time, the Colorado River began to debouch into the Salton Trough through a gap between the Chocolate and Gila Mountains (Fig. 1). The influx of fine-grained Colorado River sediment built a deltaic dam across the rift, cutting off the northwestern corner to form a lacustrine basin. The Salton Trough thus became a geographic entity distinct from the Gulf of California. For the Colorado River to enter the Salton Trough, the Bouse Embayment had to emerge above sea level and become the Colorado River valley (Fig. 2). This radical alteration of the landscape appears to have occurred quite rapidly, over a million years or less. We will see some of the evidence for this chronology (Fig. 3) during the field trip. The thick pile of Colorado River sediments in the rift not only formed a topographic divide, but also fundamentally altered the nature of crustal formation in pull-apart basins along the evolving strike-slip plate boundary (Elders and others, 1972; Elders, 1979). The apex of the bipolar delta plain has remained anchored to the North American plate by the entry point of the Colorado River (Fig. 1), while the Salton Trough has widened and its tip (at San Gorgonio Pass) has propagated toward the northwest (Fig. 2).

San Andreas Fault System

Between the middle Miocene and the some time in the Pliocene, Baja California was transferred from the North American to the Pacific Plate. Thus, the original Gulf of California rift basin evolved into the present transtensional plate boundary, accommodating the majority of the 5-6 cm/yr of dextral slip observed today between the Pacific and North American plates. This evolution is reflected in the history of the San Andreas fault system, which is best constrained in the vicinity of the San Gabriel Mountains and Ridge Basin (Crowell, 1982). Stratigraphic relationships there indicate that strike-slip faulting began in the middle Miocene (~12 Ma) on the San Gabriel fault. However, the modern San Andreas system, accommodating a major part of interplate slip, did not begin until ~5 Ma (Crowell, 1982). Since that time, both the Gulf of California and Salton Trough have widened considerably through transtensional opening (Fig. 2). Large-scale tectonic translation of the western Salton Trough relative to the North American plate is manifested by paleocurrent data from Pliocene deltaic deposits (Winker and Kidwell, 1986), which reflect a former position on the southeastern rather than the northwestern flank of the delta plain (Fig. 2).

In the Salton Trough, we infer that strike slip was first accommodated primarily on the San Andreas

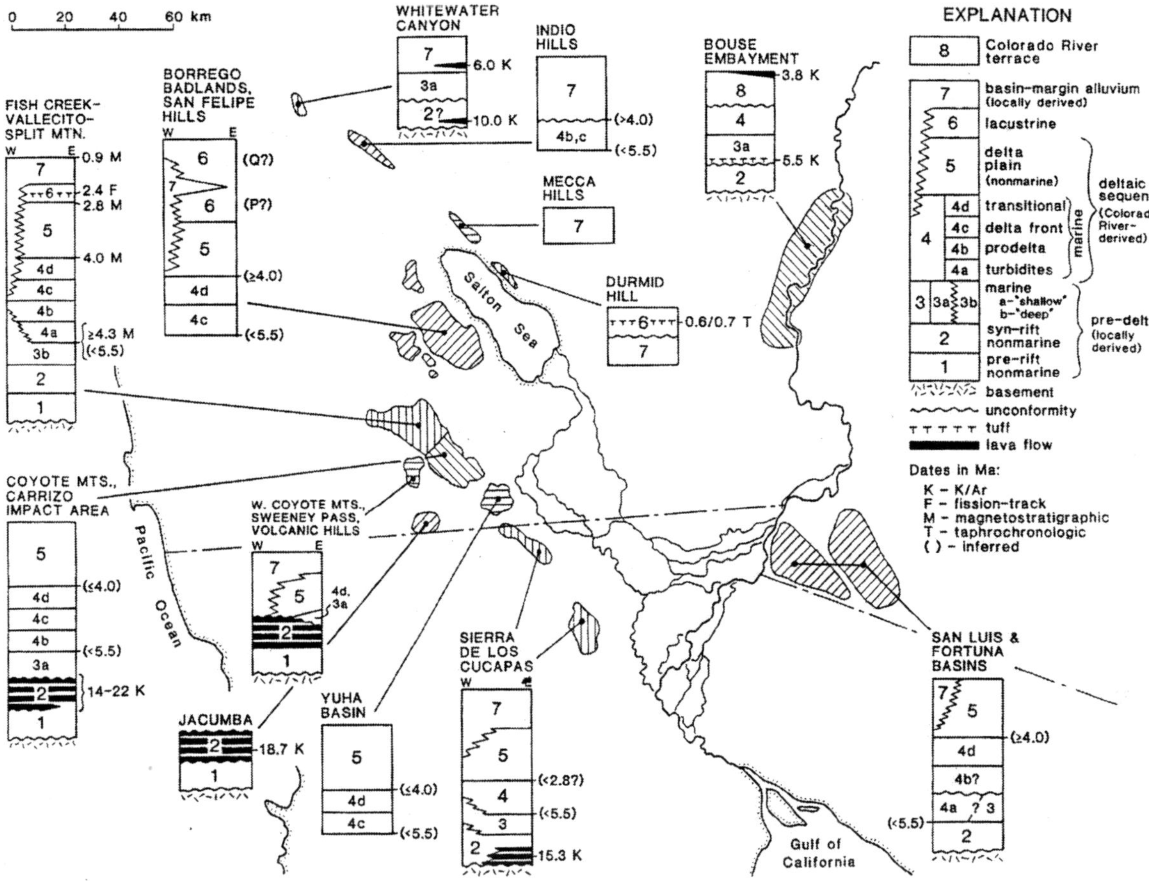

Figure 3. Neogene stratigraphic summary of Salton Trough based on informal genetic-stratigraphic units. Synonymy with formal nomenclature is discussed in Winker (1987). Sources of age dates: Fish Creek- Vallecito, Johnson and others (1983); Coyote Mts., Ruisaard (1979) and Mace (1981); Jacumba, Hawkins (1970); Whitewater Canyon, Matti and others (1985); Durmid Hill, Babcock (1974); Sierra de los Cucupas, Barnard (1968); Bouse Embayment, Shafiqullah and others (1980).

fault proper, which runs along the eastern margin of the basin (Fig. 1). Major strike-slip faults in the western Salton Trough have modern slip rates disproportionate to their small total displacement, suggesting that they are of relatively recent origin (Winker, 1987). Other manifestations of wrench tectonism in the western Salton Trough include: (1) numerous subordinate strike-slip faults of all orientations; (2) rapid Quaternary uplift (Johnson and others, 1983), responsible for the outcrops on this field trip; and (3) transpressive folding, dominated by east-west trending fold axes.

STRATIGRAPHIC NOMENCLATURE

In Neogene outcrops of the Salton Trough it is usually fairly easy to distinguish between marine and nonmarine strata. As discussed in the next section, it is also reasonably straightforward to differentiate between sediment derived from the Colorado River ("C"-suite of Winker, 1987) and sediment derived more locally ("L"-suite). These relationships are portrayed in a fence diagram of the field trip area (Fig. 4), and provide the primary basis for our stratigraphic subdivision of the Salton Trough (Fig. 3). An informal system of genetic units is convenient for summarizing basic stratigraphic relationships within the basin (Fig. 3), and sidesteps the problems of proliferating local stratigraphic names and ambiguous or conflicting usages of more widely used names.

Nevertheless, formal nomenclature based on properly defined units provides the foundation for careful mapping and stratigraphic description. In our initial studies we used the nomenclature of Woodard (1963) in the Fish Creek-Vallecito Basin, that of Christensen (1957) in the Coyote Mountains, and that of Dibblee (1954, 1984) in the rest of the western Salton Trough. In Split Mountain and the Fish Creek-Vallecito Basin, Winker (1987) employed nomenclature modified from Woodard (1963, 1974) and incorporated the facies-based mapping of Kerr (1982). Previous nomenclature has employed only two ranks,

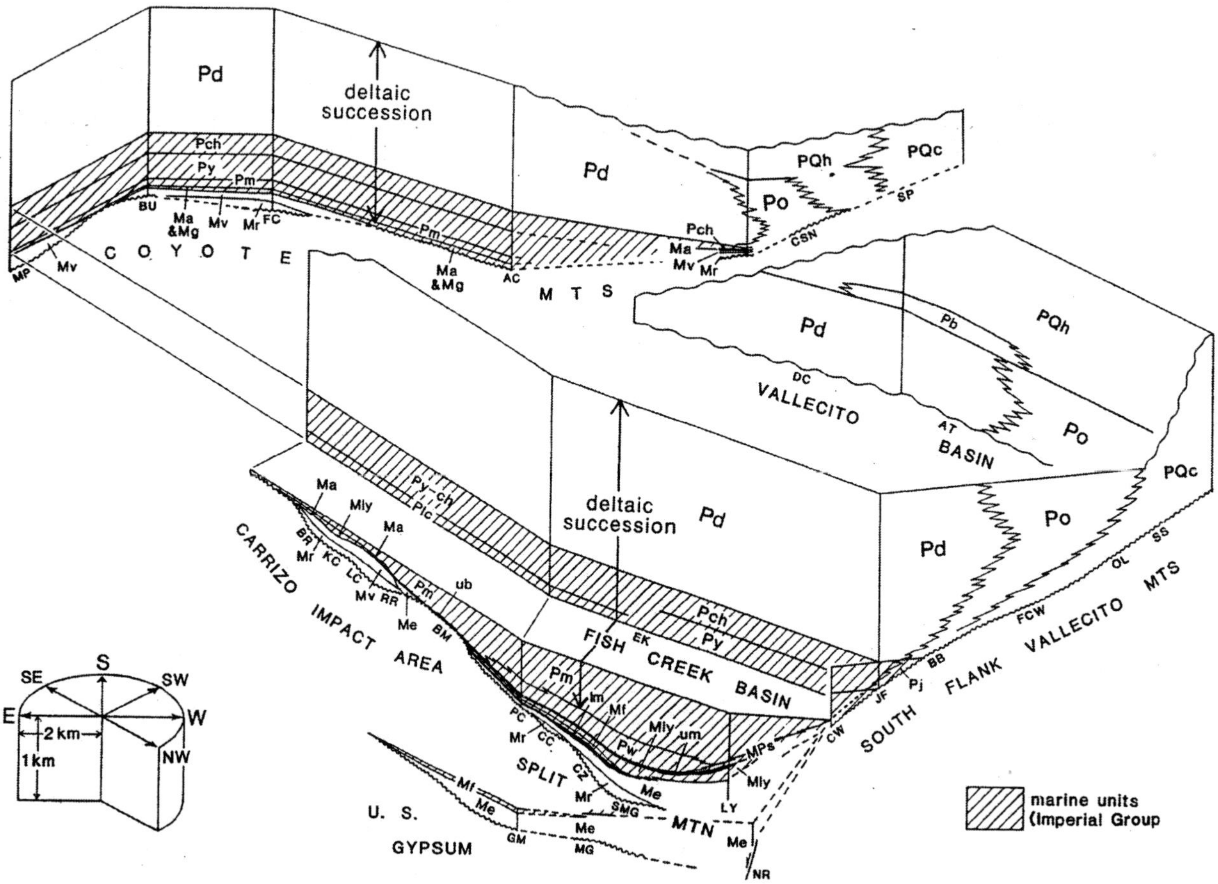
Pd
deltaic
succession
Pd
PQh
PQc
Po
SP
CSN
Pch
Py
Pm
BU
Ma
&Mg
Mv
Mr
FC
Pm
Pch
Ma
Mv
Mr
MP
Mv
C O Y O T E
Ma
&Mg
AC
M T S
Pd
Pb
PQh
DC
VALLECITO
AT
BASIN
Po
PQc
Po
Pd
deltaic
succession
Pd
Pd
Ma
Mly
Ma
Py-Ch
Plc
BR
Mr
KC
LC
Mv
RR
Pm
ub
Me
BM
CARRIZO
IMPACT
AREA
Pch
Py
EK
FISH
CREEK
BASIN
Pm
lm
Mf
PC
Mr
CC
Mly
um
Pw
MPs
CZ
Mr
SMG
Me
LY
Mly
SPLIT
MTN
Mf
Me
U. S.
GYPSUM
GM
Me
MG
Me
NR
CW
JF
Pj
BB
FCW
OL
SS
SOUTH
FLANK
VALLECITO
MTS
S
SE
SW
E
W
2 km
1 km
NW
marine units
(Imperial Group

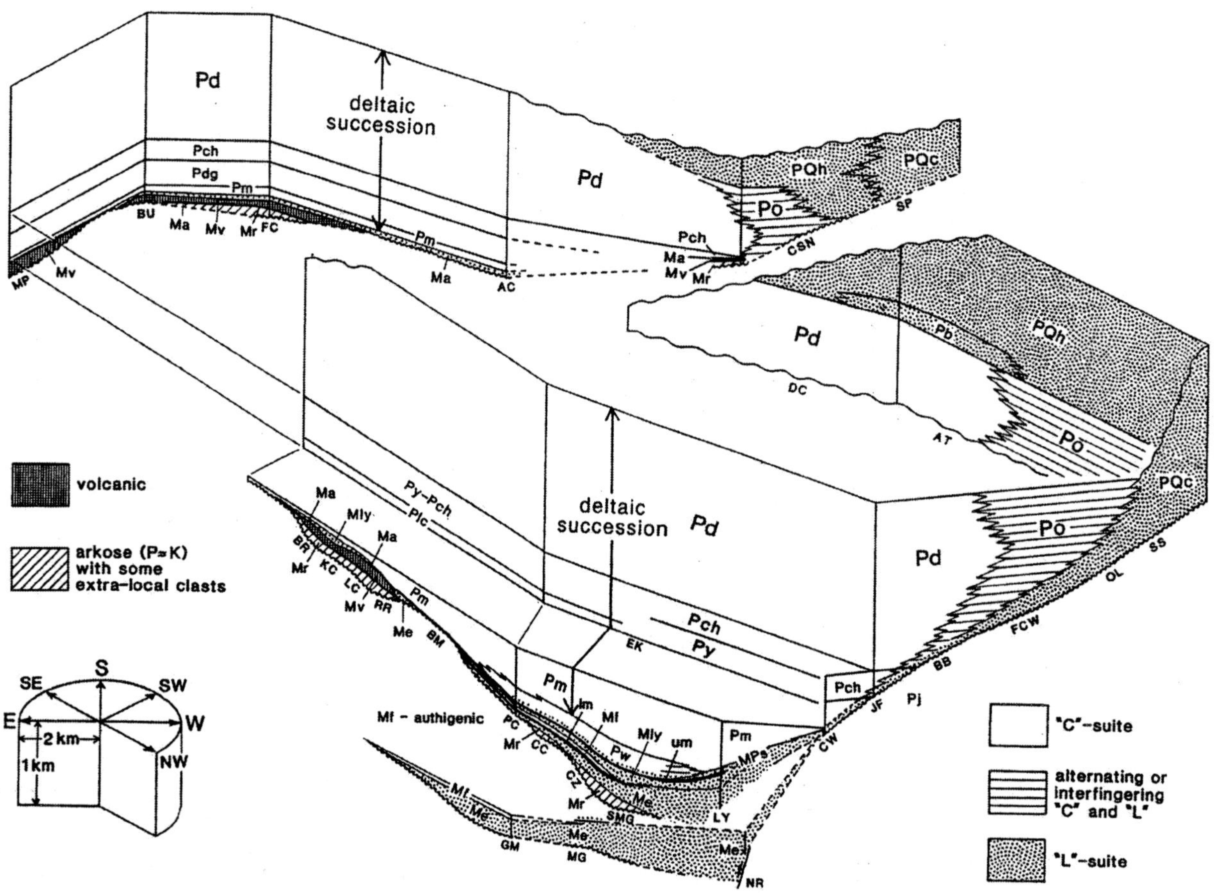
Pd
deltaic
succession
Pch
Pdg
Pm
BU
Ma
Mv
Mr
FC
Pm
MP
Mv
Ma
AC
Pd
PQh
PQc
Po
SP
CSN
Pch
Ma
Mv
Mr
Pd
Pb
PQh
DC
AT
Po
PQc
Po
volcanic
arkose (P≈K)
with some
extra-local clasts
Ma
Mly
Ma
Py-Pch
Plc
BR
Mr
KC
LC
Mv
RR
Pm
Me
BM
deltaic
succession
Pd
Pd
Pch
Py
EK
Pm
lm
Mf
Mf - authigenic
PC
Mr
CC
Mly
um
Pw
Pm
MPs
CZ
Mr
Me
SMG
LY
Mf
Me
Me
GM
MG
Me
NR
Pch
JF
Pj
BB
CW
FCW
OL
SS
S
SE
SW
E
W
2 km
1 km
NW
"C"-suite
alternating or
interfingering
"C" and "L"
"L"-suite

Figure 4 (facing page). Schematic fence diagram of stratigraphic relationships in the field trip area, showing (A) distribution of marine deposits comprising the Imperial Group, and (B) differentiation in terms of provenance. For key to stratigraphic units, see Fig. 5. Initials of some localities visited or discussed in this field trip: FC, Fossil Canyon; AC, Andrade Canyon; CSN, Canyon Sin Nombre; SP, Sweeney Pass; EK, Elephant Knees; RR, Red Rock Canyon; BR, Barrett Canyon; BM, Boundary Mountain; CZ, Crazycline Canyon; SMG, Split Mountain Gorge; LY, Lycium Canyon; CW, Coral Wash; JF, Jackson Fork; GM, U.S. Gypsum mine; MG, mouth of Gorge; NR, No Return Canyon.

formation and member. As our work progressed, we have found it convenient to add a third rank by raising three units of long standing (Split Mountain, Imperial, and Palm Spring) from formation to group. This revision is employed in this guidebook (Fig. 5).

Basal nonmarine strata in the Salton Trough comprise the Split Mountain Group of Miocene age. The Split Mountain Group (Ms) consists of volcanic rocks (primarily basalt) of the Alverson Formation (Mv), and a variety of clastic units. The Red Rock Formation (Mr) consists of relatively localized fluvial and eolian sandstone and conglomerate, apparently filling bedrock paleovalleys. The Elephant Trees (Me) and Garnet (Mg) Formations are stratigraphically higher and consist of alluvial fan conglomerates and sandstones in Split Mountain and the Coyote Mountains, respectively.

Marine strata in the Salton Trough comprise the Imperial Group (MPi), ranging in age from late Miocene to Pliocene. The Imperial Group is subdivided into: (1) evaporites of the Fish Creek Gypsum (Mf), (2) predominantly "L"-suite clastics of the Latrania Formation (MPl), and (3) predominantly "C"-suite" clastics of the Deguynos Formation (Pdg). The "L"-suite Latrania Formation is in turn subdivided into several members (Lycium, Mly; Wind Caves, Pw; Stone Wash, MPs; Andrade, Ma; Jackson Fork; Pj) reflecting lateral facies variations in the early Gulf of

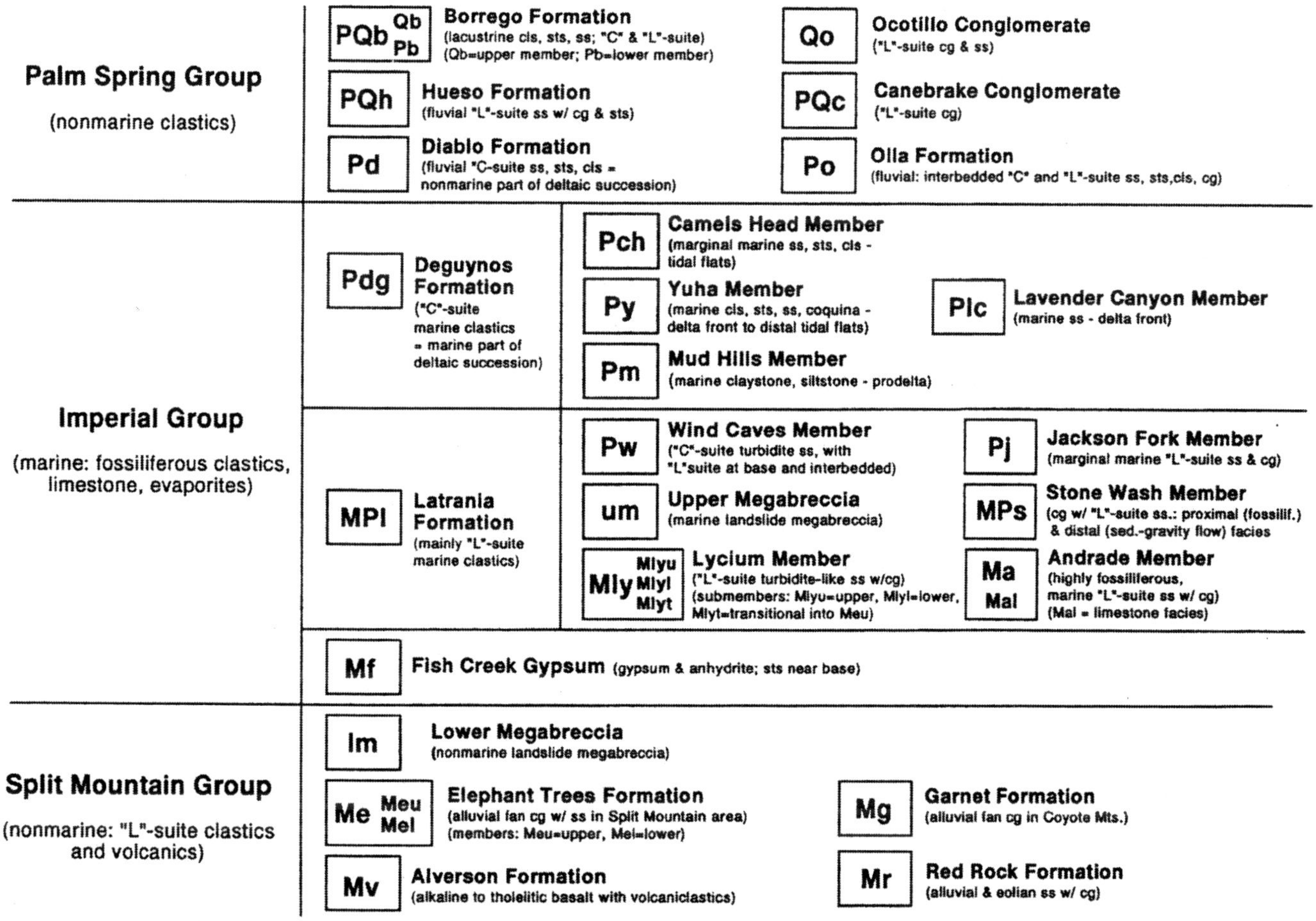

Figure 5. Stratigraphic nomenclature used in this guidebook. Abbreviations: ss = sandstone, sts = siltstone, cls = claystone, cg = conglomerate).

California. The "C"-suite Deguynos Formation is subdivided into members (Mud Hills, Pm; Lavender Canyon, Plc; Yuha, Py; Camels Head, Pch) reflecting the vertical succession of a prograding marine delta.

Post-Imperial nonmarine strata are all assigned to the Palm Spring Group of Pliocene to Pleistocene age. In the western Salton Trough the Palm Spring Group is divided into five formations. "C"-suite fluvial deposits of the Colorado River delta plain comprise the Diablo Formation (Pd). "L"-suite alluvial strata are subdivided into the Canebrake Conglomerate (PQc) and sandstones of the Hueso Formation (PQh). Where "C" and "L" alluvial deposits interfinger, they are assigned to the Olla Formation (Po). Lacustrine strata, which include both "C" and "L" clastics, are all assigned to the Borrego Formation (Pb and Qb).

While conventional lithostratigraphic units provide the foundation for mapping and regional correlation in the Salton Trough, we have not ignored sequence-stratigraphic aspects of these rocks. We have attempted to define the external geometry of lithostratigraphic bodies on a scale that would be recognizable in seismic stratigraphy. Where possible, we have also defined cyclic units and traced stratal surfaces as a basis for defining stratal geometries. However, in much of the study area this is simply not feasible, due to structural deformation, poor exposures of some of the more friable units, and above all, the nature of stratigraphic relationships in this tectonically active setting.

In vertical succession one commonly sees sudden, unique changes of lithology and paleoenvironment, rather than multiple repetitions of similar cyclic units. From a cyclostratigraphic standpoint, the most interesting parts of the Neogene basin fill are: (1) the Latrania Formation associated with the Gulf of California transgression (Days 1 and 3), and (2) the middle part of the Deguynos Formation (upper Mud Hills Member to lower Yuha Member) associated with progradation of the early Colorado River delta (Day 2). In these situations we can recognize flooding surfaces and parasequences, with retrogradational stacking in the Latrania and progradational stacking in the Deguynos. Some candidates for sequence boundaries can be identified, such as the base of the Lavender Canyon Member in the Deguynos Formation. However, we are unable to verify their regional significance or to use them for more than short-distance correlation.

Stratigraphic relationships and geometries in the region of Split Mountain, Fish Creek-Vallecito Basin, and Coyote Mountains are summarized in a schematic fence diagram (Fig. 4). This diagram was constructed to help visualize the geometry of basin fill as it existed prior to the Pleistocene (post-0.9 Ma) uplift that created the present-day topography and outcrops. It was drawn with a vertical exaggeration of 2:1 to approximate a typical aspect ratio for seismic cross sections (for example, 1"=4000', 1"=2.5 sec, and 10,000 ft/sec yields a 2:1 vertical exaggeration).

SOURCES AND DISPERSAL OF SEDIMENT

Numerous petrographic studies have described upper Cenozoic sands and sandstones in the Salton Trough and northern Gulf of California (Van Andel, 1964; Merriam and Bandy, 1965; Muffler and Doe, 1968; Van de Kamp, 1973; Kerr, 1982; Potter, 1978; Winker, 1987; Armitage, 1989; Sentianin, 1989; Nuffer, 1989, and Guthrie, 1990). These studies indicate three end-members for composition and source terrane (Fig. 6). (End-members may plot somewhat differently from study to study due to differences in methodology). First, local plutonic and metamorphic sources produce arkoses rich in plagioclase (P/F 70-85%). Second, Miocene volcanic rocks yield volcanic litharenites. Third, the Colorado River delivers lithic subarkoses to arkosic sublitharenites, with P/F of 23-65%. Colorado River sands also have secondary characteristics such as better rounding and sorting, quartz overgrowths, hematite coating, and reworked Cretaceous foraminifera which indicate recycling from sedimentary rocks (Merriam and Bandy, 1965; Winker, 1987). Mixing is common between the arkose and volcanic litharenite end-members (Fig. 6). Admixtures of Colorado River sands with the other two are relatively uncommon but are usually easy to identify because of the mixture of rounded and angular quartz grains.

Arkoses, litharenites, and their admixtures typically occur in association with conglomerates representing the lithologies of local plutonic, metamorphic and volcanic rocks. (The Red Rock Formation is an exception; it also contains extra-local clasts of silicic metavolcanics similar to the Eocene Poway and Ballena Gravels, according to Kerr, 1982, 1984). Associated siltstones are typically biotite-rich. Together, these sediments comprise the "L" suite. Typical colors are gray, olive, green, brown, white, red, orange, and pink (reddish colors occur mainly in the Miocene Red Rock, Elephant Trees, and Garnet Formations).

The "C" suite consists of Colorado River-type sandstones associated with siltstone and claystone. These lithologies occur together in two color schemes. In predominantly marine rocks, gray claystones are associated with pale yellow siltstones, and pale yellow, ochre, or brown sandstones. In predominantly nonmarine rocks, red claystones are associated with pale orange (buff) siltstones and sandstones.

Diagenesis in these sandstones is dominated by physical compaction and carbonate cementation.

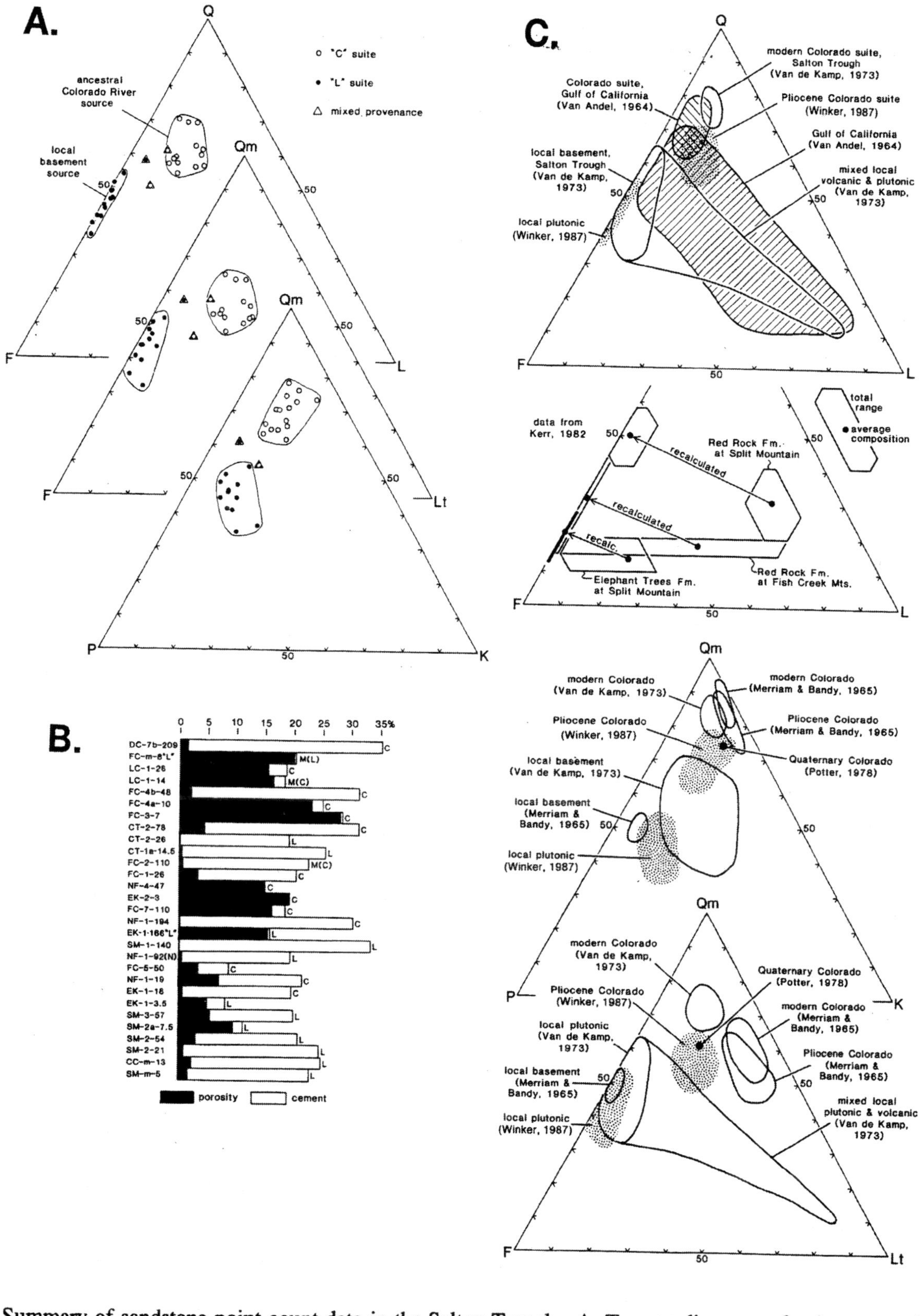

Figure 6. Summary of sandstone point-count data in the Salton Trough. A. Ternary diagrams of point- count data from Winker (1987) using the technique of Dickinson (1985). B. Summary of porosity and cement in several thin sections, plotted in approximate stratigraphic order, from Winker (1987). Most samples are keyed to cross sections (Figure 16). C. Ternary diagrams comparing point-count data from various studies. Plots for the same end-member can differ from study to study due to differences in methodology.

Hematite cementation resulting from degradation of biotite is seen in some of the "L"-suite deposits. Little evidence was seen for dissolution of feldspars. Sandstone porosities derived from point counts are highly variable, ranging from zero to 28% (Fig. 6). Porosities are generally low (<10%) in "L"-suite sandstones, but can range as high as 16%. "C"-suite sandstone porosities range from a few percent up to 28%. These numbers are somewhat biased toward low porosities because competent samples were favored over highly friable samples. Because "C"-suite sandstones are more commonly friable than "L"-suite, an unbiased sample of porosities should show an even greater disparity between "C" and "L" suite. Cement plus porosity is more similar for the two suites, ranging from 8 to 35% with a mode around 20% (Fig. 6). "C"-suite sandstones are all fine to very fine grained, but are consistently better sorted than "L"-suite sandstones. In summary, "C"-suite sandstones would have relatively good reservoir potential in their present state, while "L"-suite sandstones would make relatively poor reservoirs unless dissolution produced secondary porosity.

Paleocurrent patterns and proximal-distal relationships within the field trip area are strongly correlated with provenance. In "L"-suite units, paleocurrent modes are directed mostly toward the east or ENE (Fig. 7), away from the western basin margin. (One exception is the Red Rock Formation, where paleocurrent directions seem to be controlled more by the orientation of bedrock paleovalleys). In "C"-suite units, paleocurrent modes are directed mostly toward the south to southeast, or parallel to the paleo-basin margin (Fig. 8). With respect to the overall shape of the rift basin (Fig. 2), the "L" suite represents centripetal transport while the "C" suite represents axial transport.

STRATIGRAPHIC SUMMARY

Basement Rocks

Crystalline basement in the western Salton Trough is dominated by lower to middle Cretaceous plutonic rocks usually classified as tonalite. These granitoid rocks are rich in plagioclase and biotite, as reflected in the composition of "L"-suite sandstones and siltstones. Several distinctive lithologies occur in veins, including a "spotted tonalite" (Kerr, 1982) and tourmaline-bearing pegmatites. Similar plutonic rocks form much of the Peninsular Ranges between San Diego and the Salton Trough.

Metamorphic rocks of greenschist to amphibolite grade, including schist, gneiss, and marble, occur in the Fish Creek Mountains and the Coyote

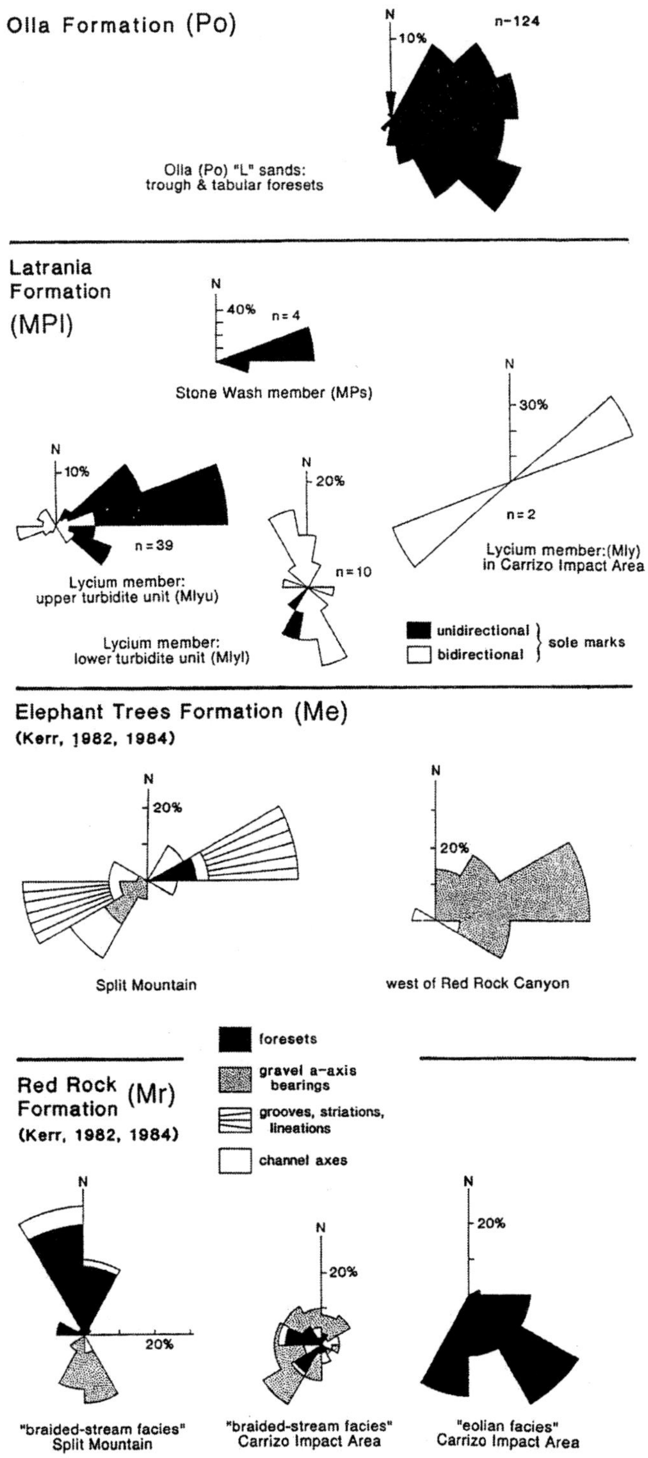

Figure 7. Paleocurrent data for various units composed of "L"-suite sediments, from Winker (1987). Except for Red Rock Member, main dispersal direction was toward the east or ENE, in contrast to south to SSE directions for the "C"-suite deltaic succession (Figure 8).

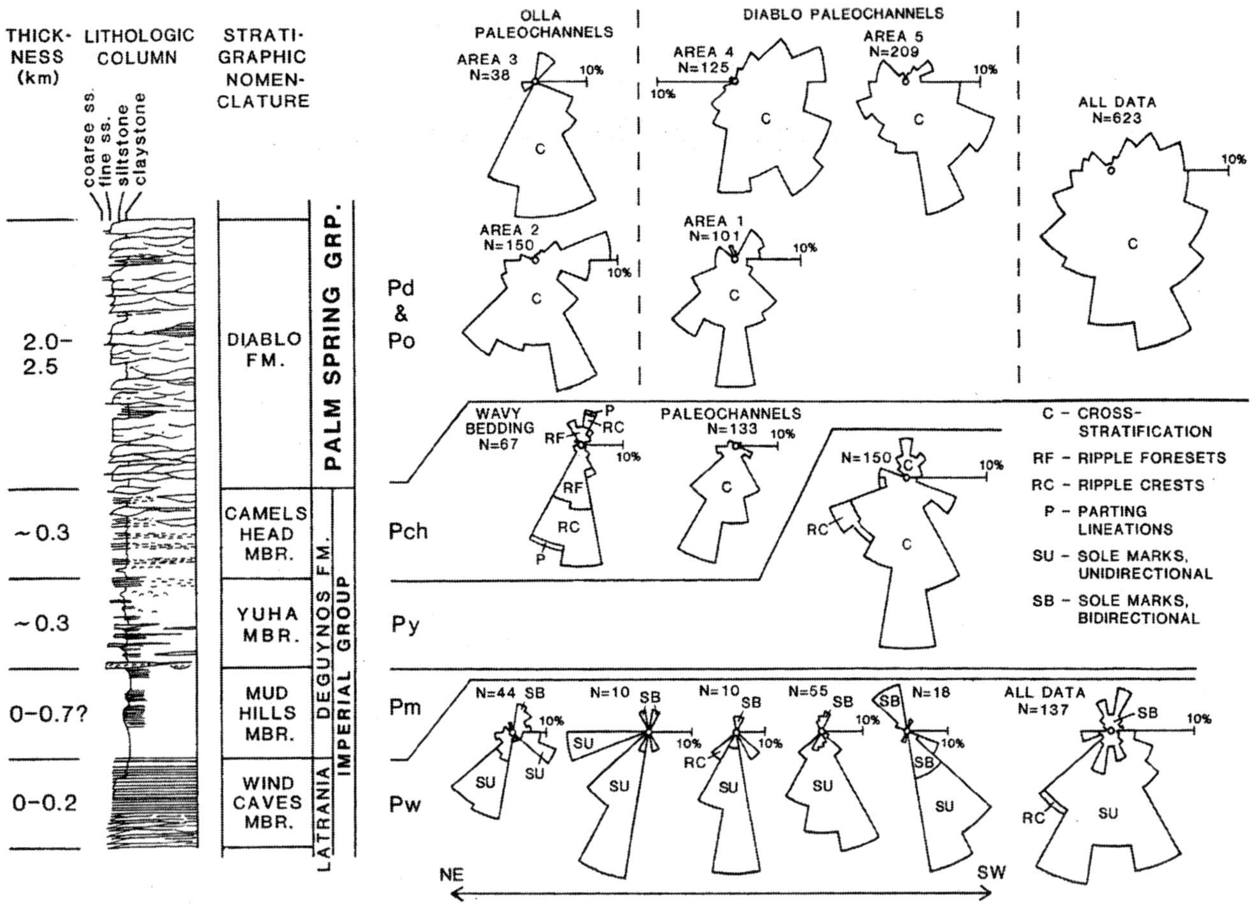

Figure 8. Paleocurrent data for the "C"-suite deltaic succession (Deguynos, Diablo, and Olla Formations) and the underlying submarine fan (Wind Caves Member).

Mountains (Day 3). Marble from the Coyote Mountains has yielded Ordovician conodonts (Miller and Dockum, 1983). Marble is of particular interest for marine paleogeography because it was unusually resistant to erosion during formation of the Gulf of California rift, and formed small islands in the late Miocene Gulf of California. Where marble is overlain directly by marine strata it is typically pockmarked by teardrop-shaped borings of pholads (a marine pelecypod) and the network galleries of boring clionid sponges (e.g. Watkins, 1990a, b). Other metamorphic and igneous rocks were also exposed locally on the seafloor based on rare occurrences of encrusting oysters on schist, corals on basalt, and barnacles on plutonic rocks.

Sheared rocks ranging from augen gness to ultramylonite comprise the Santa Rosa Cataclasite Zone (SRCZ), formed during late Cretaceous or Paleogene time. These rocks are exposed at Yaqui Ridge (optional **STOP 1-6**), as described by Simpson (1984), Schultejann (1984), and Engel and Schultejann (1984). The SCRZ provides piercing points for determining the offset of the San Jacinto fault (Sharp, 1967).

Frost and Shafiqullah (1990) reported a K-Ar date of 10.4±3 Ma from pseudotachylite on a detachment fault at Yaqui Ridge overlain by tilted conglomerate.

Clastics in Basement Paleovalleys

Basal rift-related strata in the Salton Trough are basalts and fanglomerates, commonly resting on basement. Locally, thick (up to 200 m) occurrences of sandstone, conglomeratic sandstone, and conglomerate intervene between basement rocks and unequivocally rift-related strata. We refer to these rocks collectively as the Red Rock Formation (Mr). The Red Rock Formation tends to be distinctively colored -- pink in Split Mountain, and white grading up to bright red in the Carrizo Impact Area. It consists mainly of braided-stream deposits, but also contains eolian deposits at Red Rock Canyon (Kerr, 1982, 1984). As mentioned in the previous section, Red Rock clastics differ from typical "L"-suite deposits in terms of paleocurrents (Fig. 7) and the presence of resistant, extralocal clasts of silicic metavolcanics (Kerr, 1982, 1984). In contrast to our interpretation, Kerr (1982, feather edge ~4 km to the east. In a proximal direction it terminates fairly abruptly against highly fractured plutonic basement, as seen in No Return Canyon (**STOP 1-5**). Kerr (1982, 1984) and Dean (1988) differentiated these fanglomerates into a lower, reddish unit of Miocene age, and an upper, greenish unit of Pliocene age, whereas Winker (1987) mapped them as a single unit in which color was not believed to be significant for stratigraphic correlation.

Lithology ranges from boulder conglomerate with thick debris-flow beds in Split Mountain Gorge (1-2) and No Return Canyon, to ripple-laminated sandstone and siltstone just below the Fish Creek Gypsum in the vicinity of the U.S. Gypsum quarry. Colors range from reds and browns to greens and grays. Kerr (1982, 1984) described the sedimentology of alluvial fan deposits in Split Mountain (**STOP 1-2**), recognizing: (1) debris-flow beds, (2) sheet-flood beds, and (3) clast-supported conglomerate beds. Clast-supported conglomerates appear near the top of the section at the expense of sheet-flood beds, suggesting overall progradation of the fan (Kerr, 1982, 1984).

A monomictic megabreccia referred to as the Lower Megabreccia (lm) overlies Elephant Trees fanglomerates in Split Mountain from Split Mountain Gorge to Boundary Mountain. At Split Mountain Gorge (**STOP 1-3a**) deformation of the underlying fanglomerate and internal shearing in the megabreccia indicate that it was emplaced as landslide deposit (Robinson and Threet, 1974; Kerr and Kidwell, 1991). According to Kerr (1982), the megabreccia attains a thickness of 180 m on the western flank of Boundary Mountain.

Fanglomerate and megabreccia overlie Alverson volcanics between Boundary Mountain and Red Rock Canyon (BM and RR in Fig. 10) (Ruisaard, 1979). This is the only direct indication that the Elephant Trees is younger (at least in part) than the Alverson.

In the Coyote Mountains, reddish-brown conglomerates of the Garnet Formation occupy an analogous stratigraphic position between Alverson volcanics and marine Latrania sandstones (Andrade Member). Christensen (1957) named the Garnet Formation and mapped it as Pleistocene alluvial deposits postdating the Neogene units. However, our mapping recognized that his Garnet Formation conglomerates actually underlie and grade laterally into marine Andrade sandstones, even along the crest of the Coyote Mountains (e.g. **STOPS 3-1b & c, 3-2a**). The Garnet Formation generally wedges out toward the 1984) interprets these braided stream deposits as a distal facies of the fanglomerates, rather than representing a distinctly earlier phase (as discussed in Kerr and Kidwell, 1991).

Miocene Volcanic Rocks

Miocene basalts and associated volcanic and volcaniclastic rocks comprise a discrete, mappable formation that extends discontinuously from the Carrizo Impact Area and the Coyote Mountains southward to the Jacumba area. This unit is known as Jacumba Basalt near Jacumba (Hawkins, 1970; Minch and Abbott, 1973) and as Alverson Formation (Mv) in the Coyote Mountains and Carrizo Impact Area. The Jacumba Basalt and Alverson Formation have similar geochemistry -- alkaline to tholeiitic -- and yield similar early to midle Miocene radiometric ages in the range of 14 to 22 Ma (Hawkins, 1970; Minch and Abbott, 1973; Eberly and Stanley, 1978; Ruisaard, 1979; Mace, 1981; Gjerde, 1982). These volcanics seem to be the immediate precursor to rifting rather than actual syn-rift rocks, based on their geographic distribution and stratigraphic relationships to alluvial fan deposits. The predominant Alverson lithology is amygdaloidal olivine basalt in 10-30 m thick flows with brecciated flow tops. Other lithologies include tuff, agglomerate, and volcanic sandstone and conglomerate. The Alverson is generally 50-150 m thick (Ruisaard, 1979).

Alverson basalt was exposed during inundation of the early Gulf of California, based on marine fossils encrusting its surface and sediment infiltrated into fractures in the basalt. Volcanic conglomerate is locally incorporated into fossiliferous sandstone of the Andrade Member.

Miocene Alluvial Fan and Landslide Deposits

Alluvial fan conglomerates (grading distally into sandstones) are present in the area of Split Mountain (Elephant Trees Formation, Me) and in the Coyote Mountains (Garnet Formation, Mg). These fanglomerates were shed from the west or southwest, based on paleocurrents, facies relationships, and distal thinning and pinchouts (Fig. 9). Thus, they provide the oldest clear evidence for a basin margin to the southwest.

Stratigraphic relationships in Split Mountain are summarized in Fig. 9 and 10. The Elephant Trees Formation attains a maximum thickness of over 400 m in Split Mountain just west of the gorge (Kerr, 1982) (**STOPS 1-1** and **1-2**). It wedges distally toward a

Figure 9 (facing page). Fence diagram summary of pre-Deguynos stratigraphy, facies, and paleogeographic elements. Bottom: nonmarine units (Split Mountain Group) and marine evaporites (Fish Creek Gypsum). Top: pre- deltaic marine units (Latrania Formation)

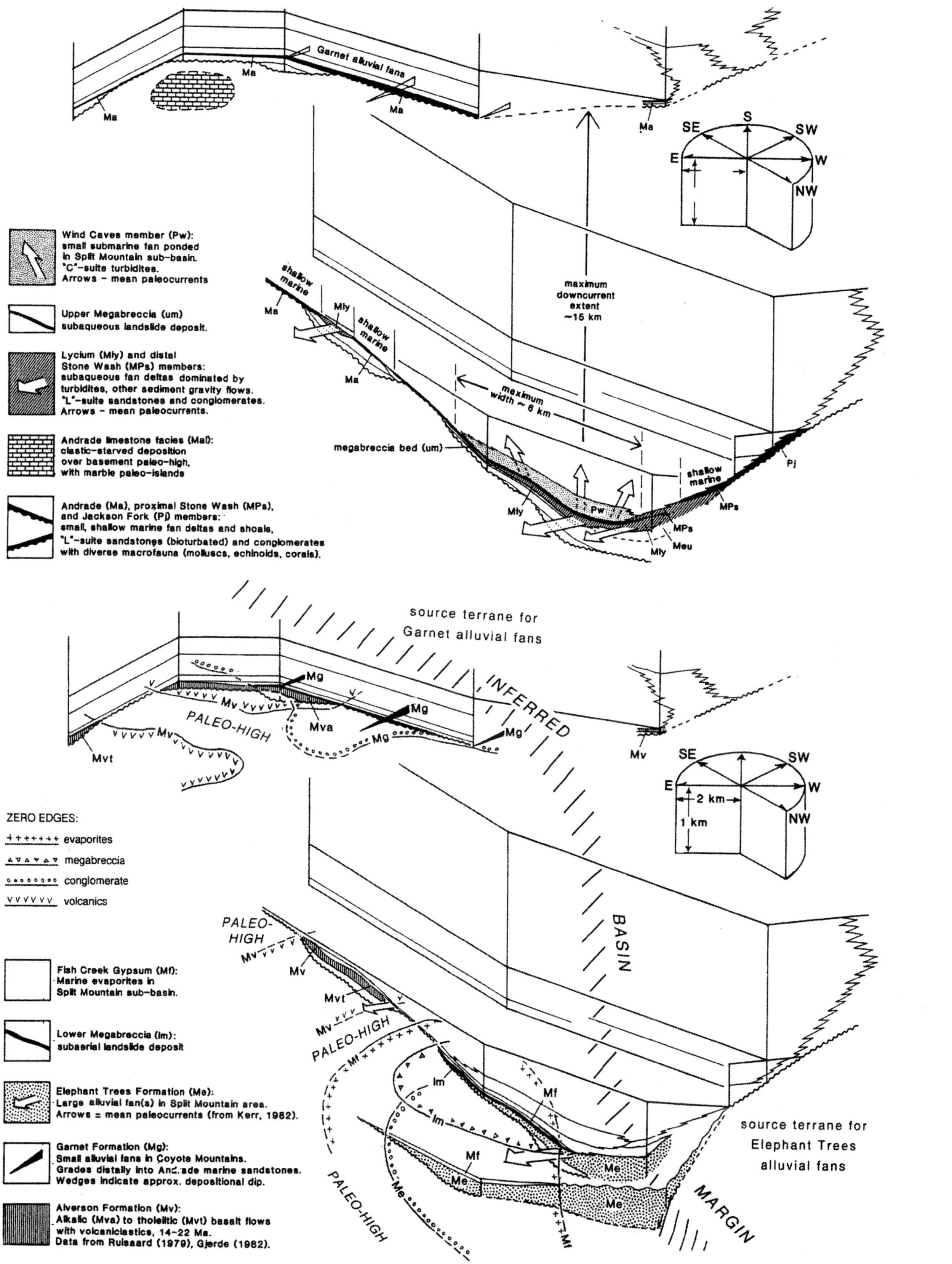
Garnet alluvial fans
Ma
SE
S
SW
E
W
NW
Wind Caves member (Pw):
small submarine fan ponded
in Split Mountain sub-basin.
"C"-suite turbidites.
Arrows - mean paleocurrents
Upper Megabreccia (um)
subaqueous landslide deposit.
Lycium (Mly) and distal
Stone Wash (MPs) members:
subaqueous fan deltas dominated by
turbidites, other sediment gravity flows.
"L"-suite sandstones and conglomerates.
Arrows - mean paleocurrents.
Andrade limestone facies (Mal):
clastic-starved deposition
over basement paleo-high,
with marble paleo-islands
Andrade (Ma), proximal Stone Wash (MPs),
and Jackson Fork (Pj) members:
small, shallow marine fan deltas and shoals,
"L"-suite sandstones (bioturbated) and conglomerates
with diverse macrofauna (molluscs, echinoids, corals).
shallow marine
Mly
maximum
downcurrent
extent
~15 km
maximum
width ~ 6 km
megabreccia bed (um)
Pw
MPs
Meu
Pj
source terrane for
Garnet alluvial fans
Mg
Mv
Mva
Mvt
PALEO-HIGH
INFERRED
BASIN
MARGIN
2 km
1 km
ZERO EDGES:
evaporites
megabreccia
conglomerate
volcanics
Fish Creek Gypsum (Mf):
Marine evaporites in
Split Mountain sub-basin.
Lower Megabreccia (lm):
subaerial landslide deposit
Elephant Trees Formation (Me):
Large alluvial fan(s) in Split Mountain area.
Arrows = mean paleocurrents (from Kerr, 1982).
Garnet Formation (Mg):
Small alluvial fans in Coyote Mountains.
Grades distally into Andrade marine sandstones.
Wedges indicate approx. depositional dip.
Alverson Formation (Mv):
Alkalic (Mva) to tholeiitic (Mvt) basalt flows
with volcaniclastics, 14-22 Ma.
Data from Ruisaard (1979), Gjerde (1982).
lm
Mf
Me
source terrane for
Elephant Trees
alluvial fans

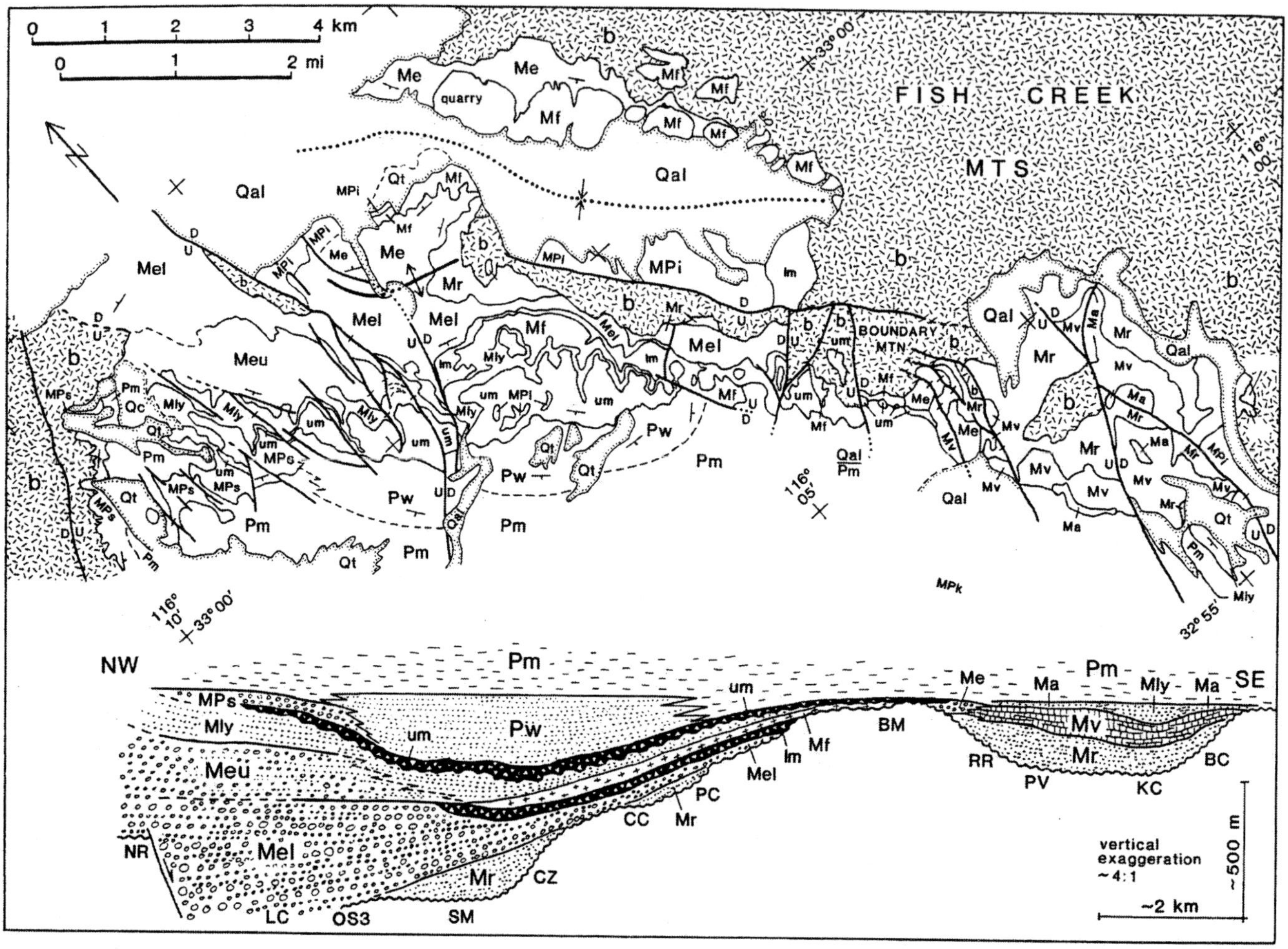

Figure 10. Geologic map and schematic cross section of Split Mountain and southwestern flank of Fish Creek Mountains in Carrizo Impact Area, showing units in Split Mountain Group, Fish Creek Gypsum, Latrania Formation.

breccia, but is overlain by full marine turbidites of the Lycium Member. It appears to occupy a paleo-low straddling the distal edge of the Elephant Trees alluvial fan (Fig. 9). Winker (1987) referred to the Fish Creek Gypsum as "pre-marine" evaporites, and suggested that the evaporite basin could have been connected to the advancing Gulf of California by a permeable sill or narrow passage.

Primary laminated gypsum also occurs as thin seams interbedded with sandstone in Gypsum Canyon near the Flatiron in the Coyote Mountains (Bell-Countryman, 1984). These gypsum-bearing sandstones of the Andrade Member yield no marine microfossils (T. Cronin, pers. comm.), whereas overlying Andrade sandstones contain a diverse marine macrofauna.

Non-Deltaic Marine Deposits

Marine deposits of the Latrania Formation contain diverse macrofauna and ichnofauna and complex internal stratigraphic relationships. Broadly speaking, east or northeast, and grades laterally and vertically into marine sandstone near the toe of the wedge. Our view of paleogeographic relationships in the Coyote Mountains (Fig. 11) contrasts with previous interpretations of the Coyote Mountains as a Neogene paleo-island rimmed by concentric bands of marine facies (e.g. Stump, 1972).

Marine Evaporites

The Fish Creek Gypsum (Mf) provides the oldest evidence for incursion of Gulf of California waters into the western Salton Trough. Dean (1988) demonstrated a marine origin for this extensive body of remarkably pure (95-99% in area of U.S. Gypsum quarry) gypsum and anhydrite on the basis of calcareous nannofossils extracted from clay seams. Prior to Dean's study, the origin of the gypsum had been uncertain because of its ambiguous stratigraphic context: it rests on nonfossiliferous, probably nonmarine strata of the Elephant Trees Formation and Lower Mega-

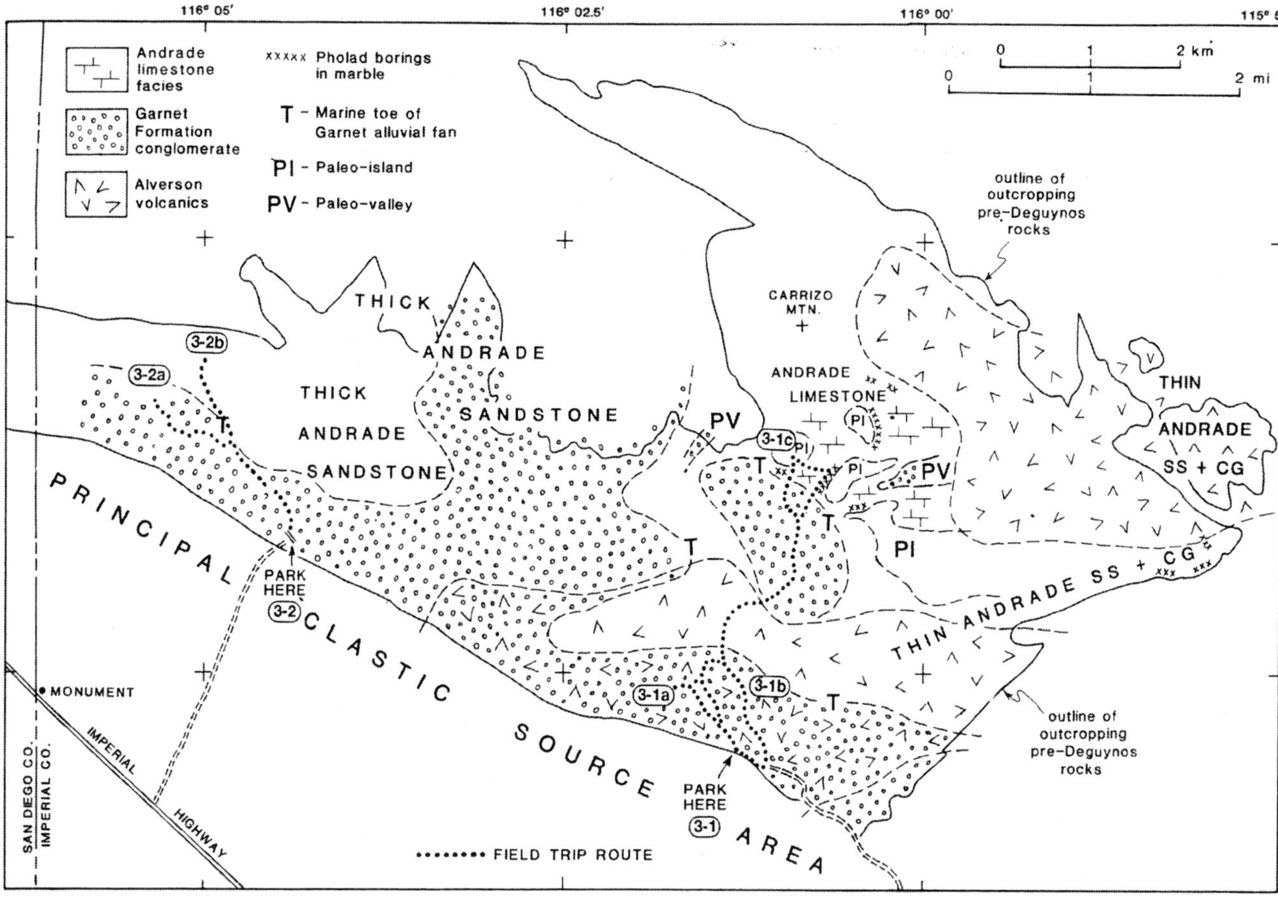

Figure 11. Late Miocene paleogeographic elements of Coyote Mountains. Andrade sandstone (Ma) covered most of this area except in areas of Andrade limestone (Mal) and paleo-islands.

base of the diatomite is not exposed, but it is overlain by the Cañon las Cuevitas Member, a shallower-marine, more clastic-rich, and more aerobic deposit. The diatomite indicates the presence of a deep-marine basin with a well-developed oxygen minimum and vigorous upwelling by the end of Miocene time (Boehm, 1984; Ingle, 1984). While this facies is not exposed in the Salton Trough, it could exist in the subsurface, raising the possibility of local source rocks for hydrocarbon generation.

Shallow Marine Facies

Fossiliferous, shallow-marine sandstones (and local limestones) of the Andrade Member (Ma) are extensively distributed in the Coyote Mountains and Carrizo Impact Area. Similar deposits occur along the flank of the Vallecito Mountains in Coral Wash (proximal Stone Wash Member) and Jackson Fork (Jackson Fork Member) (**STOP 2-6**).

The most typical aspect of the Andrade Member is massive, fossiliferous, bioturbated sandstone. Fossils typically include large pychodontid oysters and they can be divided into: (1) shallow marine deposits with a diverse macrofauna, and (2) deeper-marine deposits with abundant trace fossils, few body fossils, and sedimentary structures indicative of deposition by sediment gravity flows. Facies relationships and paleocurrent directions indicate that "L"-suite clastic sediments were shed mainly from the southwest, similar to the alluvial Elephant Trees and Garnet Formations. In fact, deposition of the Latrania is closely related to the underlying alluvial fan deposits; the intervening contact is time-transgressive and sometimes gradational. As we will see, transgression occurred by a series of flooding or back-stepping events. Between flooding surfaces both progradational and retrogradational successions can be observed.

An additional, even deeper-marine facies of the pre-deltaic Neogene Gulf of California is exposed in the Sierra San Felipe, Baja California (Fig. 1, 2), and was described as follows by Boehm (1982, 1984). Laminated diatomite of the San Felipe Diatomite Member contains a bathyal microfauna yielding biostratigraphic age of 5.5-6.0 Ma (latest Miocene). The

other epifaunal bivalves (pectinids, spondylid rock oysters), infaunal bivalves, gastropods, corals, barnacles, bryozoans, and serpulid worms. Many variations on this theme occur in the Andrade Member, especially in the Coyote Mountains. Andrade sandstones overlie unfossiliferous Garnet conglomerates and extend beyond the downdip Garnet pinchout to rest directly on basement or Alverson volcanics.

For example, on the northern flank of the Coyote Mountains, the Andrade Formation attains its maximum thickness of 60-70 m and exhibits a variety of features including: (1) bioturbated fossiliferous sandstone; (2) thick (1-2 m) *in situ* accumulations of *Pycnodonte heermanni* (a thick-valved free-living oyster) and the coral *Porites*; (3) large-scale trough crossbeds in medium-grained echinoid-bearing sandstone with infaunal echinoids; and (4) near the top, knobs of channelized fossiliferous conglomerate with basal erosional relief of 2 m. In the Fossil Canyon area (**STOP 3-1b**), Garnet conglomerate grades up and laterally into Andrade sandstone which in turn grades up and laterally into *Pycnodonte-Porites* limestone (Kidwell, 1988). In the vicinity of Marble Mountain (**STOP 3-1c**) Andrade sandstone grades laterally into thin limestone (Mal) which laps out against paleo-islands of marble with local occurrences of pholad borings. In the easternmost Coyote Mountains, where Alverson volcanics made up much of the substrate, the Andrade is typically thin and locally absent. As discussed earlier, basalt, marble, and other basement rocks were also exposed on the sea floor. Richly fossiliferous Andrade sandstone and limestone facies are also present in the Carrizo Impact Area, where they rest on Alverson volcanics.

Other fossiliferous, shallow-marine deposits of the Latrania Formation occur on the western flank of the Vallecito Mountains in the vicinity of Coral Wash and Jackson Fork. Fossiliferous deposits at Coral Wash comprise a proximal facies of the Stone Wash Member (MPs) described below. The Jackson Fork Member (Pj) of the Latrania Formation underlies the Camels Head Member (Pch) of the Deguynos Formation, and is probably a lateral equivalent as well. The Jackson Fork comprises a diverse assemblage of "L"-suite sandstones and conglomerates, including both debris-flow and sheetflood beds similar to those of the Elephant Trees. However, it also contains claystones, suggesting an age equivalence to the Deguynos Member. The Jackson Fork Member contains a fossil assemblage dominated by marine molluscs and echinoids, and trace fossils including camel tracks.

Deeper Marine Facies

Latrania Formation clastics in Split Mountain are dominated by sediment-gravity flow processes, probably in somewhat deeper water, perhaps up to 200 m (Winker, 1987). Sandstones typically occur as discrete beds with bioturbated tops, commonly with turbidite-like features such as coarse-tail grading and sole marks. Interpretation as turbidites is fairly straightforward at Split Mountain Gorge (**STOP 1-3**), where partial Bouma sequences can be recognized. Interpretation is more problematic in more proximal settings and with more conglomeratic lithologies, as at Lycium Canyon (**STOP 3-5**). Nevertheless, we interpret most of the Latrania Formation in Split Mountain as submarine sediment-gravity flow deposits.

In Split Mountain, stratigraphic relationships are laterally quite variable (Fig. 10). We recognize four members there within the Latrania Formation: Lycium Member (Mly), comprising bedded "L" sandstones; Wind Caves Member (Pw) comprising bedded "C" and "L" sandstones); Stone Wash Member (MPs), comprising amalgamated conglomerates; and the Upper Megabreccia (um). The Upper Megabreccia has a similar but slightly more extensive distribution compared to the Lower Megabreccia. The Upper Megabreccia separates the underlying Lycium Member from the overlying Wind Caves and Stone Wash Members; where the Megabreccia is absent the Lycium grades vertically and proximally into the Stone Wash. Tongues of the Upper Megabreccia protrude down into the Lycium Member, which is deformed into outcrop-scale disharmonic folds (**STOP 1-3b**). The Upper Megabreccia is polymictic, with abundant metamorphic clasts, in contrast with the monomictic plutonic-clast Lower Megabreccia. Other workers have postulated a source to the east or southeast in the Fish Creek Mountains (Pappajohn, 1980; Abbott and Rightmer, 1995). We prefer a source in now-buried basement terrain to the south or southwest, because facies relationships in the Carrizo Impact Area (based on recent field work by the authors) are inconsistent with the Fish Creek Mountains as a major source of coarse clastics.

Lycium Member sandstones are subdivided into lower and upper submembers. The lower submember (Mlyl) comprises thin-bedded turbidites that overlie the Fish Creek Gypsum, and grade proximally into a transitional member (Mlyt) with thin-bedded debris flow deposits ("green beds") at Split Mountain Gorge (**STOP 1-3a**). These grade more proximally into the upper part of the Elephant Trees Formation (Mel) (Fig. 10). The upper (Mlyu) submember is substantially thicker and crops out extensively in Split Mountain Gorge (**STOP 1-3**) and Lycium Canyon (**STOP 1-5**).

The upper Lycium grades upward and proximally (westward) into conglomerates of the Stone Wash Member (MPs). The Stone Wash is best exposed in Lycium Canyon (**STOP 1-5**), where it consists of thick, amalgamated, lenticular beds. The Stone Wash Member extends more than 2 km west of Lycium Canyon, where it laps out against plutonic basement. In Coral Wash, a typical Andrade-like assemblage of oysters, echinoids, and corals occurs on the top of the Stone Wash Member.

The Wind Caves Member overlies the Upper Megabreccia and grades laterally into the Stone Wash Member. The Wind Caves starts with basal "L"-suite turbidite sandstones similar to those of the Lycium, followed by some 200 m of "C"-suite turbidites alternating with subordinate, thin, "L"-suite turbidites. This unit represents the oldest datable occurrence of Colorado River-derived sediment in the Salton Trough. On the basis of foraminiferal biostratigraphy, K. MacDougall (1995, pers. comm.) places the Miocene-Pliocene boundary (5.1 Ma) at the base of "C"-suite turbidites. Unlike the other members of the Latrania Formation in Split Mountain, which are all related to alluvial fan-deltas, the Wind Caves Member is interpreted as a small submarine fan sourced from the mouth of Colorado River and ponded in the Split Mountain sub-basin (Fig. 9). Average bedding thickness and overall sandstone percentage decrease upsection and laterally toward the margins of the fan. This retrograde vertical succession may be related to enlargement of the fan as the sub-basin was progressively filled.

Deltaic Succession

In contrast to the underlying units, the marine Deguynos and nonmarine Diablo Formations represent a single, very large depositional system, the Colorado River delta, which spanned the width of the Salton Trough, and spread across the mosaic of environments existing in pre-deltaic times. Also unlike the underlying units, the deltaic deposits show axial rather than centripetal sediment transport (Fig. 8). The vertical succession described here (Fig. 8 and 12) is based on exposures in the Fish Creek Basin, but a similar succession occurs in the Carrizo Impact area, on the northern and eastern flanks of the Coyote Mountains, and in the Yuha Desert south of the Coyote Mountains.

Prodelta

Prodelta claystones and siltstones comprise the Mud Hills Member (Pm) (**STOPS 2-2, 2-3**). This unit is very thick in the Fish Creek Basin, roughly 700 meters (difficult to estimate due to scarcity of bedding and presence of faulting). It thins to a few tens of meters in the Carrizo Impact Area and Coyote Mountains (Fig. 4). Claystones in the Mud Hills Member are massive and mostly featureless.

Siltstones commonly exhibit "rhythmite" bedding (**STOP 2-3**). Unlike turbidites, the rhythmites do not represent discrete depositional events, but rather more continuous, traction-dominated deposition alternating between slightly coarser and finer textures. Primary physical structures are mostly obscured by bioturbation, but ripple lamination is occasionally observed. We attribute the rhythmites to annual cycles controlled by seasonal discharge of the Colorado River, although other interpretations are certainly possible.

Delta Front

Upward-coarsening parasequences representing the delta front begin near the top of the Mud Hills Member and extend up into the lower Yuha Member (Py), for a total thickness of 200-250 m (Fig. 13). A typical parasequence begins with claystone, grades up into mudstone and siltstone, commonly with rhythmite bedding, and may continue to grade up into sandstone and/or oyster-shell coquina with a sandy matrix. Fine-grained sandstones at the top of parasequences are generally trough cross-bedded, indicating southward paleocurrents (Fig. 8). Coquina beds are composed largely of single valves and fragments of the small oyster Dendostrea vespertina in a cemented sand matrix, oriented parallel or oblique to bedding, suggesting sorting and redeposition (see description by Watkins, 1990b). Articulated specimens are rare and mostly limited to the very tops of coquinas, where they may be accompanied by a modest assemblage of other mollusks, solitary corals, sharks teeth, and marine mammal bones. Coquinas exhibit both trough and low-angle crossbedding and are especially resistant to erosion, resulting in hogbacks like the Elephant Knees ridge (**STOP 2-2**). High-energy sandstones and coquinas capping parasequences are mostly discontinuous, extending laterally for one to a few kilometers. Some high-energy caps are locally sharp-based, and even channelized.

The thickest and most extensive of the high-energy sandstone caps begins at the Elephant Knees (Fig. 13) where it forms the basal Yuha Member, and extends some 10 km east into the Carrizo Impact Area where it comprises a mappable, sharp-based unit known as the Lavender Canyon Member (Plc). The Lavender Canyon ranges up to ~100 m in thickness. Trough crossbedding can be observed locally, but the Lavender Canyon Member generally appears structureless.

Parasequences in the Yuha Member maintain a fairly uniform thickness and can be readily correlated (Fig. 13). In contrast, parasequences in the Mud Hills Member thicken toward the southeast (oblique to depositional dip), and are relatively difficult to correlate without closely spaced control. Parasequence thick-

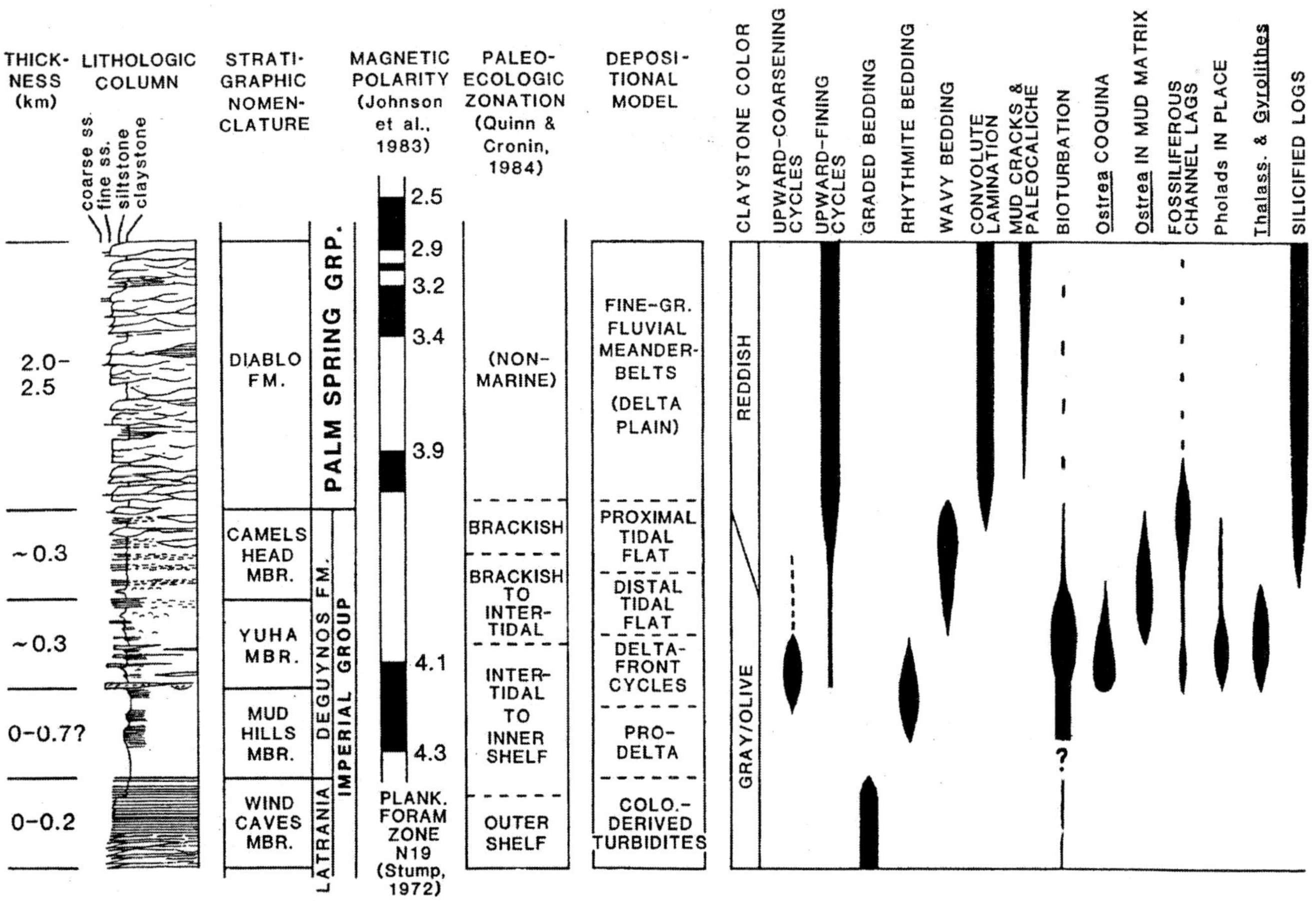

Figure 12. Summary of the deltaic succession in the Fish Creek-Vallecito Basin.

ening in the Mud Hills Member is interpreted to represent clinoform geometry developed on the leading edge of a prograding shoal-water delta platform, whereas parasequences in the Yuha Member are interpreted as smaller-scale prograding or upward-shoaling cycles developed on top of the delta platform.

Clearly, these are not the parasequences of barrier islands or a wave-dominated shoreface, and given the paleogeographic setting (Fig. 2), that type of parasequence would hardly be expected. This type of upward-coarsening cycle has not been reported or predicted in studies of the modern Colorado River delta (Thompson, 1968, 1975; Meckel, 1975). A possible analog for this type of delta front is the modern Mahakam delta front, a mixed tide- and river-dominated delta (Allen and others, 1976, 1979). The Mahakam has a broad delta platform with numerous distributary and tidal channels and channel-mouth and channel-margin shoals. The boundary between delta platform and prodelta is marked by a sharp break in slope. A series of long cores through the delta front show a consistent pattern of upward-coarsening parasequences, rather than upward-fining channel-fill cycles (Allen and others, 1976, 1979). This suggests that the channelized topography of the Mahakam delta platform was created more by accretion of shoals than by downcutting and migration of channels. If the parasequences in the Deguynos Formation are analogous, paleocurrent directions (Fig. 8) would indicate that the shoals are very strongly ebb-dominated.

Tidal Flats

Above the delta-front parasequences is a very thick (~500 m) transition zone into the nonmarine delta plain, which is represented by the Diablo Formation (Pd). This transition zone comprises the upper Yuha Member (Py) and the Camels Head (Pch) Member (defined by mixed gray and red claystones). The transition zone is distinguished from the delta front by the paucity of upward coarsening cycles and the absence of rhythmite bedding. Several features appear in the transition zone which are absent in the prodelta and delta front. The first is beds of articulated, and mutually attached Dendostrea and anomiid bivalves in a muddy matrix, interpreted as small in situ oyster biostromes. Second is an association of lenticular,

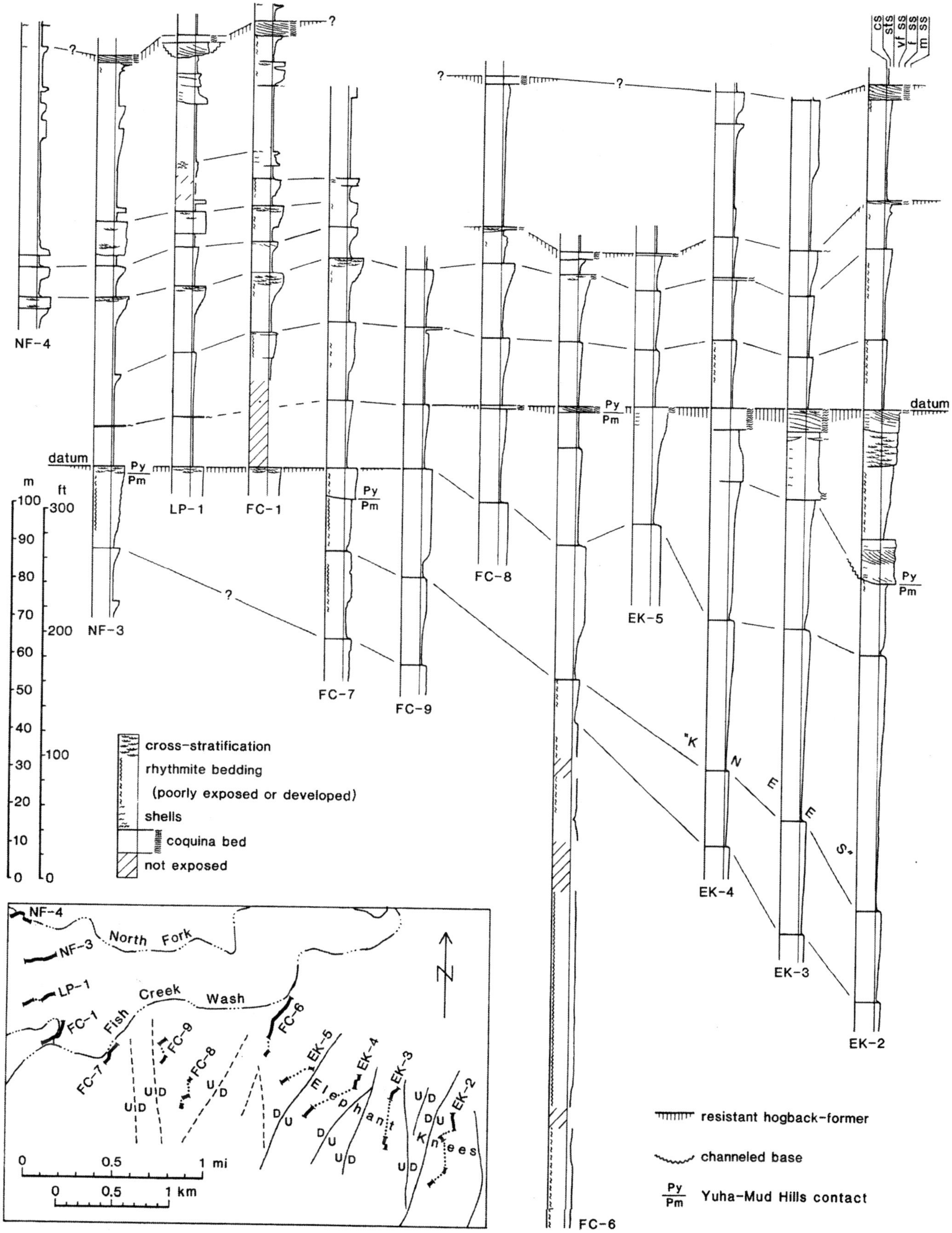

Figure 13. Cross-section of delta-front parasequences (upper Mud Hills Member and lower Yuha Member) in the Fish Creek Basin. STOP 2-2 is at EK-2 and EK-3; STOP 2-3 is at FC-6; and STOP 2-4 is at NF-3.

wavy, and flaser bedding typical of tidal flats (Reineck and Wunderlich, 1968), and referred to here as the wavy-bedding association. Third, shell material occurring as lag deposits in channel sandstones is typically encrusted with sheet-form bryozoans or serpulid worm-tubes in circumrotary colonies, indicating repeated reworking (Gyllenhaal and Kidwell, 1989). These distinctive features all suggest that the thick transition zone represents broad, channelized tidal flats between the delta plain and delta front.

Facies and features in the transition zone are vertically gradational, without obvious depositional breaks. Several vertical trends are recognized (Fig. 12). (1) Bioturbation is intense and pervasive in the lower transition zone (middle Yuha Member) but decreases up-section. (2) Oyster-shell coquinas decrease in abundance. (3) Reddish claystones begin to appear (defining the base of the Camels Head) and become increasingly prevalent up-section. (4) Sandstones increase in abundance up-section, occurring both in the wavy-bedding association and as massive, trough cross-bedded, channel sandstones. (5) Sandstones change in color up-section from pale yellow to pale orange (buff). The overlying Diablo Formation is dominated by massive, buff-colored channel sandstones and reddish claystones. Oyster shells occur locally in the Diablo as channel lags, but coquina beds, *in situ* biostromes, the wavy-bedding association, and bioturbation are all absent.

Delta Plain

The nonmarine Diablo (Pd) and Olla (Po) Formations are quite interesting from a sedimentologic standpoint (Winker, 1987), but are beyond the scope of this field trip and will be seen only from a distance. The Olla Formation represents the interfingering of Colorado River and locally derived alluvium (Fig. 4). Relative to the width of the paleo-Salton Trough of tens of kilometers (Fig. 2), the outcrop width of the Olla is fairly narrow, about 5 km. Average paleocurrents in "C"-suite channel sandstones in the Olla are south to southeast, similar to those in the Diablo Formation (Fig. 8). Average "L"-suite paleocurrents in the Olla are east-directed (Fig. 7). Thus, the Olla demonstrates the mixture of centripetal sediment transport from the basin margins with axial sediment transport from the Colorado River.

The Olla Formation occupies a fairly narrow outcrop belt in the western Salton Trough (Fig. 14). These outcrops are of considerable paleogeographic position because they provide the best information on the orientation of the basin margin during the Pliocene, post-dating the initial Miocene rifting but predating Quaternary uplift and deformation. Presumably, the Jackson Fork Member has a similar distribution and paleogeographic significance, but it crops out in only one locality whereas the Olla occurs in four discrete localities (Fig. 14). At present the Olla outcrops define a roughly north-south trend, but this distribution has certainly been affected by dextral shear during the Quaternary. Winker (1987) attempted to account for these distortions by restoring: (1) 24 km of dextral shear on the San Felipe fault system (Sharp, 1967), (2) 35° of CCW rotation in the Fish Creek-Vallecito basin (Johnson and others, 1983), and (3) ~3 km of dextral shear on the Elsinore fault, based on offset of the Canebrake-Olla-Diablo facies tract. After restoration the average orientation is NNW-SSE, bending from N-S in the north to NW-SE in the south (Fig. 14).

TECTONICS AND PALEOGEOGRAPHY

Stratigraphic relationships in the western Salton Trough demonstrate three tectonic phases, each with a different paleogeographic setting. The first phase, from the middle Miocene to earliest Pliocene, was the time of initial rifting, volcanism, and marine transgression (Split Mountain Group, Fish Creek Gypsum, Latrania Formation). During this time the major paleogeographic elements were the basin-bounding fault or faults, the uplifted basement terrain to the west providing the major source for clastic sediment, and a pattern of sub-basins and local, low-relief basement highs to the east (Fig. 9). Sub-basins include Split Mountain, the Carrizo Impact Area from Barret Canyon to Red Rock Canyon, and the north-central Coyote Mountains. Local basement paleo-highs are recognized in the eastern Coyote Mountains, the eastern Fish Creek Mountains east of Barrett Canyon, and Boundary Mountain. None of these local highs appear to have been major sources for clastic sediment.

The second phase, during the Pliocene and early Pleistocene, was characterized by initially rapid but decelerating subsidence (Johnson and others, 1983) and high sediment influx which blanketed the paleogeographic elements remaining from the first tectonic phase with a thick cover of deltaic, lacustrine, and alluvial strata (Deguynos Formation and Palm Spring Group). With the second tectonic phase, the postulated (but as yet unidentified) basin-bounding faults apparently stepped westward relative to the basin-bounding faults of the first tectonic phase. The basin-margin bajada represented by the Canebrake Conglomerate and Hueso Formation became increasingly extensive during the late Pliocene and early Pleistocene, and as "L"-suite clastics became more voluminous and finer-grained overall. This may reflect progressive dissection and drainage integration in basement terrain to the west.

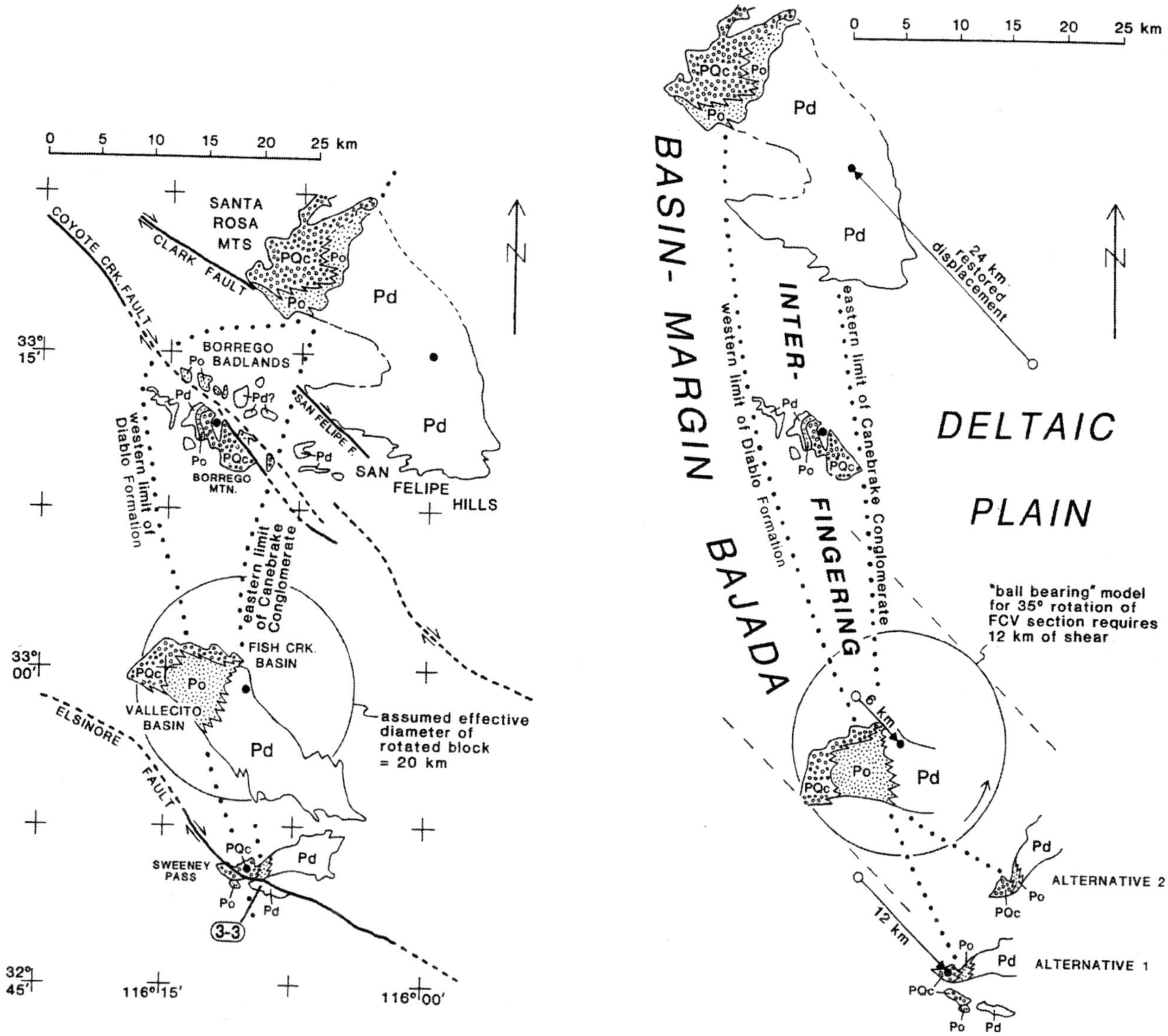

Figure 14. Distribution of the Diablo (Pd) and Olla (Po) Formations and Canebrake Conglomerate (PQc) in the western Salton Trough, in their present distribution (left) and after palinspastic restoration of Quaternary tectonic displacements (right). A similar paleogeographic relationships is thought to apply to the Deguynos Formation (marine deltaic platform) and Camels Head Member (basin-margin alluvial fans and tidal flats).

The final phase, during the Quaternary, was a time of locally rapid uplift (Johnson and others, 1983), major strike-slip faulting (Sharp, 1967), and transpressive deformation throughout the western Salton Trough. This resulted in substantial reorganization of the major topographic elements to create the modern landscape. Stratigraphically, Quaternary wrench tectonism is manifested by the regional-scale (and probably time-transgressive) angular unconformity between the relatively conformable Neogene to Pleistocene succession and Quaternary terraces and alluvium, by the eastward displacement of east-transported, locally-sourced clastics, and probably by increased mixing of "L" and "C"-suite sands. Quaternary wrench tectonism also caused regional-scale distortion of paleogeographic relationships, as suggested by our attempted palinspastic restoration of the Canebrake-Olla-Diablo facies (Fig. 14).

During all three phases the Salton Trough experienced geographic distortion through tectonic translation on a regional scale. Palinspastic reconstruction of the Salton Trough is complicated by uncertainties about total displacements and slip histories on the component faults (Fig. 1). Fig. 2 is based on a simplified kinetic model (Winker and Kidwell, 1985); Winker (1987) discussed some of the uncertainties in the analysis. Nevertheless, a few things seem clear. First, the late Miocene northern Gulf of California was considerably narrower (~50 km) than at present (~150 km). Second, the western

Salton Trough was originally farther southeast (relative to the North American plate) than at present, opposite northwestern Sonora rather than Arizona. Finally the Colorado River delta prograded southward rather than westward from the Colorado River entry point to reach the field trip area. Lacustrine sedimentation did not begin in the western Salton Trough localities until they had been transported tectonically past the topographic divide formed by the Colorado delta plain.

RESERVOIR ANALOGS FOR CLASTIC-DOMINATED MARINE RIFT BASINS

Potential reservoir-seal combinations in Neogene strata of the western Salton Trough include alluvial fan clastics overlain by marine evaporites, marine sandstones overlain by marine claystones, and alluvial sandstones overlain by lacustrine claystones. Of these, the combination of marine sandstones and claystones is most attractive due to the lateral continuity, thickness, and purity of the claystones, the typically sharp intervening contact, and the relatively good sorting of most of the sandstones.

Latrania sandstones are widespread beneath Mud Hills claystones, except where absent over paleo-highs, replaced by limestone over paleo-highs, or replaced by conglomerate near the basin margin. Of the various facies that occur at the top of the Latrania, Andrade sandstone is the most extensive in the study area. Andrade sandstones are generally well-sorted near the top; the primary concern from a reservoir standpoint would be occlusion of porosity in the "L"-suite sandstones by carbonate cement. "C"-suite sandstones in the Wind Caves Member generally show better preservation of primary porosity, reflected in their friable nature in outcrop. Here, the main concern from a reservoir standpoint would be upward decrease in bedding thickness and net/gross sandstone, making for poorer quality reservoir near the top than near the base. In summary, all facies of the Latrania are somewhat problematic as potential reservoirs, each for a different reason. Some of this lateral variability should be predictable from seismic stratigraphy on the basis of gross geometry (Fig. 4, 9).

Delta-front sandstones of the Deguynos Formation generally show better reservoir potential, due to the combination of good sorting, upward-coarsening, sharp tops, and relatively good porosity preservation in "C"-suite sandstone, again reflected in their friable nature. Where sandstone is replaced by coquina, however, porosity is generally very low due to carbonate cementation. Another reservoir-quality concern is the lateral variability of individual sandstone bodies (Fig. 13). Clearly, these tide- and river-dominated sandstone bodies are clearly much more variable than those of wave-dominated deltas. Because wave-dominated deltas would not be expected in narrow rift basins, the Deguynos Formation is probably a good analog for large marine deltas in this tectonic setting. In the subsurface, the delta-front facies should be detectable as a low-continuity, high-reflectivity seismic facies overlying oblique clinoforms. Correlating and mapping individual reservoirs would require closely spaced well control and/or a well-calibrated 3D seismic survey.

ROAD LOG

Day 1. Inception, Northern Gulf of California

Depart San Diego Convention Center to I-8 East and leave the San Diego area. Proceed across the Peninsular Ranges toward the Salton Trough. These ranges constitute a largely intact block of continental crust forming the backbone of Baja California; separation of this block from the North American plate formed the Gulf of California. As we descend from the Peninsular Ranges to the Salton Trough from Jacumba to Ocotillo (Fig. 15), we pass outcrops of Jacumba Basalt, the southern counterpart of the Alverson Formation (Mv) (**STOP 3-1a**). Weather permitting, we can see the Colorado delta plain to the southeast, and the Gulf of California beyond; the Salton Sea is to the northeast.

Refer to Fig. 15 for the field trip route. Exit at Ocotillo and turn left to cross I-8 heading north. The range in front is the Coyote Mountains (Day 3). In Ocotillo, turn right (east) onto S-80 and proceed to Plaster City. The low outcrops are of Pliocene Diablo Formation (Pd), which represents the ancient Colorado River delta plain. On the east side of Plaster City, turn left (north) on the road that parallels the narrow-gauge railroad to the U.S. Gypsum quarry ("USG" in Fig. 15). Our route here follows that of the de Anza expeditions (1774-1776) which established an overland route from Mexico to California and led to the founding of San Francisco (Lindsay and Lindsay, 1978)

The topographic high on our right is Superstition Mountain, a ridge of crystalline basement and Neogene strata pushed up along the Superstition Mountain fault, one segment of the San Felipe fault (Fig. 1). Beyond this range are the low Superstition Hills, composed of uplifted Plio-Pleistocene lacustrine Borrego Formation (Pb & Qb), and beyond them the Salton Sea, standing at 70 m below sea level. The Salton Sea was created by accidental diversion of the Colorado River in 1905-1907, and is now maintained against evaporation by irrigation runoff (Hely and others, 1966).

The range to our left is the Fish Creek Mountains, made of recently uplifted plutonic and metamorphic basement. As recently as one million years ago, these rocks were buried under ~5 km of late Cenozoic sediment represented today by the exposed Fish Creek-Vallecito section. This thickness is comparable to depths to refraction basement in the deepest parts of the Salton Trough (Fuis and Kohler, 1984). The Fish Creek Mountains have since been uplifted to their present elevation and are likely still rising. The steep slope facing us appears to be a fault scarp, though none has been mapped there.

Near the base of the escarpment is the highstand shoreline of Holocene Lake Cahuilla, last occupied ~600 years ago (Weide, 1976; Wilke, 1978; Waters, 1983). William Blake, who published the first geological description of the Salton Trough as part of the Pacific Railway Surveys (U.S. Senate Executive Documents, 1856), was much impressed by the travertine deposits and freshwater mollusc shells left by the ancient lake. Lake Cahuilla is named for the local Cahuilla Indians whose legends recalled past inundations of the Salton Trough by waters of the Colorado River. Maximum elevation of this lake stands ~12 m above sea level, controlled by the sill formed by the Colorado delta-plain. Lake Cahuilla is only the latest stage of the naturally occurring intermittent lake recorded in the Plio-Pleistocene Borrego Formation (Fig. 15).

At Split Mountain Road (paved), turn left (south) and look for signs to the Fish Creek entrance to Anza-Borrego Desert State Park. Turn right into Fish Creek Wash, where we will stop a short way in for an overview of the Split Mountain area. The first two days will be spent in Anza-Borrego, the largest state park in the United States. Collecting or altering the natural environment in any way is prohibited within park boundaries, so do not use rock hammers. Stops for Days 1 and 2 are summarized in index and geologic maps in Fig. 16.

STOP 1: Fish Creek Wash, Park Entrance

The view from Fish Creek Wash just inside the park entrance provides an overview of two manifestations of Miocene rifting -- fanglomerates of the Elephant Trees Formation (Me), and overlying "pre-marine" evaporites of the Fish Creek Gypsum (Mf). Fanglomerates resting on plutonic basement form the cliff to the WSW, and are exposed in Split Mountain Gorge in the core of the Split Mountain anticline (**STOP 1-2**). The fanglomerate attains a maximum thickness there of ~400 m (Kerr, 1982), but thins to the east and southeast, and pinches out entirely about 4 miles (6 km) southeast of here. The No Return fault (**STOP 1-5b**), which we interpret as syn-depositional with the Elephant Trees, is a little more than 2 miles (~4 km) to the west, and downthrown toward us.

The Fish Creek Gypsum is the distinctive tan-colored unit exposed on hills to the north, south, and southeast of us. Gypsum is mined at the U.S. Gyp--sum quarry just southeast of here and hauled to Plaster City. Strontium was once mined from the Fish Creek Gypsum on the hill just to our north.

Proceed up Fish Creek Wash and enter Split Mountain Gorge. The long straight stretch of the gorge follows an up-to-the-west fault with low-angle

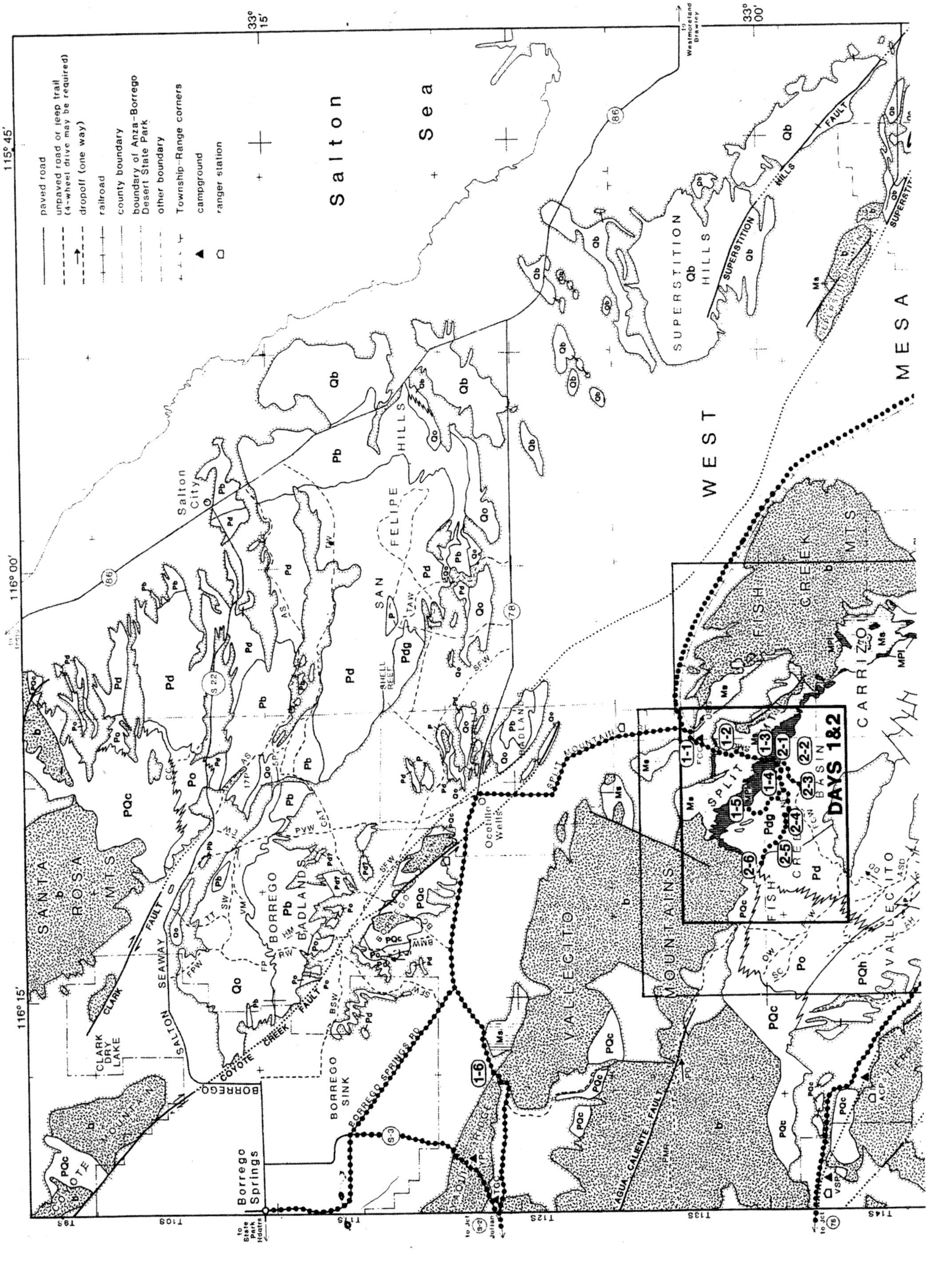
paved road
unpaved road or jeep trail (4-wheel drive may be required)
dropoff (one way)
railroad
county boundary
boundary of Anza-Borrego Desert State Park
other boundary
Township-Range corners
campground
ranger station
Salton Sea
SAN FELIPE HILLS
SUPERSTITION HILLS
SUPERSTITION HILLS FAULT
WEST MESA
FISH CREEK MTS
CARRIZO
SPLIT MOUNTAIN
FISH CREEK BASIN
DAYS 1&2
VALLECITO MOUNTAINS
BORREGO BADLANDS
BORREGO SINK
SANTA ROSA MTS
COYOTE MOUNTAIN
CLARK DRY LAKE
CLARK FAULT
COYOTE CREEK FAULT
SALTON SEAWAY
AGUA CALIENTE FAULT
Borrego Springs
Salton City
Ocotillo Wells
BORREGO SPRINGS RD
116° 15'
116° 00'
115° 45'
33° 15'
33° 00'
T9S
T10S
T11S
T12S
T13S
T14S

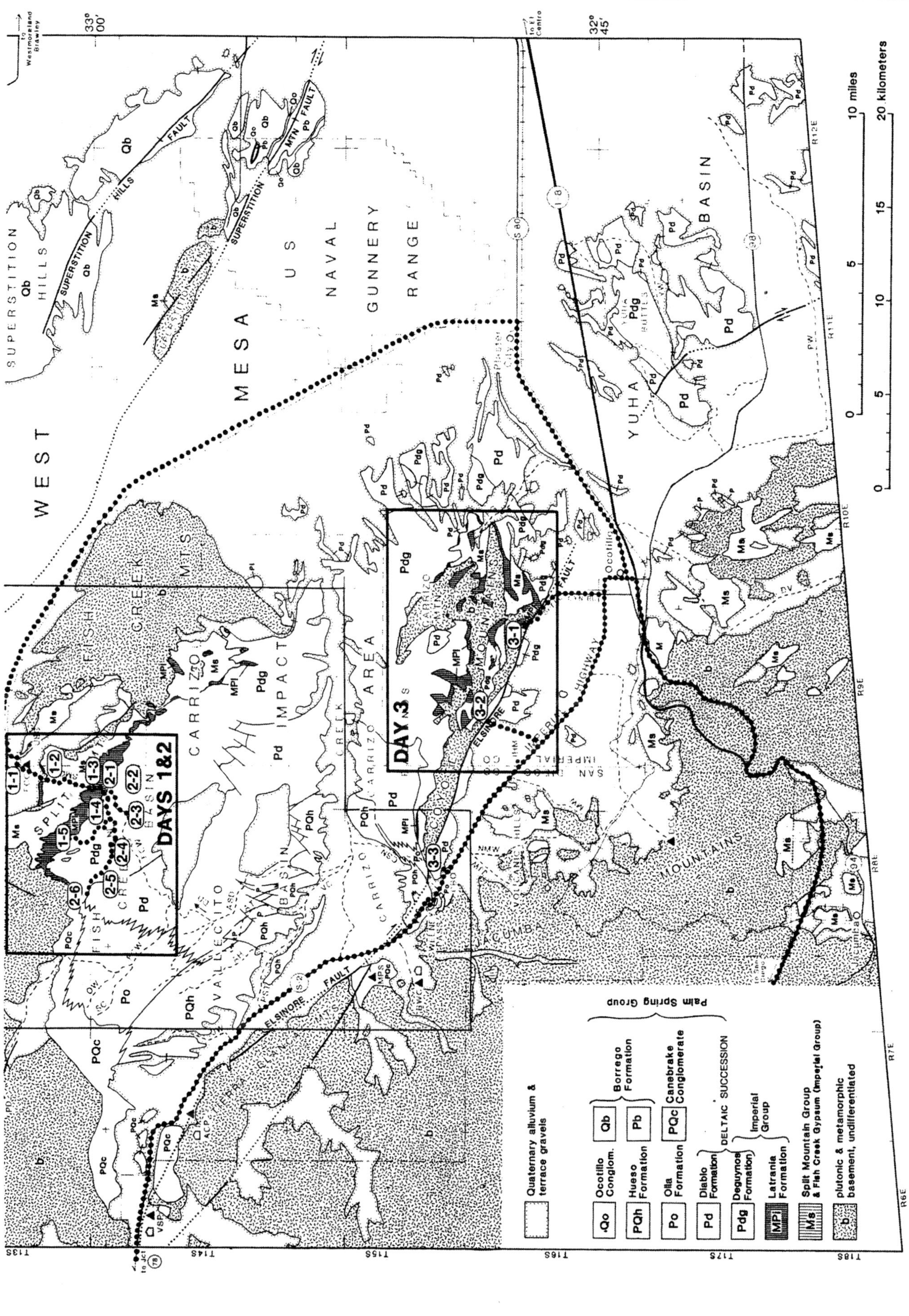

Figure 15. Geologic map and field trip route, western Salton Trough. "DAYS 1 & 2" area is shown in Figure 16; "DAY 3" area in Figure 23.

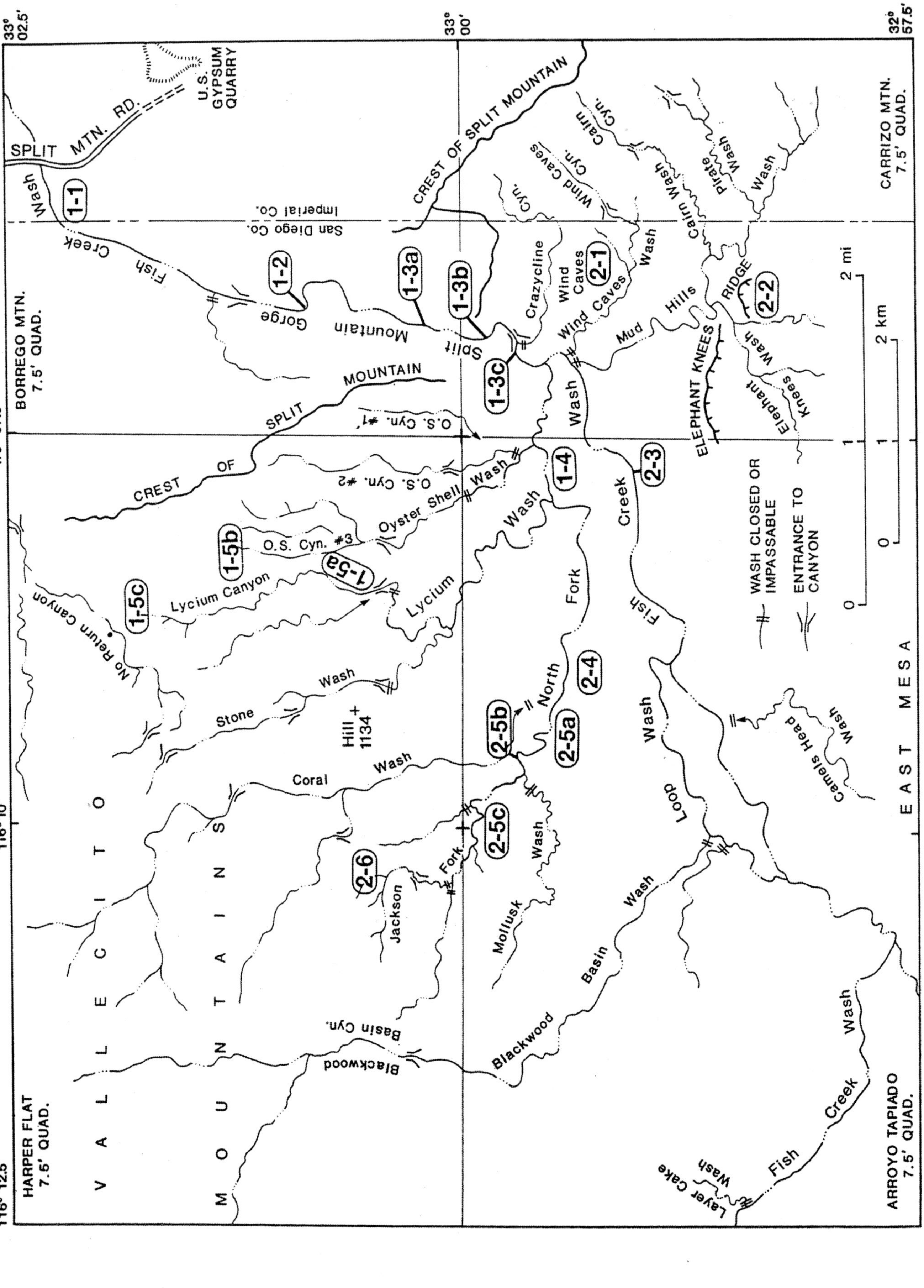
116°05′
33° 02.5′
33° 00′
32° 57.5′
116° 07.5′
116°10′
116° 12.5′
SPLIT MTN. RD.
U.S. GYPSUM QUARRY
CARRIZO MTN. 7.5′ QUAD.
BORREGO MTN. 7.5′ QUAD.
HARPER FLAT 7.5′ QUAD.
ARROYO TAPIADO 7.5′ QUAD.
Imperial Co.
San Diego Co.
CREST OF SPLIT MOUNTAIN
Fish Creek Wash
Split Mountain Gorge
Crazycline
Wind Caves
Wind Caves Cyn.
Cairn Cyn.
Cairn Wash
Pirate Wash
Mud Hills
RIDGE
ELEPHANT KNEES
Elephant Knees Wash
O.S. Cyn. #1
O.S. Cyn. #2
O.S. Cyn. #3
Oyster Shell Wash
Lycium Canyon
Lycium Wash
No Return Canyon
Stone Wash
Hill 1134
Coral Wash
North Fork
Fish Creek Wash
Loop Wash
Camels Head Wash
Jackson Fork
Mollusk Wash
Blackwood Basin Wash
Blackwood Basin Cyn.
Layer Cake Wash
Fish Creek Wash
VALLECITO MOUNTAINS
EAST MESA
1-1
1-2
1-3a
1-3b
1-3c
1-4
1-5a
1-5b
1-5c
2-1
2-2
2-3
2-4
2-5a
2-5b
2-5c
2-6
WASH CLOSED OR IMPASSABLE
ENTRANCE TO CANYON
0 1 2 mi
0 1 2 km

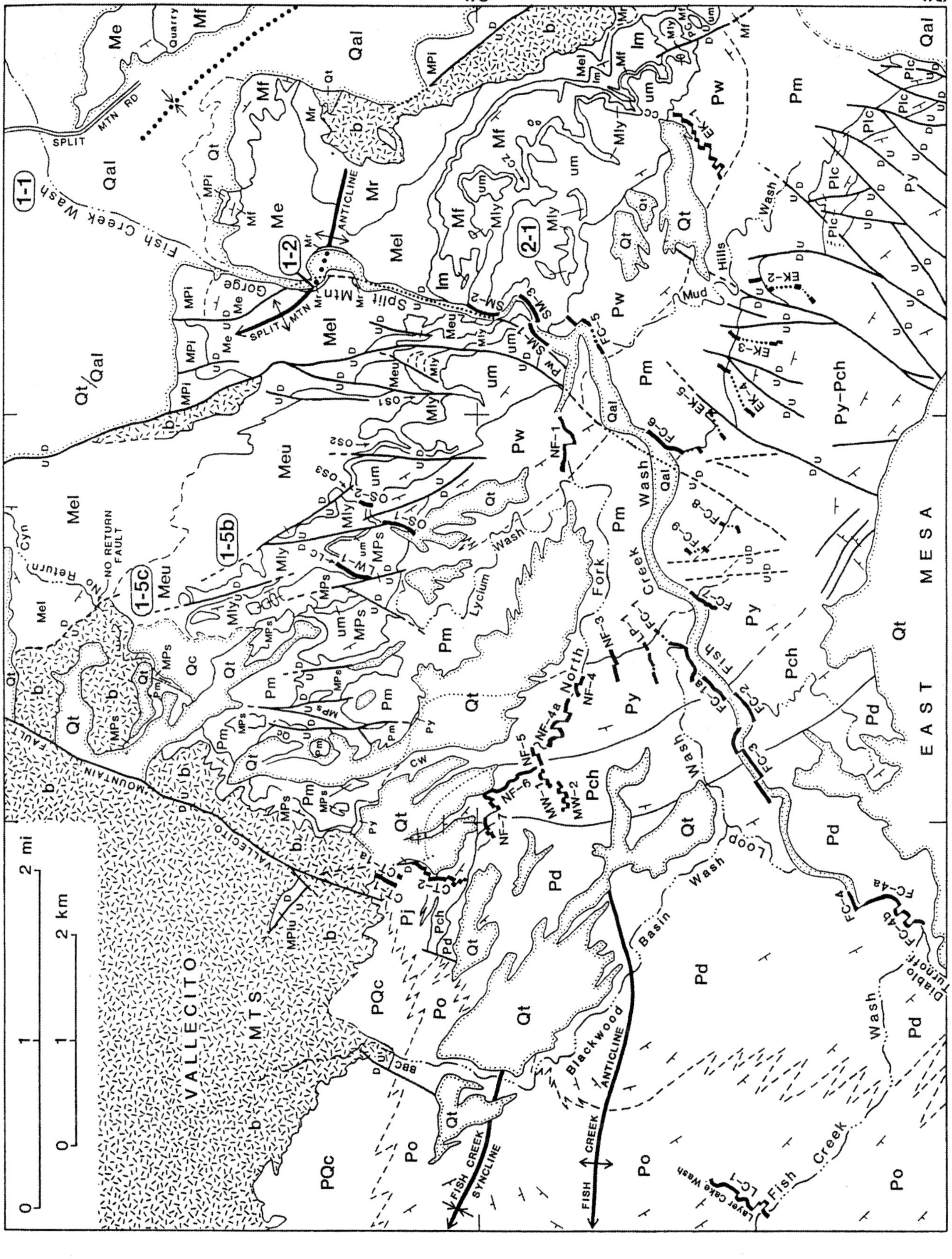

Figure 16. Index map (facing page) and geologic map of Split Mountain Gorge and Fish Creek Basin (Days 1 and 2).

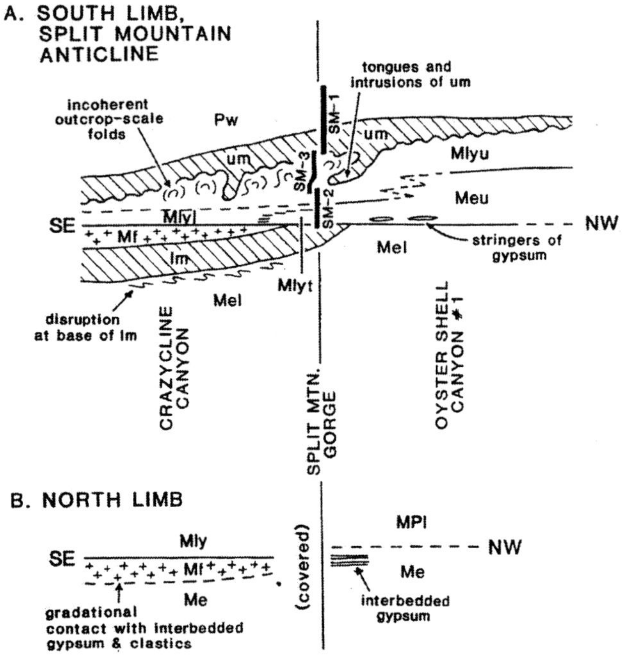

Figure 17. Details of stratigraphy in and adjacent to Split Mountain Gorge.

slickensides, and transects the Split Mountain Anticline. The stratigraphic succession differs considerably on opposing limbs of the anticline (Fig. 17). Stop in the axis of the anticline by the high, nearly vertical cliff on the right, where the gorge jogs to the left.

STOP 1-2. Split Mountain Gorge, Anticlinal Axis

Pink sandstone at the top of the Red Rock Formation is overlain by reddish conglomerates of the Elephant Trees (a.k.a. Split Mountain or Anza) Formation. There is a slight angular discordance at or near the top of the Red Rock (the contact is lithologically gradational). This outcrop demonstrates the combination of debris-flow and sheetflood deposits characteristic of the Elephant Trees Formation.

Drive up the gorge, stopping short of the massive, dark gray, ledge-forming unit on the east side of the wash, which constitutes the Lower Megabreccia. From here we will hike to the head of the gorge, stopping at points of interest along the way.

STOP 1-3a. Split Mountain Gorge, Nonmarine-Marine Contact

At the base of the Lower Megabreccia (lm), note the intense deformation of the underlying reddish flanglomerate, and the refolded shear zones within the massive bed. On the top of the bed, observe the large boulders, several meters in diameter, rafted along the top of the flow. These features are characteristic of landslide megabreccias, as discussed by Robinson and Threet (1974), Kerr (1982, 1984), and Kerr and Kidwell (1991). The megabreccia is present in a single locality on the west wall, but is otherwise absent west of the gorge.

The Fish Creek Gypsum (Mf) overlies the megabreccia but does not quite extend to the wash; its feather edge can be found with a short climb east of the gorge. Where the megabreccia is absent the basal contact of the Fish Creek Gypsum is gradational with the Elephant Trees, with gypsum and "L"-suite siltstone and sandstone interbedded.

Above the Lower Megabreccia and the feather edge of the gypsum is a succession of marine sediment gravity flow deposits summarized in a measured section (Fig. 18). These deposits constitute the Lycium Member, Upper Megabreccia, and Wind Caves Member. The Lycium Member is divided into three sub-members, all visible here (Section SM-2, Fig. 18). First, the thin greenish beds dominated by debris flows represent a lateral transition (Mlyt) between thin marine turbidites (Mlyl) which overlie the Fish Creek Gypsum in Crazycline Canyon (Fig. 17), and thin to medium, greenish to reddish beds of the upper Elephant Trees Formation (Meu), seen on the west wall of the gorge. Overlying the transitional Lycium "green beds" (Mlyt) are thin turbidite beds (Mlyl) followed the more typical medium-grained "L"-suite turbidite beds that comprise most of the Lycium Member (Mlyu).

Proceed up the gorge and look for turbidite characteristics, including coarse-tail grading, bioturbated flow tops, a variety of partial Bouma sequences, flame structures, sole marks, intraclasts, lenticular bedding geometries, and amalgamation surfaces. Beware of bedding-parallel thrust faulting and associated duplex geometries that might be mistaken for bedding geometry. Stop at the large disharmonic fold in the west wall, marked by a State Park sign.

Figure 18 (facing page). Field log of marine Lycium (Mly), upper Megabreccia (um), and lower Wind Caves (Pw) at Split Mountain Gorge. Locations are shown in Fig. 16; also see Figs. 10 and 17. Note change from "L" to "C"-suite sandstones in section SM-1 at 46.5 m.

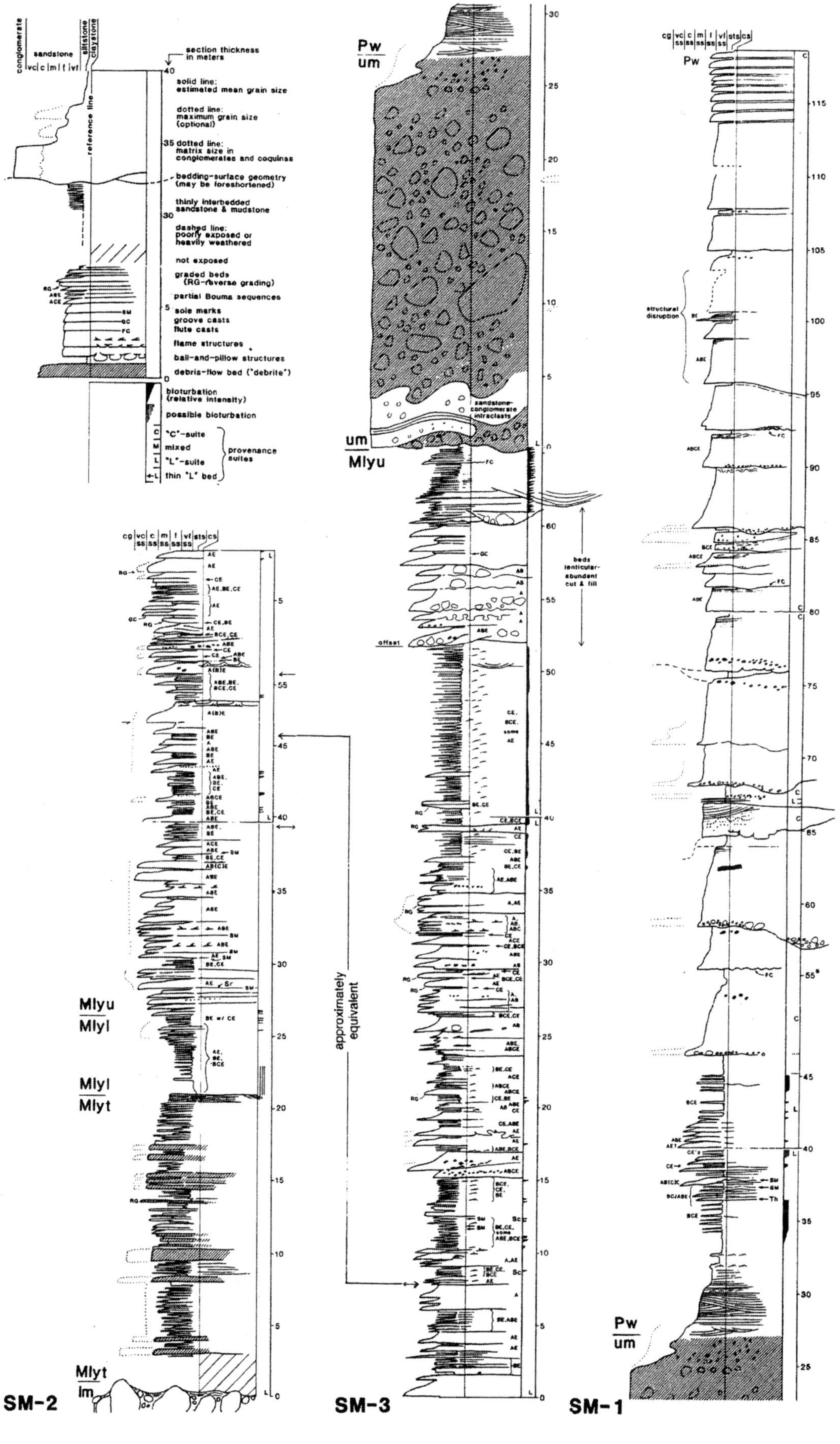
conglomerate
sandstone
siltstone
claystone
vc c m f vf
section thickness in meters
solid line: estimated mean grain size
dotted line: maximum grain size (optional)
dotted line: matrix size in conglomerates and coquinas
bedding-surface geometry (may be foreshortened)
thinly interbedded sandstone & mudstone
dashed line: poorly exposed or heavily weathered
not exposed
graded beds (RG-reverse grading)
partial Bouma sequences
sole marks
groove casts
flute casts
flame structures
ball-and-pillow structures
debris-flow bed ("debrite")
bioturbation (relative intensity)
possible bioturbation
"C"-suite
mixed
"L"-suite
thin "L" bed
provenance suites
reference line
Pw
um
Mlyu
sandstone-conglomerate intraclasts
offset
beds lenticular-abundant cut & fill
approximately equivalent
Mlyl
Mlyt
Im
structural disruption
SM-2
SM-3
SM-1

STOP 1-3b. Outcrop-scale Fold

Interpretation of this outcrop is complicated by the Split Mountain fault, which cuts into the west wall where the gorge bends. Here, the polymictic Upper Megabreccia (um) cuts down into Lycium turbidite sandstones and has "shouldered" them aside, creating the large fold. Tongues and intrusions of the Upper Megabreccia and outcrop-scale folds in the underlying Lycium turbidite sandstones are abundant in Crazycline Canyon and also occur in Oyster Shell Canyon. Fold axes show a preferred NE-SW alignment, but no preferred direction of overturning (Fig. 19). We consequently think the fold axes may be aligned with the transport direction of the megabreccia rather than transverse to it as commonly supposed. Note the "crackle breccia" or jigsaw-puzzle texture of the megabreccia, indicating pulverization of initially larger clasts during transport (Abbott and Rightmer, 1995).

Continue up the gorge on foot. Sole marks in the Lycium turbidites are poorly exposed in the gorge, but trains of intraclasts give a sense of transport toward the east. Discussion of larger-scale bedding geometry will be deferred to Lycium Canyon (**STOP 1-5**), where it is better preserved.

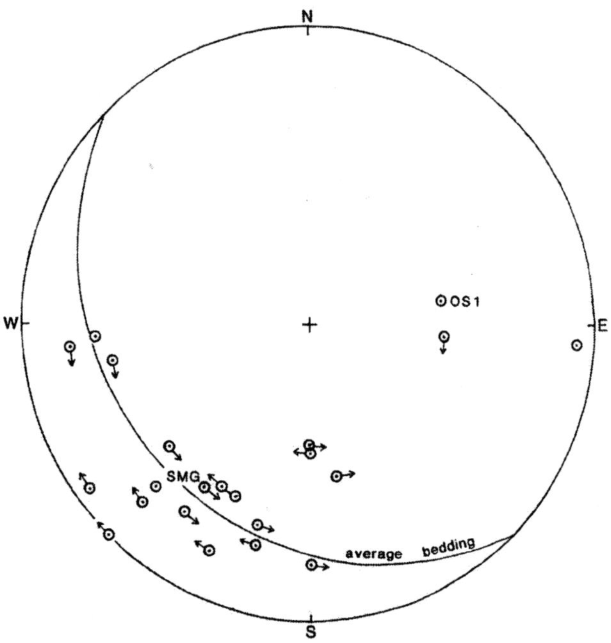

Figure 19. Stereonet of fold axes in Lycium Member (Mly) where deformed by emplacement of Upper Megabreccia (um). SMG = Split Mountain Gorge (**STOP 1-3b**); OS1 = Oyster Shell Canyon #1; other data from Crazycline Canyon (Figure 15). Arrows indicate vergence or sense of overturning.

STOP 1-3c. Head of Split Mountain Gorge

The succession here is summarized in measured sections SM-3 and SM-1 (Fig. 18). Here we can see medium and thin-bedded "L"-suite sandstones of the Lycium Member (Mly) overlain by the Upper Megabreccia (um). Near the contact the Lycium exhibits channel-like lenticular geometries and boulders and cobbles floating in a sandstone matrix. The megabrecca contains large inclusions of Lycium sandstones and exhibits the crackle breccia texture. It grades up into laminated sandstone which may represent the same flow event -- a turbiditic top to a subaqueous landslide deposit.

About 20 m above the megabreccia is another lithologic change that is less striking in outcrop but that has considerable regional significance. This is not only the first occurrence of "C"-suite sandstone in this outcrop, it is also the oldest datable Colorado River-derived sediment in the Salton Trough. K. MacDougall (1995, pers. comm.) places the Miocene-Pliocene boundary (5.1 Ma) essentially at this lithologic boundary on the basis of foraminiferal biostratigraphy. Note that the "C"-suite sandstones are more friable and thicker-bedded (~5 m) than the "L"-suite beds we have seen so far. Although it is not obvious here, paleoçurrents also change, from mainly ENE below to mainly south above (Fig. 7, 8, 9). From here, bedding thickness decreases both up-section and laterally (compare **STOP 1-4**); this appears to be in the axis of the Wind Caves ponded submarine fan (Fig. 9).

From here we will drive to an unspectacular but relatively complete and undeformed exposure of the Wind Caves Member in North Fork (**STOP 1-4**), one of two where the entire Wind Caves section is completely exposed. The route to **STOP 1-4** parallels the dip slope of the Upper Megabreccia, offset by a series of up-to-the-west faults. The exposure begins just past the entrance to Oyster Shell Wash (a misnomer!) (Fig. 16).

STOP 1-4. North Fork

The purpose of this stop is to view the upward-thinning succession characteristic of the Wind Caves Member, and the alternation of "C" and "L"-suite sandstones. Upward thinning in the Wind Caves submarine fan may represent a change from more channelized fan-lobes near the base to more sheet-like deposits near the top, reflecting net retrogradation and spreading of the fan as it filled the Split Mountain sub-basin (Fig. 9).

Proceed a short distance up North Fork, then enter Lycium Canyon on the left. This wash follows a long, winding course through the lower Mud Hills

Member (Pm) of the Deguynos Formation (Pdg), a thick, monotonous succession of slightly silty claystones representing the prodelta of the Colorado River delta (Day 2). We will see a few stray beds of "L"-suite turbidites within the Mud Hills Member. Proceed up Lycium Wash as far as possible, and park vehicles.

STOP 1-5a. Lycium Canyon

This stop, summarized by the measured section in Fig. 20, demonstrates a more proximal counterpart of the "L"-suite marine sandstones observed in Split Mountain Gorge (**STOP 1-3**). It involves a lengthy hike of about 2 miles (3 km) up the canyon to the No Return Canyon overlook, and back.

The dip slope here is formed not by the Upper Megabreccia as at the last two stops, but by the conglomeratic Stone Wash Member (MPs). The Stone Wash is overlain with sharp contact by the Mud Hills Member (there are isolated turbidite beds within the Mud Hills in this area). The Stone Wash Member is the lateral equivalent (more proximal in a facies sense) of the lower, "L"-suite part of the Wind Caves Member. The upper "C"-suite Wind Caves has lapped out and/or shaled out between here and the last stop; in other words 200 m of section has disappeared about 3 km. The Stone Wash Member laps out against plutonic basement 2 km west of here.

Enter Lycium Canyon through the narrow, slanting cut in the top of the Stone Wash Member and proceed down-section (upstream). Marine indicators are absent in the upper part of the Stone Wash. Near the top, note the wide variety of clast sizes, mostly poorly sorted. There are relatively few discrete flow units or interbed contacts, suggesting that the flow units are highly amalgamated. The strongest evidence for a sediment-gravity flow origin is in bioturbated bed tops and sole marks near the base of the Stone Wash Member.

Note also that the conglomerate is polymictic, with metamorphic clasts. The nearby Vallecito Mountains consist of exclusively plutonic basement. This lends support to the argument that paleotopography included basement exposures that are now buried -- in this case below Fish Creek Basin. Farther northwest, closer to the Vallecitos, the Stone Wash Member becomes monomictic.

At the base of the Stone Wash (by definition) is the westernmost occurrence of the Upper Megabreccia (um). Below that is the Lycium Member (again by definition) but here the upper Lycium is very similar to the Stone Wash (Fig. 20). Proceeding down-section, the Lycium becomes less conglomeratic, then entirely sandstone. Overall, the Lycium-Stone Wash succession gives the impression of a prograding fan delta (Fig. 20).

Look in the Lycium Member for evidence of a peculiar bedding geometry of "retrograde sigmoid bedsets" which was first recognized by M.B. Edwards (1986, pers. comm.) at this location (Fig. 21). The

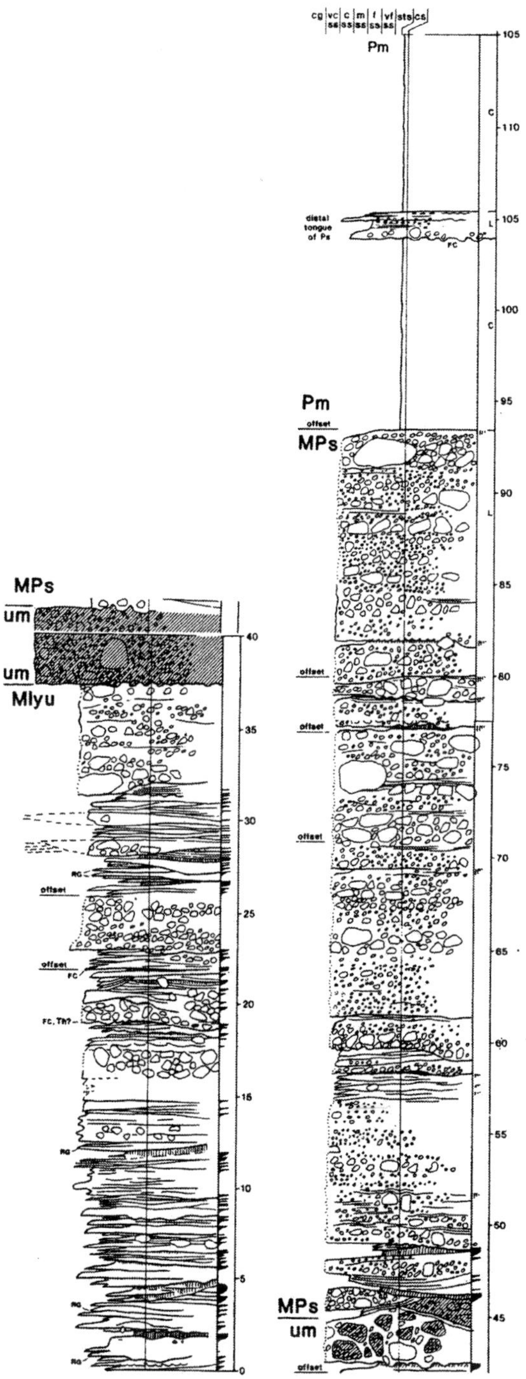

Figure 20. Field log of marine Lycium Member (Mly), Upper Megabreccia (um), and Stone Wash Member (MPs) at Lycium Canyon. Location is shown in Figure 17; also see Figure 10. Key to symbols in Figure 18.

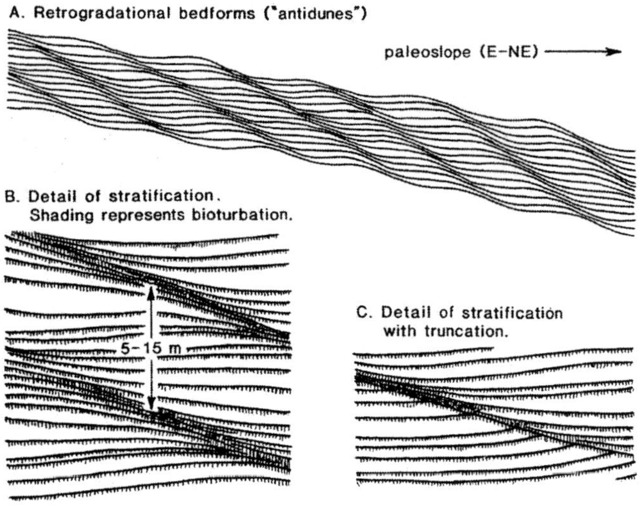

Figure 21. Schematic representation and interpretation of retrograde sigmoid bedsets in Lycium Member, with vertical exaggeration.

true geometry is best observed by examining dip-oriented (i.e. E to NE) exposures and trace out individual beds. We perceive the beds as bundled into sigmoidal packages with thinner beds near the top and base and thicker beds in the middle, and sometimes an erosional surface at the base of the bundle. Bed contacts are accentuated by bioturbation; where beds are thin the bioturbation may coalesce to obscure the bedding. We infer that these sigmoid bundles were created by a bedform with several meters of relief that migrated in a consistently up-current direction.

The Lycium Member rests with sharp contact on the upper member of the conglomeratic Elephant Trees Formation (Meu), which we last saw in the west wall of Split Mountain Gorge (**STOP 1-3a**). From here, proceed up Lycium Canyon to the overlook of No Return Canyon.

STOP 1-5b. No Return Canyon Overlook

At the eastern edge of the canyon visible from this overlook and extending down to the mouth of the canyon is a thick section of olive-gray lower Elephant Trees fanglomerate (Mel) in a very proximal facies, dipping generally toward the west. Boulders up to 2 m diameter are common, in debris-flow beds up to 10 m thick. The outcrop directly in front (NW) of us across the canyon is strewn with boulders, but without any evidence of bedding and with clear evidence of igneous dikes in place. Hence, we are viewing bedded conglomerates abutting basement, presumably in a fault contact, although the inferred fault cannot be precisely located. The plutonic basement appears to be highly fractured. In fact, the change from sedimentary to igneous rocks is difficult to recognize up close, when hiking up the canyon from the Elephant Trees area.

This outcrop illustrates some of the difficulties in recognizing and mapping major faults associated with the Miocene extensional esisode that created the early northern Gulf of California. It is evident that a thick section of fanglomerate is replaced immediately to the west by older plutonic rocks, suggesting a fault contact, but the fault or faults themselves are not clear.

Retrace route through Lycium Canyon and return to vehicles. Retrace route out of the State Park via North Fork and Fish Creek, and turn left (north) onto Split Mountain Road (paved); follow this road to Ocotillo Wells. In Ocotillo Wells, turn left at the "T" onto State Route 78. Seven miles west on Route 78 we either turn right onto Borrego Springs Road and take the short cut to Borrego Springs. Alternatively, if time permits, we can continue another 3 miles on Route 78 to the edge of Yaqui Ridge.

STOP 1-6. Yaqui Ridge. (optional).

At the road cut is a poor exposure of conglomerate just south of a better exposure of plutonic basement. The contact between conglomerate and basement is better exposed in gullies cut into the pediment just east of us. The contact is a detachment fault containing pseudotachylite that has been dated by K-Ar on feldspar concentrates at 10.4±3 Ma (Frost and Shafiqullah, 1990). This would make place the detachment faulting within the age brackets for fanglomerates of the Elephant Trees Formation.

Retrace the route to Borrego Springs Road, and turn left toward Borrego Springs for overnight accommodations.

Day 2. Deltaic Succession

Refer to Fig. 15 for the field trip route. From Borrego Springs, head southeast on Borrego Springs Road. Turn left at the T to S-78. In Ocotillo Wells turn right onto Split Mountain Road. At the entrance to Anza-Borrego State Park, turn right off pavement and return to Split Mountain Gorge, proceed up to the head of the gorge, and turn left into Fish Creek Wash. Watch on the left for the head of Wind Caves trail. Park the vehicles and hike the short trail up to the Wind Caves (Fig. 16).

STOP 2-1. Wind Caves

These caves were produced by weathering of arkosic "L"-suite turbidite sandstones of the basal Wind Caves Member (Pw). The dip slope here is formed by the Upper Megabreccia (um). This is a good vantage point from which to visualize local paleogeography of the late Miocene Gulf of California (Fig. 9) prior to the arrival of the Colorado River delta (Fig. 12).

Day 2 will focus on the Deguynos Formation, which comprises the marine portion of the deltaic succession. The next three stops are devoted to the upper Mud Hills and lower Yuha Members, representing the delta front facies (Fig. 13). In the facing slope of the large hogback to the southwest (Elephant Knees Ridge), the grayer colors represent claystone and the yellower colors represent siltstones, while sandstones and coquina mostly have a brown desert varnish. Note the leftward thickening in the facing slope, interpreted as clinoform geometry below the lowest sandstone and coquina beds (Fig. 13). Return to vehicles and proceed the short distance up Fish Creek Wash to the entrance of Mud Hills Wash (closed to vehicles).

STOP 2-2. Elephant Knees

Hike up Mud Hills Wash to the split in the Elephant Knees ridge at measured section EK-3 (Fig. 16). Cut banks of Mud Hills Wash show sandstone beds at the top of the Wind Caves Member, and thin siltstone beds within claystone of the Mud Hills Member. The Mud Hills Member contains foraminifera and a sparse aragonitic fauna of deposit-feeding bivalves.

Section EK-2 shows seven complete delta-front parasequences (Fig. 13). The "knees" on the north-facing slope consist of claystone at the base of the second exposed parasequence. Sandstone and oyster-shell coquina cap the third parasequence; the resistant coquina forms the dip slope of the hogback. At section EK-2, the coquina is 6 m thick and composed of densely packed *Dendostrea* valves in very large-scale, south-directed cross-sets.

Hike along the base of the dip slope to section EK-3 (Fig. 16). In this short distance the coquina has been replaced by sandstone with trough crossbedding indicating unimodal, south-directed paleocurrents. In addition, the third parasequence of section EK-2 has split into two. Some of the parasequences here show hints of rhythmite-type bedding, which we will observe at **STOP 2-3**. Sandstones of the basal Yuha Member at section EK-3 grade into the Lavender Canyon Member east of here, in the Carrizo Impact Area (Fig. 16).

Return to vehicles and continue up Fish Creek Wash to the strikingly ribbed cliff on the left (south) side of the wash.

STOP 2-3. Rhythmites, Fish Creek Wash.

This remarkable outcrop of rhythmically bedded siltstones is part of the upper Mud Hills Member, below the first clear parasequence (Fig. 13). Similar rhythmic bedding is commonly observed within the silty portion of parasequences. The ~10 cm cycles are symmetric, without sharp contacts, and are due to subtle textural differences which account for slight differences in resistance to erosion. In this outcrop the alternation is between clayey and sandy siltstones; elsewhere, rhythmites with finer and coarser textures are also observed. Rhythmic cycles are typically bioturbated, but finer-scale bedding and ripple lamination are locally preserved. Thickness of rhythmite cycles ranges from about 5 cm to as much as 30 cm.

We interpret the rhythmites as annual cycles controlled by seasonal discharge of the ancestral Colorado River. The modern Colorado River has a strongly seasonal discharge dominated by meltwater runoff from the Rocky Mountains. This interpretation implies rock accumulation rates within the rhythmites (50-300 mm/yr) one to two orders of magnitude greater than average rates of 3-4 mm/yr for this part of the succession based on paleomagnetism (Johnson and others, 1983). This is not unreasonable if sedimentation on the delta front occurred by rapid progradational episodes alternating with prolonged periods of relatively slow sedimentation (non-rhythmic claystones).

Backtrack down Fish Creek Wash and turn left into North Fork. Proceed up North Fork through the very thick succession of Mud Hills Member. Park the vehicles at **STOP 2-4**, and climb out the south bank of the wash and approach The Notch (Section NF-3 in Fig. 13) from below.

STOP 2-4. The Notch.

Measured section NF-3 contains two complete parasequences below the resistant cross-bedded sandstone, and at least six parasequences above (Fig. 13). This second parasequence has well-exposed rhythmite bedding in the siltstone portion and is capped by trough cross-bedded sandstone. Paleocurrent directions in this sandstone are unimodal toward the south, consistent with what we observed at **STOP 1-2**. The third parasequence is manifested by a thin lag of oyster shells; this has been physically correlated with the thick coquina beds at the Elephant Knees hogback (**STOP 2-2**). Proceed up-section by four more parasequences, where we see a sandstone unit that is sharp-based. This is laterally equivalent to a gradationally-based sandstone at section FC-7. The sharp base exhibits abundant trace fossils (*Thalassinoides* and *Gyrolithes*) and molds of *in situ* pholad bivalves. We interpet such occurrences of pholads as evidence that the clay was quite stiff due to exhumation of previously compacted clay, perhaps at the base of tidal channels.

Return to North Fork wash. Depending on time, the next stop (**2-5**) can be conducted either by hiking or driving slowly up the wash and stopping at specific points of interest. Refer to the measured sections in Fig. 22 and the base map in Fig. 16.

STOP 2-5a. Upper Yuha Member, North Fork

This stop corresponds to section NF-4a at 160 m. This is the first good example of the wavy bedding association of lenticular, wavy, and flaser bedding commonly observed on tidal flats (Reineck and Wunderlich, 1986). Occurrences of the wavy bedding association persist to the top of the Camels Head Member (Section NF-7 at 140 m). Also look for *Thalassinoides*, *Gyrolithes*, *in situ* pholads, and oyster shells in a muddy matrix.

STOP 2-5b. Upper Yuha Member, North Fork

This stop begins at the base of section NF-5. The steep cut bank demonstrates non-parallel stratal geometry in the upper Yuha Member. We envision considerable paleorelief on the delta platform due to a complex of channels and shoals. A short distance along the wash is a relatively thick accumulation of oyster shells in a muddy matrix. We interpret these as *in situ* oyster concentrations located either on tidal flats or within tidal channels. Note the large number of articulated specimens attached to others-of-kind in small life-positioned clusters, and the relative sparsity of layers composed of fragmented, reworked skeletal debris. We believe that this part of the delta served as a source of the oyster debris that forms the thick cross-bedded coquinas in the subtidal delta front (**STOPS 2-2** and **2-4**).

Continue up the wash, and observe the first occurrence of reddish claystone that defines the base of the Camels Head Member (Section NF-5 at 73 m). Also observe the increasing abundance of thick, cross-bedded, sharp-based sandstones with fossil material (interpreted as tidally influenced point bar deposits), and repeated occurrences of the wavy bedding association, interpreted as tidal flat deposits.

STOP 2-5c. Camels Head Member, North Fork

This stop demonstrates the occurrence of transported shell material within a channel sandstone of the Camels Head Member. Also note the wavy-bedding association, this time associated with reddish claystones. Bioturbation at this point is minor and burrowed horizons are uncommon. In the base of one sandstone bed observe concretion-like lumps that are actual mollusc shells encrusted by multiple concentric layers of sheet-form bryozoans, indicative of repeated reworking, presumably in a tidal channel. Cross-sections of elongate specimens reveal that many of these circumrotary colonies started as overgrowths of snail shells that had been occupied by hermit crabs (Gyllenhaal and Kidwell, 1989). Modern examples can be found along both shores of the northern Gulf of California.

Continue up North Fork to the confluence with Jackson Fork. Proceed on foot up Jackson Fork.

STOP 2-6. Jackson Fork

Walk toward the Vallecito Mountains (and back down-section), and look for locally-derived "L"-suite sandstones and conglomerates interbedded with the typical "C"-suite Camels Head lithologies. Some of the "L"-suite beds are of debris-flow origin, indicated by their dark color (presence of clay matrix), poor sorting, and absence of lamination. Some lithic clasts in the "L" suite are encrusted with serpulid worm tubes or solitary corals, indicating deposition in marine waters. Within the Camels Head Member look for : (1) beds of articulated oyster shells in a clay-silt matrix, similar to those of **STOP 2-5**; (2) mud cracks; (3) a variety of burrows; (4) casts of *in situ* pholads in claystone, suggestive of erosion to expose stiff, compacted clay; and (5) the wavy bedding association. A succession of purely "L"-suite sandstone and conglomerate comprises the Jackson Fork Member (Pj) of the Latrania Formation. This Member is younger than the Latrania Formation in other localities; the Latrania-Deguynos contact is time-transgressive. The Jackson Fork Member is interpreted as a mixture of debris-flow

Figure 22 (facing page). Field log of deltaic Deguynos Formation (from uppermost Mud Hills Member, Pm) to lowermost Diablo Formation (Pd). These logs comprise a continuous composite section.

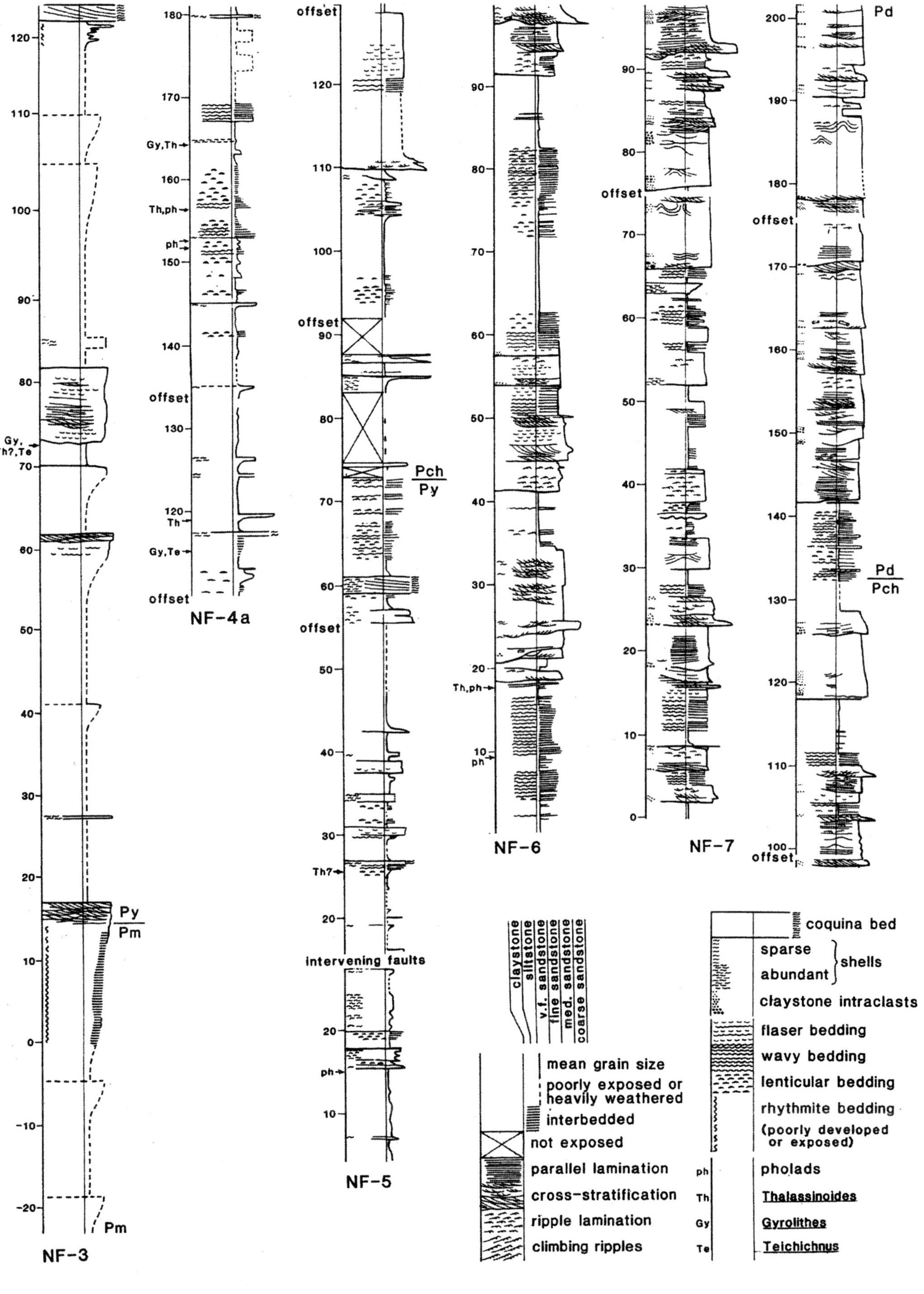
NF-3
Py
Pm
Gy, Th?,Te
NF-4a
offset
Gy,Th
Th,ph
ph
Th
Gy,Te
NF-5
offset
Pch
Py
Th?
intervening faults
ph
NF-6
Th,ph
ph
NF-7
offset
Pd
Pch
claystone
siltstone
v.f. sandstone
fine sandstone
med. sandstone
coarse sandstone
mean grain size
poorly exposed or heavily weathered
interbedded
not exposed
parallel lamination
cross-stratification
ripple lamination
climbing ripples
coquina bed
sparse
abundant
shells
claystone intraclasts
flaser bedding
wavy bedding
lenticular bedding
rhythmite bedding
(poorly developed or exposed)
ph pholads
Th Thalassinoides
Gy Gyrolithes
Te Teichichnus

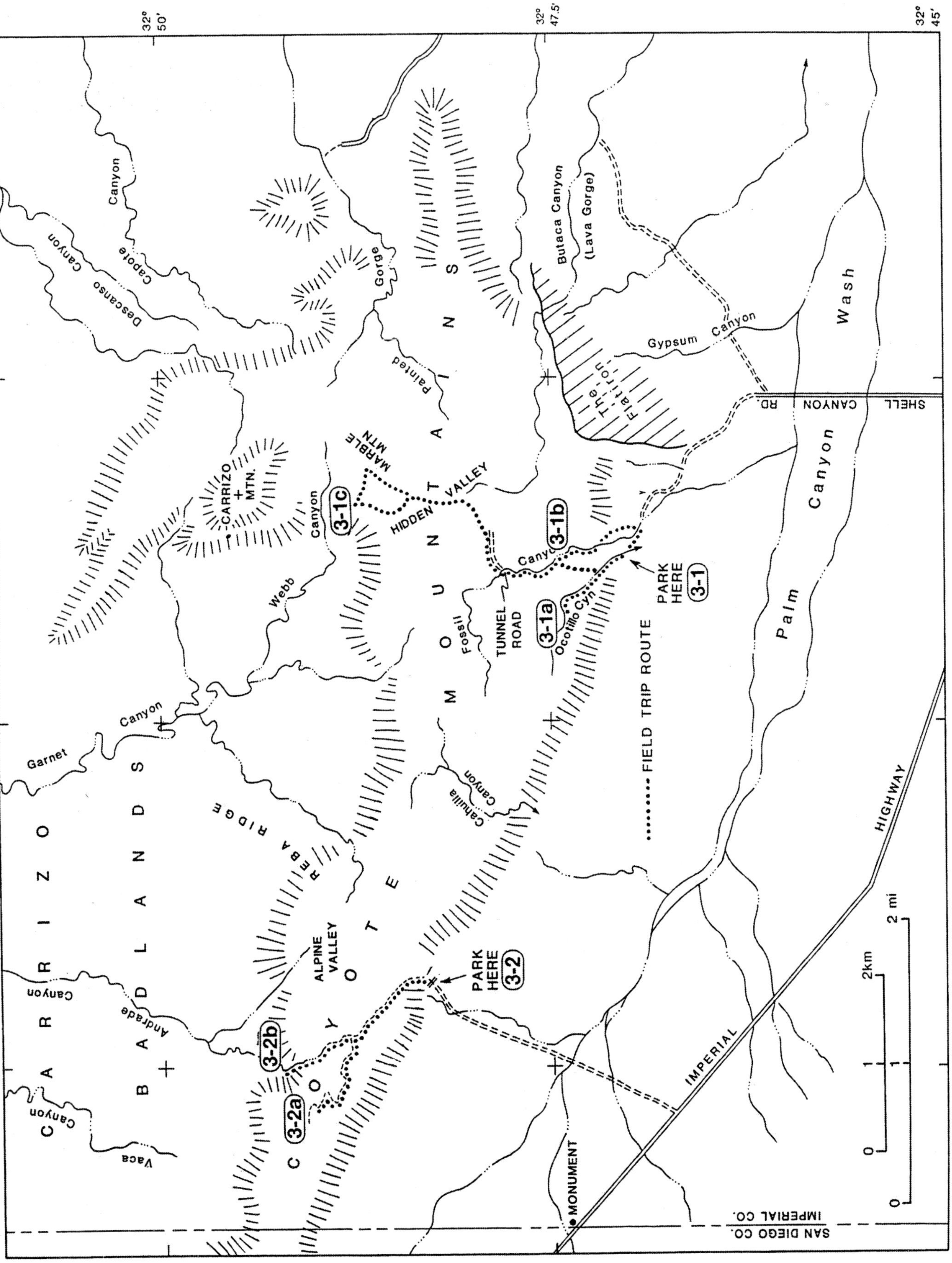

CARRIZO BADLANDS
COYOTE MOUNTAINS
CARRIZO MTN.
MARBLE MTN
HIDDEN VALLEY
ALPINE VALLEY
REBA RIDGE
The Flatiron
Butaca Canyon (Lava Gorge)
Gypsum Canyon
Painted Gorge
Capote Canyon
Descanso Canyon
Webb Canyon
Garnet Canyon
Andrade Canyon
Vaca Canyon
Fossil Canyon
Cahuilla Canyon
Ocotillo Cyn
TUNNEL ROAD
SHELL CANYON RD.
Palm Canyon Wash
IMPERIAL HIGHWAY
MONUMENT
SAN DIEGO CO.
IMPERIAL CO.
PARK HERE 3-1
PARK HERE 3-2
3-1a
3-1b
3-1c
3-2a
3-2b
FIELD TRIP ROUTE
2 mi
2km
116° 00′
116° 02.5′
116° 05′
115° 57.5′
32° 50′
32° 47.5′
32° 45′

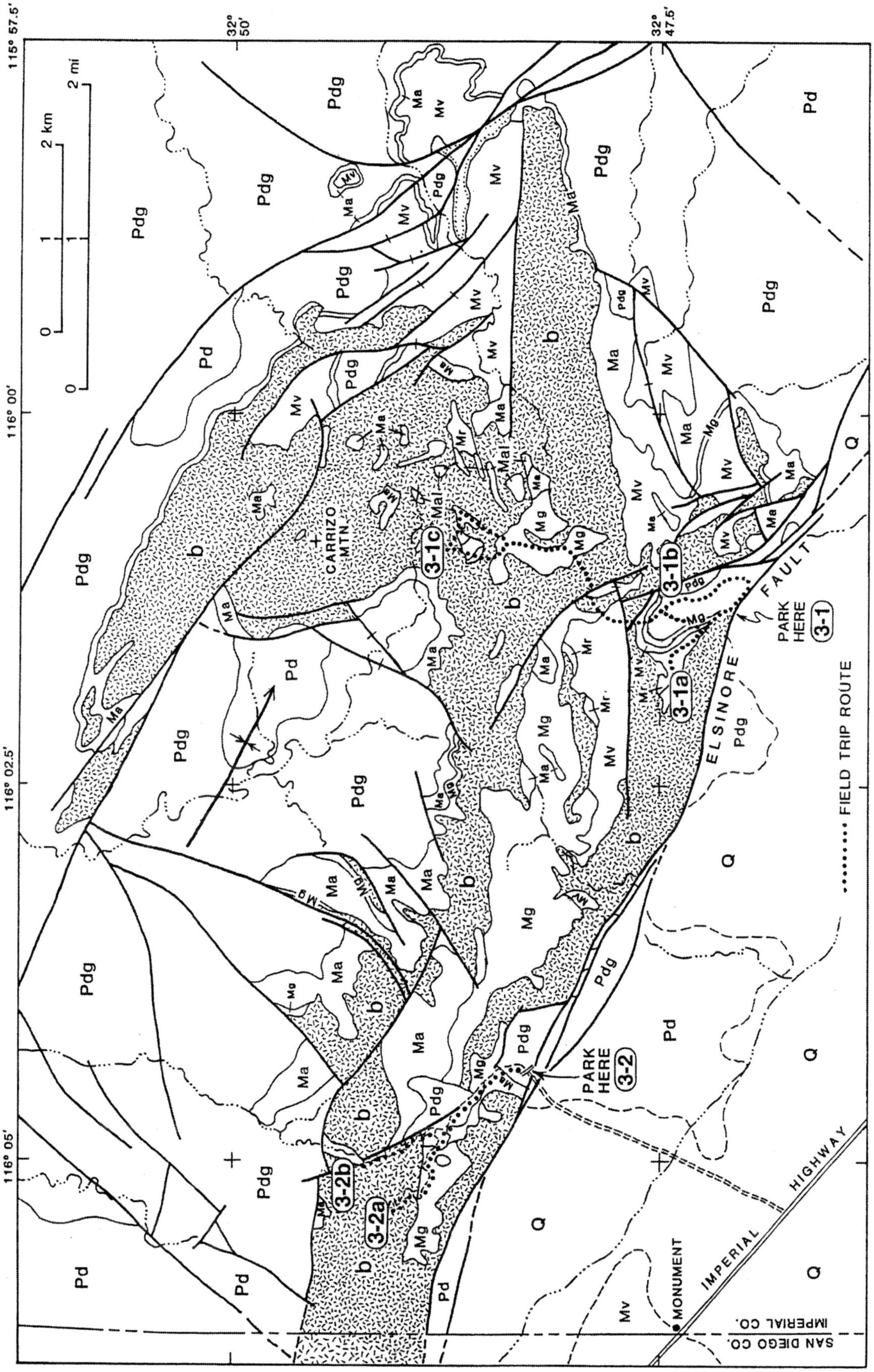

Figure 23. Index map (facing page) and geologic map of Coyote Mountains. Location is shown in Figure 15 ("DAY 3").

and sheetflood deposits typical of an alluvial fan, but with marine indicators. Look for evidence of: (1) burrows, particularly *Ophiomorpha*; (2) cloven hoof-prints of camels, 10-15 cm across, on the bases of sandstone beds; and (3) shell beds rich in shallow marine gastropods (turritellids, naticids, cerithiids, limpets, cowries), bivalves (cardiids, crassatellids, venerids, pectinids, anomiids), echinoids (sand dollars) and balanid barnacles.

After rounding a broad bend to the left, proceed up a fossiliferous dip slope toward the exposure of plutonic basement on the flank of the Vallecito Mountains. Conglomerate of the Jackson Fork Member appears to be in nonconformable contact with basement. The same lap-out relationship can be seen all along the southeastern flank of the Vallecito Mountains, in the Stone Wash Member, Jackson Fork Member, and Canebrake Conglomerate. Return to vehicles and retrace route down North Fork and Split Mountain Gorge. As on Day 1, turn left onto Split Mountain Road, then left onto State Route 78, then right onto Borrego Springs Road.

Day 3. Inception of Gulf of California, Part 2.

Referring to Fig. 15, leave Borrego Springs on Borrego Springs Road, but this time turn right (south) at Casa del Zorro onto Route S-3, and proceed across Yaqui Ridge (optional **STOP 1-6**) to the junction with State Route 78 (off the map). Turn right onto Route 78 and continue west to the junction with Route S-2 (Imperial Highway), turn left (southeast) onto S-2, and proceed toward Ocotillo. After passing the Vallecito State Park (VSP in Fig. 15) campground on the right, look for outcrops of the light gray Canebrake Conglomerate (PQc), a widespread Plio-Pleistocene unit of alluvium derived from the paleo-basin margin. Just past the Bow Willow Campground (BWC in Fig. 15), roadcuts in the switchbacks at Sweeney Pass expose conglomeratic sandstones of the Hueso Formation (PQh). Pass the Canyon Sin Nombre overlook, where we return for the last stop **(3-3)** of the field trip.

The range to the left (north) is the Coyote Mountains, a long, narrow, asymmetric basement-cored uplift. Its southern edge is marked by the active Elsinore strike-slip fault, with at least 3 km of strike slip based on offset of the Olla-Diablo contact, and a significant dip-slip component. Basement in the Coyotes is predominantly metamorphic, mainly schist and marble intruded by small plutons and pegmatites. Despite the rather low elevation (735 m maximum), the Coyote Mountains are quite rugged and access is rather limited, making for slow mapping. Patterns of faulting (predominantly strike-slip) within the Coyotes are complex and confusing; a fault that appears to be major at one point may seem to disappear or be truncated by another fault a short distance away. The extent of paleogeographic distortion due strike-slip faulting and possible block rotations is still under study.

On the outskirts of Ocotillo, turn left onto Shell Canyon Road and follow it north toward the base of the Coyotes and veer left toward the mouth of Alverson (Fossil) Canyon. Park vehicles at the mouth of Fossil Canyon (westernmost point, where road turns north into gated canyon) and proceed on foot ~100 m west to the next canyon, Ocotillo Canyon. Refer to Fig. 23 for **STOPS 3-1** and **3-2** (alternate).

STOP 3-1a. Ocotillo Canyon

Park the vehicles and hike up Ocotillo Canyon to view a thick vertical succession in a west-facing outcrop. Three units here comprise the Split Mountain Group: (1) The Red Rock Formation (Mr) consists of reddish conglomerate resting on metamorphic basement; (2) the Alverson Formation (Mv) consists of basalt flows and volcaniclastics; and (3) the Garnet Formation (Mg) consists of tan conglomerates with basalt clasts in addition to basement-derived clasts. Ocotillo Canyon is the only locality in the Coyote Mountains where all three units can be seen in succession; elsewhere in the Coyotes only two, one, or none of these units intervene between basement and marine Imperial strata (Latrania and Deguynos Formations).

From here, backtrack down Ocotillo Canyon until it is possible to climb out the east wall via a narrow gulch, then walk north up this gulch to observe in the west-facing wall the vertical and lateral transition of Garnet Formation conglomerate into fossiliferous sandstones of the Andrade Member. Climb onto the top of the east wall and walk NNE obliquely down the dip slope to the floor of Fossil (Alverson) Canyon. In recent years this dip slope has been heavily mined by fossil collectors, but we will still see abundant pycnodontid oysters and other mollusks. It is possible to trace individual cobble layers and shell beds from the gulch to **STOP 3-1b**, revealing a set of low-angle clinoforms in thin sandy parasequences (Fig. 24). Shell lags mantle the flooding surfaces; proximal infaunal bivalve-dominated sands change distally into an oyster-coral-dominated limestone at **STOP 3-1b.**

STOP 3-1b. Fossil Canyon

This exposure of Andrade Member sandstone is typical of Andrade occurrences at or near the toes of Garnet fanglomerate wedges, which can be observed in elsewhere in the Coyote Mountains at Hidden

Figure 23. Index map and geologic map of Coyote Mountains. Location is shown in Figure 15 ("DAY 3").

Valley (**STOP 3-1c**), Cahuilla Canyon, upper Fossil Canyon, and west of Andrade Canyon (**STOP 3-2a**). These sandstones are richly fossiliferous and have an almost nodular appearance due to bioturbation and carbonate cementation. At the northern end of the narrow gorge, at wash level, is the typical 2-3 meter thick, densely packed shell bed that forms a distal limestone facies within the Andrade Member (described in Kidwell, 1988). Dominated by large *Pycnodontid heermanni*, this facies includes small subspherical to dumbbell shaped *Porites* corals (corallite diameters 1-2 mm) and large, 10-20 cm massive and palmate heads of the corals *Solenastrea* and *Siderastrea*. Although coral-bearing portions of the Andrade Member are often referred to as "reefs", note that boundstone fabrics are modest to non-existent and that there are no flanking beds or other evidence of significant topographic relief on the seafloor. These pycnodontid oysters were unattached epifauna. Corals encrust single bivalve shells rather than forming interlocking frameworks of skeletons. We thus interpret these bioclastic limestones as coral-bearing shell-grounds rather than reefs.

Proceed on foot up Fossil Canyon and down-section through Garnet Formation conglomerates, greenish volcanic sandstones of the Alverson Formation, and Alverson basalt. Cross a fault into metamorphic basement, then back into Alverson. Note the absence of Red Rock Formation. Turn right up a steep mining road known as Tunnel Road. Here, densely fossiliferous Andrade limestone, much like that at **STOP 3-1b**, sits directly on Alverson volcanics; Garnet conglomerate is absent. At the top of Tunnel Road, walk north and cross a fault back into metamorphic basement which forms the eastern wall of Fossil Canyon, and continue east up a very steep, boulder-filled gulley (to the left, i.e. north, of Tunnel Road).

At the top of the canyon wall, proceed northeast over a low ridge into Hidden Valley. Here, reddish Garnet Conglomerate rests directly on basement; both Red Rock and Alverson are now absent. The low, rolling hills and limited exposure are typical of the Garnet Formation in the Coyote Mountains. Continue down to the sandy wash and follow it downstream. Marine Andrade sandstone overlies the Garnet in small patches. The drainage we are following steepens down the flank of the Coyote Mountains and joins the Garnet Canyon drainage. Before it does so, stop at an outcrop on the right (north) side of the wash.

STOP 3-1c. Garnet Wash Overlook

This outcrop displays several key stratigraphic relationships in a small area which are representative of this part of the Coyote Mountains. First, reddish conglomerate of the Garnet Formation rests directly on basement; Red Rock and Alverson are again absent. Within the space of this outcrop, the Garnet itself pinches out toward the east to where Andrade rests directly on basement. Andrade sandstone overlies the Garnet but grades upward and eastward into an arenaceous limestone, also within the space of this outcrop. To the north, this limestone facies of the Andrade Member laps out against a hill composed of marble. Two additional lines of evidence suggest that

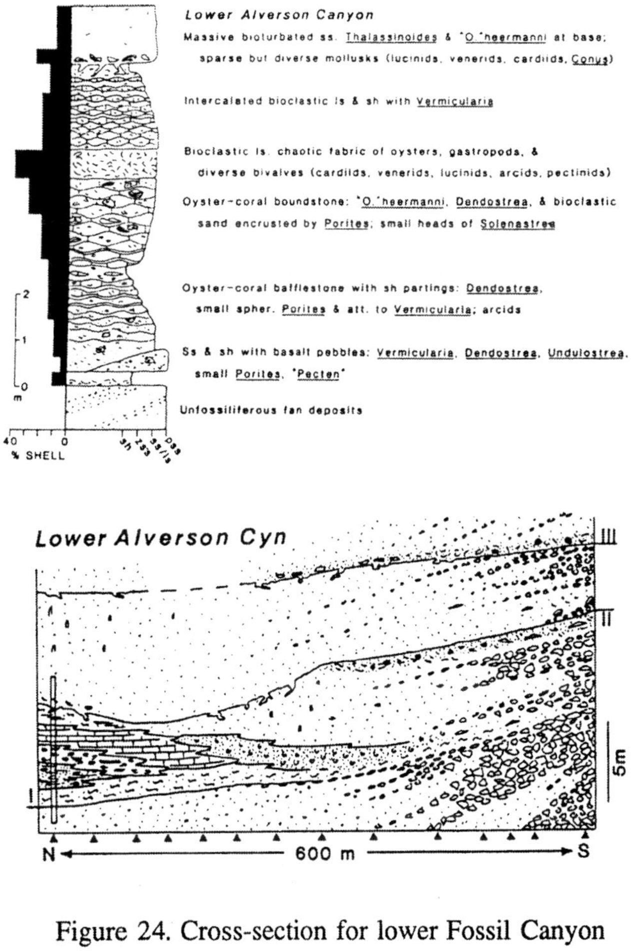

Figure 24. Cross-section for lower Fossil Canyon (**STOP 3-1b**) based on measured sections and individual bed correlations, from Kidwell (1988).

this marble hill represents a paleo-island in the Andrade sea. First, the Andrade limestone contains a bed of marble boulders, presumably shed from a very nearby source. Second, borings of marine pholad bivalves can be found in the marble adjacent to the contact with Andrade and in some boulders within the Andrade.

Climb this marble hill and paleo-island for an overlook of Garnet Wash and Coyote Mountains. To the east are other paleo-islands which were identified in similar fashion, including Marble Mountain immediately to the northeast, which we will pass on the return hike. The Andrade flanking the marble paleo-islands consists of arenaceous, sparsely fossiliferous limestone similar to this outcrop. The limestone facies of the Andrade occurs distally to the distal pinchout of wedging Garnet Formation. These marine toes of Garnet Formation can be identified at several places in the Coyote Mountains (Fig. 11).

From the Garnet Wash overlook, proceed to the southeast across the intervening valley toward Marble Mountain to the jeep trail. Observe how Andrade limestone laps out against the base of Marble Mountain, as we saw at the overlook outcrop. Also note pholad borings in marble basement and marble-clast conglomerate in the Andrade. On the opposite side of Marble Mountain is another paleo-low with Andrade limestone and similar lap-out relationships, and beyond that another paleo-high. Return to Hidden Valley via the jeep trail and backtrack down to Fossil Canyon.

Downstream of **STOP 3-1b**, in the walls of Fossil Canyon, observe the rapid facies change from oyster-coral limestone and sandstone in the north to conglomerate in the south. We are looking at the lateral change from Andrade Member into Garnet Formation, and on a broader scale the time-transgressive nature of the contact between the marine Imperial and nonmarine Split Mountain Groups.

Return to vehicles and retrace the route back down Fossil Canyon Road toward Ocotillo, and turn right onto Imperial Highway (Route S-2). Drive 6 miles west on Imperial Highway, then turn right onto an unpaved road toward Andrade Canyon.

STOP 3-2. Andrade Canyon (optional)

Continue on foot up the old mining road. Along the way note exposures of fossiliferous Andrade sandstone (especially on your left, onlapping basement) and overlying muddy strata of the Deguynos Formation (marine deltaic succession). Approximately one mile up the old mining road, continue west on a faint, steep jeep trail (rather than turning north into the wash) and follow it for approximately one-half mile. This trail crosses the edge of an outcrop of Garnet Formation conglomerate, as described at **STOP 3-1**. This occurrence is an eastward-thinning wedge overlain by Andrade sandstone at its eastern toe. As the trail crosses onto the basement complex of schist and marble, watch for pholad borings.

From this vantage point we also have a good overview of the Andrade Member in Alpine Valley to our east. In Alpine Valley, Andrade sandstone is thicker than we have seen before and generally rests directly on basement. At Reba Ridge, about 2 miles east of here but hidden from view, the Andrade attains a thickness of 60 m. In this case, the Andrade seems to thicken distally from the toe of the Garnet Formation alluvial fan that we are standing on. In contrast, at Hidden Valley, Andrade at the toe of that Garnet Formation alluvial fan becomes calcareous distally and laps out against paleo-islands.

Backtrack down the steep trail to the main wash, and then walk north (downstream) in this wash to Andrade Canyon proper on the northern flank of the Coyotes Mountains (**STOP 3-2b**). Observe the buttress nonconformity of the thin conglomerate and thick sandstone on basement in the floor of the gorge, and the rapid facies change from conglomerate to sandstone northward away from basement.

Return to vehicles and drive back to the Imperial Highway. Turn right and drive approximately 5 miles to the Canyon Sin Nombre overlook on the right, just before Sweeney Pass. Park vehicles in the parking lot.

STOP 3-3. Canyon Sin Nombre Overlook

This stop affords an overlook of Split Mountain (north) and the Carrizo Impact Area (northeast) in the distance, summarized in Fig. 10. In the foreground we can also observe exposures of the nonmarine Palm Spring Group in the Vallecito Basin and Carrizo Badlands (Fig. 15). Immediately in front of us are variegated beds of the Olla Formation (Po), representing the interfingering of "C"-suite fluvial deposits (buff and reddish colors) with "L"-suite alluvium (gray and greenish colors). The Olla grades laterally and toward the right into the "C"-suite Diablo Formation (Pd), and upward and to our left into the "L"-suite Hueso Formation (PQh). The Hueso-Olla-Diablo facies tract is offset approximately 3 km dextrally by the Elsinore fault. The Olla Formation occurs at several localities in the western Salton Trough, which define a SSE-NNW orientation of the paleo-basin margin after palinspastic reconstruction (Fig. 14).

Return to vehicles and backtrack along the Imperial Highway to Ocotillo, turn right toward I-8 west, and return to San Diego.

ACKNOWLEDGMENTS

Research supported by NSF grants EAR 84-07740, 86-06254, and a NSF Presidential Young Investigator award.

REFERENCES CITED

Abbott, P.L., and D.A.Rightmer, 1995, Pliocene sturzstrom from the Fish Creek Mountains, Anza-Borrego Desert State Park, southern California: Geological Society of America Abstracts with Programs, v. 27, p. A-278.

Allen, G. P., Laurier, D., and Thouvenin, J., 1976, Sediment distribution patterns in the modern Mahakam delta, Proc. Indonesian Petroleum Association, 5th Annual Convention.

Allen, G. P., Laurier, D., and Thouvenin, J., 1979, Étude sédimentologique du delta de la Mahakam, Total (Compagnie Française des Pétroles) Notes et Mémoires, no. 15, 156 p.

Armitage, A. L., 1989, Composition of Holocene Colorado River sand, an example of mixed-provenance sand derived from multiple tectonic elements of the Cordilleran continental margin [M. S. thesis]: San Diego State University, 169 p.

Babcock, E. A., 1974, Geology of the northeast margin of the Salton Trough, Salton Sea, California: Geological Society of America Bulletin, v. 85, p. 321-332.

Barnard, F. L., 1968, Structural geology of the Sierra de los Cucapas, northeastern Baja California, Mexico, and Imperial County, California [Ph. D. dissertation]: University of Colorado, 155 p.

Bell-Countryman, P., 1984, Environments of deposition of the Pliocene Imperial Formation, Coyote Mountains, southwest Salton Trough, *in* Rigsby, C. A., ed., The Imperial Basin - Tectonics, Sedimentation and Thermal Aspects: Pacific Section, Society of Economic Paleontologists and Mineralogists, p. 45-70.

Blake, W. P., 1853, *in* Williamson, Lt. R. S., 1856, Reports of exploration and surveys to ascertain the most practicable and economical route for a railroad from the Mississippi River to the Pacific Ocean, v. 5: Senate Executive Document no. 78, U. S. 33rd Congress, 2nd Session.

Boehm, M. C. F., 1982, Biostratigraphy, lithostratigraphy, and paleoenvironments of the Miocene-Pliocene San Felipe marine sequence, Baja California norte, Mexico [M. S. thesis]: Stanford University, 326 p.

Boehm, M. C., 1984, An overview of the lithostratigraphy, biostratigraphy, and paleoenvironments of the late Neogene San Felipe marine sequence, Baja California, Mexico, *in* Frizzell, V. A., ed., Geology of the Baja California Peninsula: Pacific Section, Society of Economic Paleontologists and Mineralogists, v. 39, p. 253-265.

Buising, A. V., 1989, Depositional and tectonic evolution of the northern proto-Gulf of California and lower Colorado River, as documented in the Mio-Pliocene Bouse Formation and bracketing units, southeastern California and western Arizona [Ph.D. dissertation]: Santa Barbara, University of California, 196 p.

Christensen, A. D., 1957, Part of the geology of the Coyote Mountain area, Imperial County, California [M. A. thesis]: Los Angeles, University of California, 158 p.

Crowell, J. C., 1982, The tectonics of Ridge Basin, southern California, *in* Crowell, J. C., and Link, M. H., eds., Geologic history of Ridge Basin, southern California: Pacific Section, Society of Economic Paleontologists and Mineralogists, p. 25-41.

Crowell, J. C., and Sylvester, A. G., 1979a, Excursion guide, *in* Crowell, J. C., and Sylvester, A. G., eds., Tectonics of the juncture between the San Andreas Fault system and the Salton Trough, southeastern California: Geological Society of America, Annual Meeting, Guidebook, p. 141-167.

Dean, M. A., 1988, Genesis, mineralogy and stratigraphy of the Neogene Fish Creek Gypsum, southwestern Salton Trough, California [M. S. thesis]: San Diego State University, 150 p.

Dibblee, T. W., 1954, Geology of the Imperial Valley region, California, *in* Jahns, R. H., ed., Geology of California, Chapter 2: Geology of the Natural Provinces: California Division of Mines, Bulletin 170, p. 21-28.

Dibblee, T. W., 1984, Stratigraphy and tectonics of the San Felipe Hills, Borrego Badlands, Superstition Hills, and vicinity, *in* Rigsby, C. A., ed., The Imperial Basin - Tectonics, Sedimentation and Thermal Aspects: Pacific Section, Society of Economic Paleontologists and Mineralogists, v. 40, p. 31-44.

Dickinson, W. R., 1986, Interpreting provenance relations from detrital modes of sandstones, *in* Zuffa, G. G., ed, Provenance of Arenites: D. Reidel, p. 333-361.

Eberly, L. D., and Stanley, T. B., 1978, Cenozoic stratigraphy and geologic history of southwestern Arizona: Geological Society of America Bulletin, v. 89, p. 921-940.

Elders, W. A., Rex, R. W., Meidav, T., Robinson, P. T., and Biehler, S., 1972, Crustal spreading in southern California: Science, v. 178, p. 15-24.

Elders, W. A., 1979, The geological background of the geothermal fields of the Salton Trough, *in* Elders, W. A., ed., Geology and geothermics of the Salton Trough: Geological Society of America Annual Meeting 92, Field Trip 7, Guidebook, p. 1-19.

Engel, A. E. J., and Schultejann, P. A., 1984, Late Mesozoic and Cenozoic tectonic history of south central California: Tectonics, v. 3, p. 659-675.

Frost, E. G., and Shafiqullah, M., 1990, Pre-San Andreas opening of the Salton Trough as an extensional basin: K-Ar ages on regional detachment faults along the western margin of the Salton Trough: Geological Society of America Abstracts with Programs, v. 21, no. 5, p. 81.

Fuis, G. S., and Kohler, W. M., 1984, Crustal structure and tectonics of the Imperial Valley region, California, *in* Rigsby, C. A., ed., The Imperial Basin - Tectonics, Sedimentation, and Thermal Aspects: Pacific Section, Society of Economic Paleontologists and Mineralogists, v. 40, p. 1-13.

Gjerde, M. W., 1982, Petrography and geochemistry of the Alverson Formation, Imperial County, California [M. S. thesis]: San Diego State University, 85 p.

Guthrie, L. L., 1990, An internally standardized study of Cenozoic sand and sandstone compositions, Salton Basin, southern California: implications for rift basin evolution with emphasis on the Palm Spring and Imperial Formations [M. S. thesis]: San Diego State University, 184 p.

Gyllenhaal, E.D., and S.M. Kidwell, 1989, Growth histories of subspherical bryozoan colonies: Ordovician and Pliocene evidence for paleocurrent regimes and commensalism: Geological Society of America Abstracts with Programs, v. 21, p. A112.

Hawkins, J. W., 1970, Petrology and possible tectonic significance of late Cenozoic volcanic rocks, southern California and Baja California: Geological Society of America Bulletin, v. 81, p. 3323-3338.

Hely, A. G., Hughes, G. H., and Irelan, B., 1966, Hydrologic regimen of Salton Sea, California: U. S. Geological Survey Professional Paper 486-C, 32 p.

Ingle, J. C., 1984, Neogene marine stratigraphy and history of the Gulf of Califotnia (abstract): Pacific Section, Society of Economic Paleontologists and Mineralogists, 59th Annual Meeting, p. 76-77.

Johnson, N. M., Officer, C. B., Opdyke, N. D., Woodard, G. D., Zeitler, P. K., and Lindsay, E. H., 1983, Rates of late Cenozoic tectonism in the Vallecito-Fish Creek Basin, western Imperial Valley, California: Geology, v. 11, p. 664-667.

Kerr, D. R., 1982, Early Neogene continental sedimentation, western Salton Trough, California [M. S. thesis]: San Diego State University, 138 p.

Kerr, D. R., 1984, Early Neogene continental sedimentation in the Vallecito and Fish Creek Mountains, western Salton Trough, California: Sedimentary Geology, v. 38, p. 217-246.

Kerr, D. R., and Kidwell, S. M., 1991, Late Cenozoic sedimentation and tectonics, western Salton Trough, California, *in* Walawender, M. J., and Hanan, B. B., eds., Geological excursions in southern California and Mexico, Geological Society of America Annual Meeting, Guidebook, p. 397-416.

Kidwell, S.M., 1988, Taphonomic comparison of passive and active margins: Neogene shell beds of the Atlantic coastal plain and northern Gulf of California: Palaeogeography Palaeoclimatology Palaeoecology v. 63, p. 201-224.

Lindsay, L., and Lindsay, D., 1978, The Anza-Borrego Desert region: Berkeley, Wilderness Press, 165 p.

Lyle, M., and G.E. Ness, 1991, The opening of the southern Gulf of California, *in* Dauphin, J.P., and Simoneit, B.R.T., eds., The Gulf and Peninsular Province of the California: American Association of Petroleum Geologists Memoir 47, p. 403-423.

Mace, N. W., 1981, A paleomagnetic study of the Miocene Alverson Volcanics of the Coyote Mountains, western Salton Trough, California [M. S. thesis]: San Diego State University, 142 p.

Matti, J. C., Morton, D. M., and Cox, B. F., 1985, Distribution and geologic relations of fault systems in the vicinity of the central Transverse Ranges: U. S. Geological Survey, Open-File Report 85-365, 27 p.

Meckel, L. D., 1975, Holocene sand bodies in the Colorado Delta area, northern Gulf of California, *in* Broussard, M. L., ed., Deltas: Models for Exploration: Houston, Texas, Houston Geological Society, p. 239-265.

Merriam, R., and Bandy, O. L., 1965, Source of upper Cenozoic sediments in Colorado delta region: Journal of Sedimentary Petrology, v. 35, p. 911-916.

Metzger, D. G., 1968, The Bouse Formation (Pliocene) of the Parker-Blythe-Cibola area, Arizona and California: U. S. Geological Survey, Professional Paper 500-D, p. 126-136.

Miller, R. H., and Dockum, M. S., 1983, Ordovician conodonts from metamorphosed carbonates of the Salton Trough, California: Geology, v. 11, p. 410-412.

Minch, J. A., and Abbott, P. L., 1973, Post-batholithic geology of the Jacumba area, southeastern San Diego County, California: San Diego Society of Natural History Transactions, v. 17, p. 129-136.

Muffler, L. J. P., and Doe, B. R., 1968, Composition and mean age of detritus of the Colorado River delta in the Salton Trough, southeastern California: Journal of Sedimentary Petrology, v. 38, p. 384-399.

Nuffer, C. D., 1989, Compositional variation of Holocene sand deposited in a transtensional rift system, Salton Basin, California [M. S. thesis]: San Diego State University, 101 p.

Pappajohn, S., 1980, Description of Neogene marine section at Split Mountain, easternmost San Diego County, California [M. S. thesis]: San Diego State University, 77 p.

Potter, P. E., 1978, Petrology and chemistry of modern big river sands: Journal of Geology, v. 86, p. 423-449.

Quinn, H. A., and Cronin, T. M., 1984, Micropaleontology and depositional environments of the Imperial and Palm Spring Formations, Imperial Valley, California, *in* Rigsby, C. A., ed., The Imperial Basin - tectonics, sedimentation, and thermal aspects: Pacific Section, Society of Economic Paleontologists and Mineralogists, v. 40, p. 71-85.

Reineck, H.-E., and Wunderlich, F., 1968, Classification and origin of flaser and lenticular bedding: Sedimentology, v. 11, p. 99-104.

Robinson, J. W., and Threet, J. L., 1974, Geology of the Split Mountain area, Anza-Borrego Desert State Park, eastern San Diego Sounty, California, *in* Hart, M. W., and others, eds., Recent Geologic and Hydrologic Studies, Eastern San Diego County and Adjacent Areas: San Diego Association of Geologists, Guidebook, 101 p.

Ruisaard, C. I., 1979, Stratigraphy of the Miocene Alverson Formation, Imperial County, California [M. S. thesis]: San Diego State University, 125 p.

Sawlan, M. G., 1991, Magmatic evolution of the Gulf of California rift, *in* Dauphin, J. P., and Simoneit, B. R. T., eds., The Gulf and Peninsular Province of the California: American Association of Petroleum Geologists Memoir 47, 301-369.

Sawlan, M. G., and Smith, J. G., 1984, Petrologic characteristics, age and tectonic setting of Neogene volcanic rocks in northern Baja California Sur, Mexico, *in* Frizzell, V. A., ed., Geology of the Baja California Peninsula: Pacific Section, Society of Economic Paleontologists and Mineralogists, v. 39, p. 237-251.

Schultejann, P. A., 1984, The Yaqui Ridge antiform and detachment fault: mid-Cenozoic extensional terrane west of the San Andreas fault: Tectonics, v. 3, p. 677-691.

Sentianin, B. A., 1989, Petrology and provenance of sandstones in the Pleistocene Borrego Formation, Salton Basin, California, 79 p.

Shafiqullah, M., Damon, P. E., Lynch, D. J., Reynolds, S. J., Rehrig, W. A., and Raymond, R. H., 1980, K-Ar geochronology and geologic history of southwestern Arizona and adjacent areas: Arizona Geological Society Digest, v. 12, p. 201-259.

Sharp, R. V., 1967, San Jacinto fault zone in the Peninsular Ranges of southern California: Geological Society of America Bulletin, v. 78, p. 705-730.

Simpson, C., 1984, Borrego Springs-Santa Rosa mylonite zone: a Late Cretaceous west-directed thrust in southern California: Geology, v. 12, p. 8-11.

Smith, J.T., 1991, Cenozoic marine mollusks and paleogeography of the Gulf of California, *in* Dauphin, J. P., and Simoneit, B. R. T., eds., The Gulf and Peninsular Province of California: American Association of Petroleum Geologists Memoir 47, p. 637-666.

Smith, P. B., 1960, Fossil foraminifera from the southeastern Californian deserts: U. S. Geological Survey Professional Paper 400-B, p. B278-B279.

Smith, P. B., 1970, New evidence for a Pliocene marine embayment along the lower Colorado River area, California and Arizona: Geological Society of America Bulletin, v. 81, p. 1411-1420.

Stock, J. M., and Hodges, K. V., 1989, Pre-Pliocene extension around the Gulf of California and the transfer of Baja California to the Pacific Plate: Tectonics, v. 8, p. 99-115.

Stump, T. E., 1972, Stratigraphy and paleontology of the Imperial Formation in the western Colorado Desert [M. S. thesis]: San Diego, California State University, 132 p.

Thompson, R. W., 1968, Tidal flat sedimentation on the Colorado River delta, northwestern Gulf of California: Geological Society of America Bulletin, Memoir 107, 133 p.

Thompson, R. W., 1975, Tidal flat sediments of the Colorado delta, northwestern Gulf of California, *in* Ginsburg, R. N., ed., Tidal Deposits: New York, Springer-Verlag, p. 57-65.

Van Andel, Tj. H., 1964, Recent marine sediments of Gulf of California, *in* Van Andel, Tj. H., and Shor, G. G., eds., Marine Geology of the Gulf of California: American Association of Petroleum Geologists, Memoir 3, p. 216-310.

Van de Kamp, P. C., 1973, Holocene continental sedimentation in the Salton Basin, California: a reconnaissance: Geological Society of America Bulletin, v. 84, p. 827-848.

Waters, M. R., 1983, Late Holocene lacustrine chronology and archeology of ancient Lake Cahuilla, California: Quaternary Research, v. 19, p. 373-387.

Watkins, R., 1990a, Paleoecology of a Pliocene rocky shoreline, Salton Trough region, California: Palaios, v. 5, p. 167-175.

Watkins, R., 1990b. Pliocene channel deposits of oyster shells in the Salton Trough region, California. Palaeogeography Palaeoclimatology Palaeoecology v. 79, p. 249-262.

Weide, D. L., 1976, Regional environmental history of the Yuha Desert, *in* Wilke, P. J., ed., Background to Prehistory of the Yuha Desert Region: Ramona, California, Ballena Press, p. 9-20 and 95-97.

Wilke, P. J., 1978, Late prehistoric human ecology at Lake Cahuilla, Coachella Valley, California: University of California Archeological Research Facility Contributions, no. 38, 168 p.

Winker, C. D., and Kidwell, S. M., 1986, Paleocurrent evidence for lateral displacement of the Pliocene Colorado River delta by the San Andreas fault system, southeastern California: Geology, v. 14, p. 788-791.

Winker, C. D., 1987, Neogene stratigraphy of the Fish Creek-Vallecito section, southern California: implications for early history of the northern Gulf of California and Colorado Delta [Ph.D. dissertation]: Tucson, University of Arizona, 494 p.

Winterer, J. I., 1975, Biostratigraphy of Bouse Formation: a Pliocene Gulf of California deposit in California, Arizona, and Nevada [M. S. thesis]: Long Beach, California State University, 132 p.

Woodard, G. D., 1963, The Cenozoic succession of the west Colorado Desert, San Diego and Imperial Counties, southern California [Ph. D. dissertation]: Berkeley, University of California, 173 p.

Woodard, G. D., 1974, Redefinition of Cenozoic stratigraphic column in Split Mountain Gorge, Imperial Valley, California: American Association of Petroleum Geologists Bulletin, v. 58, p. 521-539.

Is the Fluoride Content Too High in the Drinking Water at San Diego County's Agua Caliente Regional Park?

Carole L. Ziegler
Southwestern College, 900 Otay Lakes Road, Chula Vista, California 91910-7299
cziegler@swccd.edu

ABSTRACT

Addition of fluoride to public drinking waters has been a subject of controversy since it was first shown that fluoridated waters prevent tooth decay. The fluoride content of the ground water at Agua Caliente Regional Park located in San Diego County, California, is higher than the 1.6 parts per million (ppm) standard for public drinking water systems in San Diego County. Studies in communities where the fluoride level is as high as 10 ppm have shown no adverse effects. However, since the fluoride level at Agua Caliente is greater than 5 ppm it may be detrimental to people whose health has been compromised by disease.

Since the time of the Haawii Indians, whose name means "water in rock", humans have benefited from the waters found at Agua Caliente Regional Park located in San Diego County, California. The area is traversed by the Elsinore Fault and due to tectonic processes the rock is highly fractured. Water that possibly comes from the Laguna Mountains is able to percolate down into the ground where it is then heated and returns to the surface. Mike Walawender of SDSU suggests that this water may be traveling quite deep due to the temperatures of the waters as they exit from springs.

Today the San Diego County Park System operates a campground at Agua Caliente Regional Park that utilizes the water from the springs for recreational, perceived medical, and drinking water purposes. An indoor therapy spa is maintained at 102°F by mechanical heating. There is also an outdoor wading pool that is maintained at a temperature of 90°F with no additional energy use due to the exiting temperature of a nearby spring. This spring is located behind some outdoor showers near the pool and has an exiting temperature of 98°F. This is the warmest of the springs in the area with most springs not any cooler than 88°F. There is one spring that has a temperature of 76°F so perhaps the source water for this spring does not circulate as deep. Ranger Mark Stockdale notes that the drinking water well at his home nearby in Canebrake registers 105°F at 500 feet below the surface.

At the entrance to the therapy pool, there is a sign posted warning that the fluoride content is high and drinking the water may be detrimental to one's health. Because Agua Caliente Regional Park falls under the description of having a public drinking water system, the waters must be tested for certain constituents. Some constituents, such as fluoride and zinc, need only be tested once. Nitrites are tested every three years and nitrates are tested every year. The waters are tested four times per year for *E. coli*. Water sampled on January 29, 1991 showed a fluoride level of 5.56 milligrams per liter (mg/L) or parts per million (ppm) and water sampled on November 15, 1999, showed a fluoride level of 5.1 ppm. These levels are considered high. San Diego County standards for public drinking water systems should be less than 1.6 ppm. The fluoride is possibly being brought in by ground water from other sources, or deep seated circulation of ground water is dissolving the fluoride out of the two types of micaceous granites associated with the rocks of the La Posta Pluton. Interesting to note is that the spring waters at Vallecito County Park which is also in the area shows a much lower fluoride content at 0.56 ppm.

But is the fluoride content too high? It has been shown that communities with fluoridated water have anywhere from 15 to 40% less tooth decay than communities without fluoridated water. Bone is composed of calcium hydroxyapatite $Ca_5(PO_4)_3(OH)$, and teeth are mostly composed of another related hexagonal hydroxyapatite. When fluoride is ingested it replaces a hydroxyl group in these calcium phosphates forming fluorapatite, $Ca_{10}(PO_4)_6(F,OH)_2$. This fluorapatite is an essential ingredient in strong bone formation. When fluoride is present in saliva it acts as a bacteriacide. It also aids in the buffering action of saliva which neutralizes the acids associated with plaque. These acids cause demineralization of teeth. The buffering action allows remineralization of the teeth at which time fluorapatite is deposited in the tooth enamel. Fluorapatite is more resistant to demineralization than hydroxyapatite. Furthermore, fluorapatite surfaces have fewer defects, such as pits and fissures, which can provide sites for tooth decay to begin.

The ideal level of fluoride in drinking water for the prevention of tooth decay and strengthening of bones has been determined to be 1 ppm. With this in mind the American Academy of Pediatric Dentistry has recommended that water containing less than 0.3 ppm fluoride be fluoridated to a level of 1 ppm.

Some adverse effects have been noted when someone is receiving too much fluoride. The most common effect which, in general, is not harmful is fluorosis causing mottling or staining of the teeth. In some cases, however, it may be an indication that a person is ingesting too much fluoride and that it has reached toxic levels. Fluoride toxicity can lead to weight loss and anemia. Another interesting side effect is the increase in bone mass of the spine when fluoride is used to treat osteoporosis. The size and structure difference of fluorapatite from hydroxyapatite causes a decrease in mechanical efficiency, so even though fractures of the spine are reduced, fractures of the hip, ankle, knee and foot may increase. The administration of fluoride compounds for medical purposes is often the reason for development of these and other side effects. Serious adverse effects from naturally fluoridated waters are not generally noted until the drinking water contains greater than 10 ppm fluoride.

Since the waters at Agua Caliente contain just over 5 ppm fluoride it would appear that, in general, using the water at Agua Caliente for drinking purposes does not present a problem. However, some people visit Agua Caliente because they have some medical problems and they believe in the therapeutic value of mineralized hot spring water. They may have diseases that are exacerbated by the ingestion of extra fluoride. Fluoride that is not taken up in bone and teeth is filtered through the kidney. People with compromised kidney function should avoid ingesting extra amounts of fluoride. Some medications may enhance fluoride activity in the body, or affect the body's ability to filter out fluoride. These people should consider bringing their own water for drinking purposes if they intend to spend any length of time at Agua Caliente.

Constituent Results from Testing for Water Quality at Agua Caliente Regional Park

Constituent	Level, ppm	Maximum Acceptable Level For Drinking Water, ppm
Fluoride	**5.1, 5.56**	1.6
Chloride	120.2	500
Sulfate	58.9	500
Nitrate	<0.10	45
Nitrite	<0.05	10
Copper	0.05	1.0
Iron	0.1	0.1
Manganese	0.03	0.05
Zinc	0.01	5.0
Foaming Agents	0.5	0.5
Total Dissolved Solids	478	1000

Note 1: One of my earliest encounters with geochemistry was my mother telling me how she never had a cavity until well into her twenties due to the naturally high fluoride content in the drinking water where she was raised in Arkansas. This in turn sparked my interest many years later to look into what the effects might be with high ingestion of fluoride when visiting Agua Caliente.

Note 2: Due to the interest in the water chemistry by the San Diego Association of Geologists and myself in particular David Martinez, Manager of San Diego County Parks, is requesting that fluoride be retested when normal testing of the waters at Agua Caliente occurs in January 2007.

REFERENCES

Annual Water Quality Reports for Agua Caliente County Park, 1998, 2000 and 2004 provided by David Martinez, San Diego County Parks Manager.

Calonge, N. "Prevention of dental caries in preschool children: recommendations and rationale", *American Journal of Preventive Medicine*, 2004, May: 26 (4):326-9.

"Clinical guideline on fluoride therapy", Chicago (IL): American Academy of Pediatric Dentistry, 2003. 2 p.

Lindsay, Diana, *Anza-Borrego A to Z: People, Places and Things*, 2000, Sunbelt Publications, Inc., San Diego, California.

"Oral Health Resources", US Centers for Disease Control (CDC), available at www.cdc.gov/oralhealth/data_systems/index.htm.

"Recommendations for using fluoride to prevent and control dental caries in the United States", Centers for Disease Control and Prevention, MMWR Recomm. Rep 2001, August 17:50 (RR-14):1-42.

Stockdale, Mark, Ranger, Agua Caliente County Park – personal communication.

Walawender, Michael J., *The Peninsular Ranges, A Geologic Guide to San Diego's Back Country*, 2000, Kendall/Hunt Publishing Company, Dubuque, Iowa.

Walawender, Michael J. – personal communication.

Ziegler, M.D. Michael G. – personal communication.

Morphology, Mineralogy, and Geochemistry of Oolitic Manganite-rich Nodules from the Upper Miocene Fish Creek Gypsum, Imperial County, California

David G. Peterson Jr.*, National Gypsum Company, Technology Innovation Center, 5901 Carnegie Boulevard, Charlotte, North Carolina 28209,* dgpeterson@nationalgypsum.com

George T. Jefferson*, Colorado Desert District Stout Research Center, California Department of Parks and Recreation, 200 Palm Canyon Drive, Borrego Springs, California 92004,* gjefferson@parks.ca.gov

Bruce A. Hudgens*, United States Gypsum Company Research Center, 700 North Highway 45, Libertyville, Illinois 60048*

ABSTRACT

Two fist-sized oolitic manganite-rich nodules were recovered from a previously excavated part of the United States Gypsum Company quarry. This large excavation lies within the upper Miocene Fish Creek Gypsum deposit (Dibblee, 1954), located on the west side of the Salton Trough in Imperial County, southern California. Both nodules are roughly ellipsoidal. The larger one is prolate measuring 14.4 cm along its major axis and 7.5 cm along its minor axis (diameter). The other is oblate with a 10 cm major axis (diameter) and 7 cm minor axis.

The origin of the Fish Creek Gypsum, previously considered a sabkha-like surface evaporite of marine brines (Sharpe and Cork, 1995) or evaporation from a restricted shallow-marine embayment (Dean, 1996), has been reevaluated (Peterson and Jefferson, 1997; Escamilla, 2003). An alternative hypothesis suggests that the Fish Creek Gypsum is an alteration product of anhydrite that precipitated from hydrothermal fluids in a subaqueous, marine environment of shallow to moderate depths (Peterson and Jefferson, 1997).

The nodules are composed of 25 to 75 percent steel gray and black manganite ooliths, <1.0 mm in diameter, with 26 to 52 percent gypsum-filled interstices. Both nodules were found isolated on the surface (not *in situ*) in an area within the quarry that contains deposits of dark gray to black gypsum and anhydrite enriched with pyrolusite. Pyrolusite is common in the Fish Creek Gypsum and found throughout the gypsum and anhydrite in dull sooty, earthy masses. The Fish Creek Gypsum nodules are similar in composition and structure to oolitic manganese ores of the Tshiaturi type (Baumann, 1976), suggesting they may have formed in a shallow marine environment (Force et al., 1999), with a hydrothermal fluid source for the manganese.

INTRODUCTION

Two oolitic manganite-rich nodules were discovered in the United States Gypsum (USG) quarry within the Fish Creek Gypsum, an extensive gypsum ($CaSO_4 \cdot 2H_2O$) and anhydrite ($CaSO_4$) deposit in the Fish Creek Mountains, Imperial County, California (Figure 1). Nodule "A" was

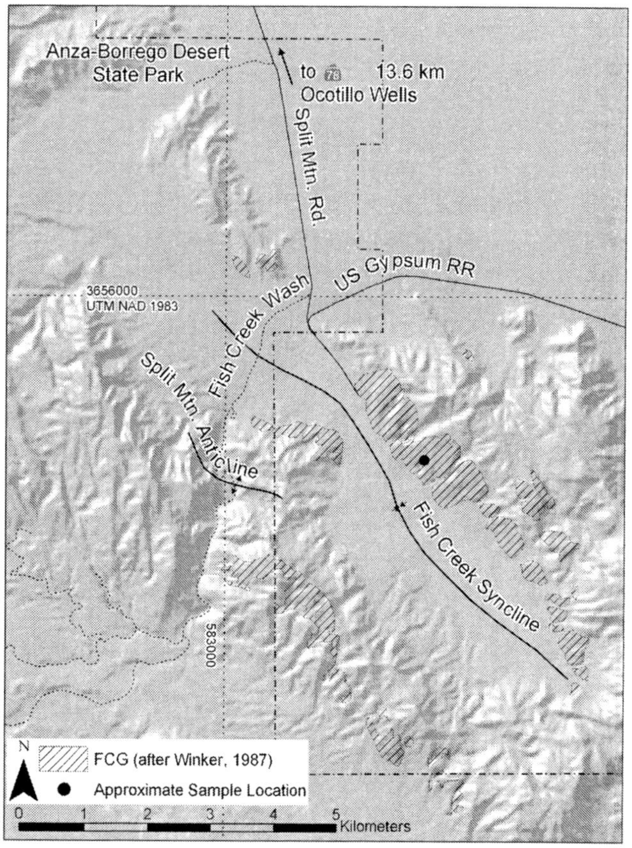

Figure 1. Vicinity Map, showing the approximate location where the two manganite-rich nodules were collected, in the Fish Creek Gypsum deposit at the United States Gypsum quarry, Imperial County, California.

discovered in the fall of 1997 and a second, smaller nodule "B" was found in the same area in the fall of 2000. Both nodules were recovered not *in situ* from a previously excavated bench above the floor of the USG quarry in an area of dark gray to black gypsum and anhydrite enriched with pyrolusite (MnO_2).

Pyrolusite is an oxidation product of weathered manganese minerals such as manganite [MnO(OH)]. Pyrolusite is common in the Fish Creek Gypsum (FCG) and found throughout the gypsum and anhydrite in dull, sooty, earthy masses. Samples of gypsum from the FCG contain as much as 13 percent pyrolusite. One black selenitic gypsum specimen contained 2 percent anhydrite, 3 percent calcite ($CaCO_3$), 1 percent manganese sulfide, and 2 percent manganese dioxide. Dendritic pyrolusite is frequently exposed on the surface of the anhydrite within the FCG deposit.

BACKGROUND

Deep-sea manganese nodules were first recovered during the 1873 *Challenger* expedition southwest of the Canary Islands in the Atlantic Ocean. Manganese nodules generally form slowly at the sediment-water interface in deep ocean basins with low sedimentation rates, and are often found near spreading centers and hydrothermal activity. Nodules differ in morphology, mineralogy, and chemical composition (see Table 1). Minor amounts of copper, nickel, zinc, cobalt, and rare earth elements have been found in manganese nodules. Variations are a result of the different sources of constituent elements, and local depositional settings (Baturin and Yemel'yanov, 1991). Individual nodules are usually potato-sized and shaped (Rogers, 1995), earthy brown to bluish black lumps of ferromanganese oxides layered concentrically around one or more nuclei of detrital materials (Usui et al., 1993).

GEOLOGIC SETTING

Divergent motion along the contact of the Pacific and North American plates has caused the Salton Trough and Gulf of California to open up by oblique rifting and extension over the past approximately 12 to 14 Ma (Winker and Kidwell, 1996). Development of the Trough has been controlled by northwest-southeast trending, right lateral, transform faults. The Fish Creek Gypsum is late Miocene, about 6.0 to 5.5 Ma. This formation is located on the western margin of the Salton Trough, 12 km south of Ocotillo Wells, and adjacent to Anza-Borrego Desert State Park. The deposit lies between the northwest-southeast trending Elsinore and San Jacinto fault zones, and west of an active spreading center near the south end of the Salton Sea (Elders et al., 1986). Late Pleistocene extrusive volcanic and geothermal activity is centered at the southeastern edge of the Salton Sea along an active spreading center. This surface activity likely is the result of the shallow emplacement of sheet-like dikes along "leaky" transform faults (Sharp and Cork, 1995).

Situated on the east flank of the Fish Creek Mountains, the Fish Creek Gypsum United States Gypsum Company quarry is on the northeast limb of the Fish Creek syncline, east of the Split Mountain anticline (Sharp and Cork, 1995) (see Figure 1). The approximately 80 m thick and laterally extensive, 5.8 km along strike, gypsum deposit and an associated celestite ($SrSO_4$) deposit occur as lens-shaped bodies stratified within marine sediments at the base of the Latrania Formation, the lowest member of the Imperial Group (Hanna, 1926; Winker, 1987; Dean, 1988; Remeika, 1998). At this location, the FCG stratigraphically overlies granodiorite basement (USG core sample analyses), the Elephant Trees Conglomerate (Winker and Kidwell, 1996), and the Split Mountain Sturzstrom (Kerr and Abbott, 1996).

Several hypotheses have been proposed for the origin of this large gypsum and anhydrite deposit. Sharp and Cork

Table 1. Percent metal content of manganese nodules.

	Pacific	Indian O.	Atlantic	Overall maximum	Overall minimum
Mn	17.2	14.9	13.6	34	5.41
Fe	11.8	14.6	15.5	26.32	4.36
Ni	0.63	0.38	0.33	2	0.13
Co	0.36	0.31	0.24	2.57	0.045
Cu	0.36	0.17	0.16	2.5	0.028

Notes: Table modified from Seibold and Berger (1996, Table 10.2).

(1995) described a scenario in which the Fish Creek Gypsum formed above the high-tide line in a hyperarid environment near the margin of the proto-Gulf of California:

> Evaporite minerals were deposited in a coastal sabkha environment along the margin of a sub-basin or embayment of the Salton Trough. The sabkha is formed in a hot, dry climate with high evaporation rates. The seawater is 'pumped' laterally through the wedge of tidal flat sediments by evaporative reflux. High evaporation rates draw the pore water to the surface of the sabkha by capillary action. Diagenic minerals, such as gypsum, anhydrite, and carbonates, are precipitated when the pore water becomes saturated. Anhydrite is precipitated in the supratidal zone (above the mean high water level). Anhydrite and gypsum grows interstitially by pushing aside the soft, fine-grained sediments. Upon burial, the gypsum may react with pore water to form nodules of anhydrite crystals. Post-depositional burial and high crustal heat flow rates converted all of the gypsum to anhydrite. Recent uplift and exposure allowed groundwater to circulate through joints, fractures, and faults, converting the anhydrite to gypsum. (Sharp and Cork, 1995).

Dean (1996) proposed formation of the FCG primarily by evaporation in a partially isolated shallow-marine environment where limited mixing occurred with the larger Gulf of California:

> [The Fish Creek] Gypsum is interpreted to have precipitated in a restricted shallow-water basin. During early diagenesis and deep burial the gypsum was altered to anhydrite. Within the past one million years, the FCG was uplifted, exhumed, then partially gypsified by the downward percolation of meteoric waters.
>
> The FCG is one of several thick Mio-Pliocene calcium sulfate deposits occurring between non-marine and marine strata located along the Neogene eastern [*sic*] coastline of the Gulf of California. It appears that these evaporites formed in response to the extensional-rift tectonics associated with the opening of the Gulf of California. (Dean, 1996, p. 123).

Contrary to these previous models, the massive amounts of calcium sulfate, strontium sulfate, and anhydrite in the FCG are herein interpreted as the result of the alteration of anhydrite precipitated from the localized low- and high-temperature hydrothermal vent fluids and hydrothermal mound building in a subaqueous setting.

Along the west side of the Fish Creek syncline, local interfingering of the base of the FCG with conglomeratic, coarse-grained sandstones in the top of the underlying Elephant Trees Conglomerate, suggests moderate water depths to nearshore depositional environments on a relatively steep slope, not an evaporative tidal flat. Also on the west limb of the Fish Creek syncline, invertebrates of the Typanites ichnofauna (Jefferson et al., 2006) are found at the base of the Latrania Formation, about 1 km along strike from where the FCG pinches out between the base of the Latrania Formation and the top of the Elephant Trees Conglomerate. This stratigraphic relationship suggests that the base of the FCG and the lowest deposits of the Latrania are roughly coeval, both having been deposited on top of the Elephant Trees Conglomerate. These organisms indicate that nearshore, well oxygenated tropical marine conditions were present when the gypsum was deposited. Furthermore, if late Miocene paleoclimatic conditions were similar to those evidenced by the early Pliocene local Carrizo paleoflora (Remeika, 2006), then the estimated 500 to 625 mm of annual precipitation precludes a dry sabkha-like environment.

Within the USG quarry, bedding is 0.5 to 1.0 m-thick, massive, and commonly occurs in large, 4 to 6 m diameter, mound-shaped structures. The gypsum and anhydrite occur in local white, black, gray, and red, originally vertical structures that are 2 to 5 m in diameter. These 20+ m tall structures may represent chimney accumulations that emanated from the bottom of the deposit. Anhydrite occurs throughout the FCG as "finger-like" structures that rise from the base of the FCG where it is locally in contact with granodiorite basement. Anhydrite is a well-documented major constituent of hydrothermal vent black and white smokers, and an integral part of chimney development and mound formation in seafloor hydrothermal systems (Butterfield et al., 1990).

The FCG exhibits a significant absence of calcite, dolomite, and halite. These minerals are usually associated with calcium sulfate that has precipitated as a surface evaporite. Within the Latrania Formation, and on strike to the northwest of the gypsum deposit, is a large lens-shaped body of celestite (Durrell, 1953). Deposits of celestite generally are not associated with evaporite minerals. Furthermore, the elevated levels of trace elements such as copper, gold, silicon, iron, manganese, strontium, silver, barium, and lead within the FCG are much higher than in evaporites derived from normal seawater. Such levels are comparable to the trace metals in minerals precipitated from localized hydrothermal vent fluids in modern submarine hydrothermal mound deposits (Butterfield et al., 1990; Peterson and Jefferson, 1997).

Euhedral biotite and phlogopite mica crystals, interpreted to be authigenic, occur imbedded within single crystals of anhydrite, gypsum, and celestite, and metallic sulfides have been observed and detected in gypsum samples. In geothermal wells near the Salton Sea, authigenic biotite is presently forming from hydrothermal brines in an anhydrite-rich, zinc-bearing zone at depths of 2485 to 2955 m and under temperatures of 300° to 325° C (Shearer, 1988).

Recent work by Escamilla (2003) supports the idea that hydrothermal activity was involved in the formation of the FCG. Escamilla documented structures and textures in samples, high concentrations of some elements, and the presence of euhedral biotite, celestite, chlorite, epidote, and hematite that indicate a hydrothermal influence on the FCG.

PHYSICAL CHARACTERISTICS

Nodule "A" is shaped roughly prolate ellipsoidal with a 14.4 cm major axis and a minor axis (diameter) of approximately 7.5 cm. It weighs 1,025 g. The subrounded nodule is encrusted on one side (herein assumed to be the top) with a layer of white gypsum, which varies in thickness from <1.0 to 8 mm. The white gypsum is covered in areas with light and dark gray, and earthy brown minerals. The sides have a predominately dark gray, <0.3 mm thick mineral layer that covers a <0.5 mm thick layer of the white gypsum. The assumed bottom is relatively flat and is also covered with a <0.3 mm thick, dark gray mineral layer, that is over a <0.5 mm thick layer of white gypsum (Photograph 1).

Photograph 1. Oolitic manganite nodule "A". Scale is 10 cm. Videograph by D. Peterson.

The core is earthy brown and dark gray to black (see detail Photograph 2), consisting of steel gray and black sub-metallic manganite ooliths <1.0 mm in diameter that are cemented together with a white gypsum matrix (see detail Photograph 3).

Photograph 2. Cross section through the core of oolitic manganite nodule "A". Scale is 10 cm. The core is earthy brown and dark gray to black. Videograph by D. Peterson.

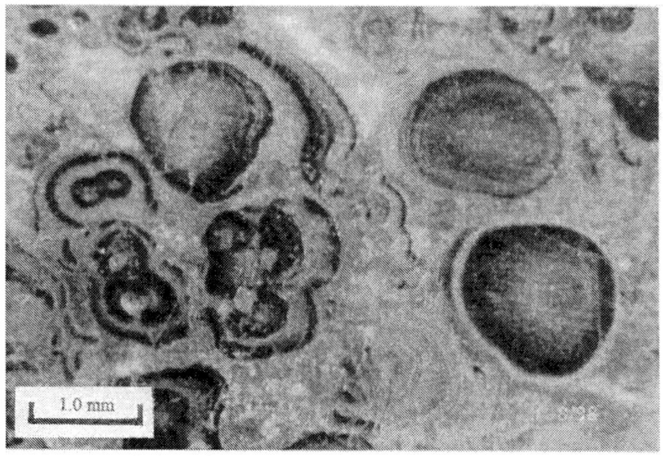

Photograph 3. Cross section through the core of oolitic manganite nodule "A". Steel gray and black, concentrically banded manganite ooliths, < 1.0 mm in diameter, with gypsum interstices were taken at 20x magnification. Videograph by D. Peterson.

Nodule "B" is shaped roughly oblate spheroidal with a major axis (diameter) of approximately 10 cm and a minor axis of 7.0 cm. It weighs 634.3 g. The cross section of the core viewed with a stereo microscope reveals identical dark gray manganite ooliths, <1.0 mm in diameter, cemented together with gypsum as seen in nodule "A" (Photograph 4).

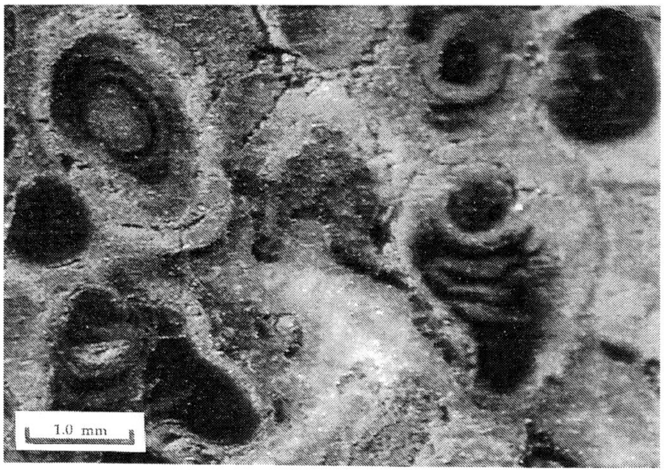

Photograph 4. Cross section of oolitic manganite nodule "B". Steel gray and black concentrically banded manganite ooliths, < 1.0 mm in diameter, with gypsum interstices were taken at 20x magnification. Video micrograph by D. Peterson.

ANALYTICAL METHODS

Polished cross sections through the cores of nodules "A" and "B" were examined under 40x magnification, reflected light, dark-field microscopy.

Sectioned samples were taken through the smaller diameters of each nodule. These samples were divided into thirds (Figure 2) for elemental and mineral composition analyses.

The mineralogy of both nodules was determined using a Scintag X-1 X-ray diffractometer (XRD) at USG's Research and Technology Center (RTC). The SiroQuant quantitative X-ray diffraction computer program was used for the analyses. The X-ray diffraction patterns were recorded, the mineral phases identified and the concentrations calculated employing the Riqas quantitative X-ray diffraction computer program (Table 2).

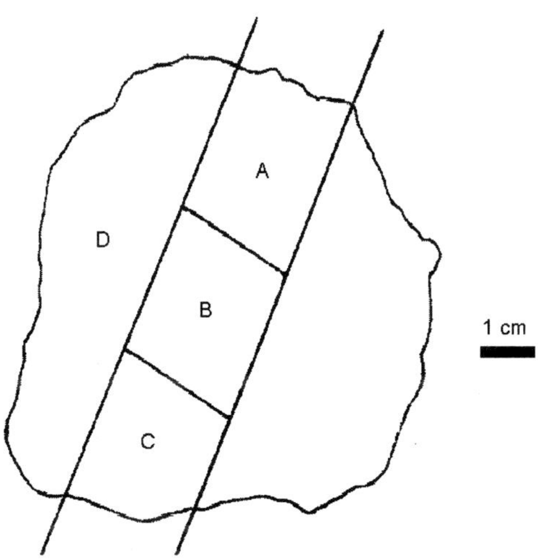

Figure 2. Oolitic manganite nodule "A" analytic sample placement on core-diameter cross section from the approximate middle and divided into thirds: (A) top third, (B) middle third, and (C) bottom third.

Sub-samples were removed from each third of nodule "A" and sent to Construction Technology Laboratories, Inc. (CTL), and analyzed by X-ray fluorescence spectroscopy (XRF). Semi-quantitative XRF analyses were also run using the Uni-Quant II computer program at USG's RTC.

A portion of nodule "A" was coated with carbon for scanning electron microscope/energy dispersive X-ray analysis (SEM/EDX). The SEM used at USG's RTC is a JOEL JSM-840A. A sample taken from the left side of the middle section of nodule "A" (Figure 3) was photographed under SEM at 20 KeV at 100x magnification, and, using electron imaging backscatter, an elemental analysis was performed.

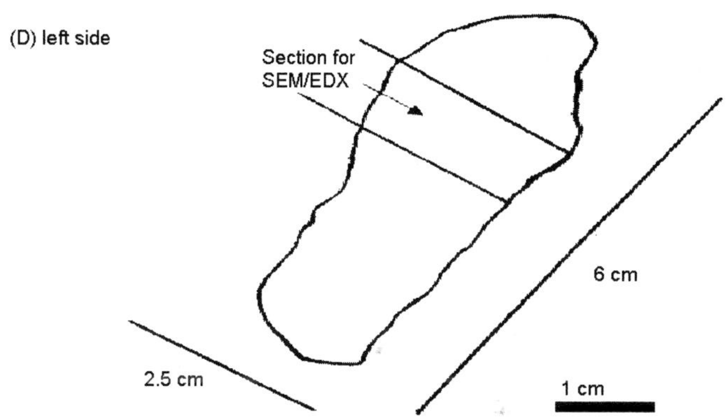

Figure 3. SEM/EDX trace element analysis of a sample of nodule "A" through diameter cross section D (left side).

The section samples were also analyzed by graphite-furnace atomic absorption spectrophotometry (GFAAS) for sodium, magnesium, aluminum, silicon, phosphorus, chlorine, potassium, calcium, titanium, vanadium, chromium, manganese, iron, cobalt, nickel, copper, zinc, arsenic, selenium, strontium, molybdenum, cadmium, tin, barium, and lead.

The Penta-Pycnometer at USG was used to measure specific gravity and calculate density, in grams per cubic centimeter (g/cc) of crushed samples from the core of nodule "A".

Samples of these two nodules are presently housed at the Colorado Desert District Stout Research Center, and are available for further study and analyses.

Table 2. Mineralogy of oolitic manganite nodule "A" using XRD analysis.

	upper third (A)	Middle third (B)	lower third (C)
Gypsum ($CaSO_4 \cdot 2H_2O$)	51.4%	27.3%	25.8%
Manganite [MnO(OH)]	48.6%	72.8%	74.2%

Notes: Results are in percent by volume. Analysis performed on cross section (divided into thirds) through middle diameter of nodule. Upper and lower portions assumed.

ANALYTICAL RESULTS

The constituent minerals of nodule "A" are 49 to 74 percent by volume (%v) manganite with 26 to 51 %v gypsum as determined by X-ray diffraction (Table 2). The density of nodule "A" is 3.41 g/cc. Manganite has a density of 4.3 g/cc and gypsum has a density of 2.32 g/cc. These measurements suggest a composition of approximately 70 percent by weight (%wt) manganite and 30 %wt gypsum.

Nodule "A" contains 721 to 4297 parts per million by weight (ppm) magnesium, 474 to 2120 ppm aluminum, 23 to 38 ppm titanium, 297 to 1348 ppm iron, and 398 to 1636 ppm strontium (Table 3). Sample D (Figure 3) of nodule "A" contains 79 ppm copper, 28 ppm zinc, 197 ppm barium, and 85 ppm uranium.

The following elements were identified from the middle section of nodule "A" using SEM electron imaging backscatter: manganese, calcium, potassium, chlorine, sulfur, silicon, aluminum, magnesium, oxygen, and carbon. Samples of nodule "A" sent to the CTL for X-ray fluorescence yielded the following elements: manganese, calcium, sulfur, bromine, dysprosium, and rhodium.

Table 3. Elemental analysis of oolitic manganite nodule "A".

Element	Upper third (A)	Middle third (B)	Lower third (C)
Na	118	61	68
Mg	4297	721	896
Al	2120	573	474
Ca	105,125	61,822	66,822
Ti	38	36	23
V	4	2	3
Cr	<2.0	<2.0	<2.0
Fe	1,348	333	297
Co	1	1	<0.5
Ni	2	1	2
Cu	5	2	3
Zn	<5.0	<5.0	<5.0
As	2	1	2
Se	<1.0	<1.0	<1.0
Sr	1636	462	398
Mo	2	2	2

Notes: Analysis performed using graphite furnace atomic absorption spectrophotometry. Results reported in parts per million by weight (ppm). Samples taken from cross section (divided into thirds) of the diameter of nodule "A". Upper and lower portions assumed.

Table 4. Mineralogy of oolitic manganite nodule "B" using XRD analysis.

Mineral	Upper third (A)	Middle third (B)	Lower third (C)
Manganite [MnO(OH)]	35.2%	24.6%	50.4%
Gypsum $CaSO_4 \cdot 2H_2O$	51.8%	44.3%	37.2%
Anhydrite $CaSO_4$	3.6%	0.9%	1.3%
Dolomite $CaMg(CO_3)_2$	1.9%	11.8%	1.7%
Ankerite [$CaFe(CO_3)_2$]	4.5%	11.2%	6.3%
Rhodochrosite ($MnCO_3$)	3.0%	7.2%	3.1%

Notes: Results in percent by volume. Analysis performed on cross section (divided into thirds) through middle diameter of nodule. Upper and lower portions assumed.

The core of nodule "B" analyzed by XRD (Table 4), contained, in volume-percent, 1 to 4 percent anhydrite, 2 to 12 percent dolomite, 3 to 7 percent rhodochrosite, 5 to 11 percent ankerite, 25 to 50 percent manganite, 37 to 52 percent gypsum, and 25 percent to 50 percent manganite.

DISCUSSION

The cores of the FCG nodules are composed of 26 to 75 percent steel gray and black manganite ooliths with 26 to 52 percent gypsum interstices. The nodules are unusual in their morphology, mineralogy, and geochemistry in comparison to typical ocean floor manganese nodules (Baturin and Yemel'yanov, 1991; Rogers, 1995). Most seafloor manganese nodules are layered concretionary deposits (Usui et al., 1997) that do not exhibit the oolitic structure of the FCG specimens.

The elemental composition of seafloor manganese nodules typically ranges from 8 to 50 percent manganese, 2 to 27 percent iron, 0.2 to 2 percent nickel, 0.03 to 1.6 percent copper, 0.01 to 2.3 percent cobalt, as well as lesser amounts of rare earth elements (Baturin and Yemel'yanov, 1991). The FCG oolitic nodules are high in manganese and very low in iron content, unlike the composition of most seafloor manganese nodules.

However, the existence of oolitic manganese-rich nodules has been reported. The FCG nodules are similar in textural characteristics to oolitic iron ores of the minette type, or oolitic manganese ores of the Tshiaturi type (Baumann, 1976). The mineralogy of the FCG samples best fits the Tshiaturi type of nodule. Manganese deposits that contain pisolitic and oolitic textures and/or nodules are generally thought to precipitate as a result of the mixing of anoxic mineral-laden waters with oxygen in shallow marine, lagoon-like settings (Gültekin, 1998; Force et al., 1999). In such shallow-marine environments, microbial metabolism may promote the precipitation of manganese oncoids and pisoliths (Gutzmer et al., 2002). Baumann (1976) observed that "In nearshore areas first pyrolusite (MnO_2) will form. The pyrolusite zone is followed by a zone with manganite ($Mn_2O_3 \cdot H_2O$) and, remotest from the coast, a zone occurs with rhodochrosite ($MnCO_3$)." These zones presumably correlate with the amount of dissolved oxygen and/or water depth. The abundance of manganite and the minor amount of rhodochrosite (nodule "B" only), suggest that the FCG nodules probably formed in shallow to moderate marine depths of less than 300 m (Force et al., 1999). A hydrothermal vent fluid source for the metallic constituents is indicated.

ACKNOWLEDGMENTS

We thank the following who have contributed to this paper: Anne Sturz, with the Marine Science and Environmental Studies Department at the University of San Diego, for her technical review and comments; David Escamilla, with the Marine Sciences and Environmental Studies Department at the University of San Diego, for his trace element analysis of FCG gypsum and anhydrite samples; Leo Macarthur at the USG quarry for his cooperation, assistance, and photographs; Art Struss at the USG RTC for the SEM/EDX analysis of the trace elements found in the macronodule cores; Scott Cimaglio at the USG RTC for the trace element analysis by GFAAS and electrothermal atomic absorption spectrophotometry; Laura Couche-Powers at CTL for providing the petrographic study; and JoAnne Delles at CTL for providing the X-ray fluorescence spectroscopy results.

DEDICATION

This paper is dedicated to Bruce Hudgens (1945-2004). He was a cat-lover, generous teacher and mentor, sky-gazer, and a very highly respected analyst. Bruce is missed by his co-workers and was a truly valuable resource. The authors are grateful for the many contributions that their colleague, Bruce Hudgens, made to this paper.

REFERENCES CITED

Baumann, L., 1976, Introduction to Ore Deposits. Halstead Press, a division of John Wiley and Sons, New York, pp. 91-96.

Baturin, G.N., and Ye.M.Yemel'yanov, 1991, Geochemistry of South Atlantic manganese nodules. Translated from Geokhimiya, no. 11, pp. 1620-1628.

Butterfield, D.A., G.J. Massoth, R.E. McDuff, J.E. Luton, and M.D. Lilley, 1990, Geochemistry of hydrothermal fluids from ASHES axial seamount hydrothermal emissions study vent field, Juan de Fuca Ridge; subseafloor boiling and subsequent fluid-rock interaction. Journal of Geophysical Research, v. 95, no. B8, pp. 12,895-12,921.

Dean, M., 1996, Neogene Fish Creek Gypsum and Associated Stratigraphy and Paleontology, Southwestern Salton Trough, California. *In* Sturzstroms and Detachment Faults, Anza-Borrego Desert State Park, California, P.L. Abbott and D.C. Seymour editors, South Coast Geological Society, Annual Field Trip Guide Book No. 24, pp. 123-148.

Dean, W.E., and A.L. Tung, 1974, Trace and minor elements in anhydrite and halite, Supai Formation (Permian), East-Central Arizona. *In* Fourth Symposium on Salt, v. 1-2, pp. 287-301, The Northern Ohio Geological Society, Inc., Cleveland, Ohio.

Dibblee, T.W., Jr., 1954, Geology of the Imperial Valley Region, California. *In* Geology of Southern California, R.H. Jahns editor, California Division of Mines and Geology Bulletin, v. 170, no. 2, pp. 21-81.

Durrell, C., 1953, Celestite Deposits near Ocotillo, San Diego County, California. California Division of Mines and Geology Special Report 32, pp. 5-7.

Elders, W.A., H.C. Lewis, and J.M. Mehegan, 1986, Saline brines and ore-forming processes in the geothermal systems of the Salton Trough. *In* Geology of the Imperial Valley, South Coast Geological Society, Annual Field Trip Guidebook No. 14, pp. 134-143.

Escamilla, D., 2003, A reevaluation of the origin of the Fish Creek Gypsum, Imperial County, California. Master of Science Thesis, Marine Sciences, University of San Diego, San Diego, California, 175 p.

Force, E.R., S. Paradis, and G.J. Simandi, 1999, Sedimentary manganese. *In* Selected British Columbia Mineral Deposit Profiles v. 3, Industrial Minerals, G.J. Simandi, Z.D. Hora, and D.V. Lefbure editors, British Columbia Ministry of Energy and Mines, 5 p.

Frost, E.G., S.C. Suitt, and M. Fattahipour, 1996, Emerging perspectives of the Salton trough region with an emphasis on extensional faulting and its implications for later San Andreas deformation. *In* Sturzstroms and Detachment Faults, Anza-Borrego Desert State Park, California, P.L. Abbot and D.C. Seymour editors, South Coast Geological Society, Annual Field Trip Guide Book No. 24, pp. 81-122.

Gültekin, A.H., 1998, Geochemistry and origin of the Oligocene Binkilic manganese deposit, Trace Basin, Turkey. Turkish Journal of Earth Sciences, v. 7, pp. 11-23.

Gutzmer, J., M.O. Schafer, and N.J. Beukes, 2002, Red bed-hosted oncolitic manganese ore of the Paleoproterozoic Soutpansberg Group, Bronkhorstfontein, South Africa. Economic Geology, v. 97, no. 6, pp. 1151-1166.

Hanna, G.D., 1926, Paleontology of Coyote Mountain, Imperial County, California. Proceedings of the California Academy of Sciences, 4th Series, v. 14, pp. 427-503.

Jefferson, G.T., T.A. Deméré, and N.S. Rugh, 2006, Appendix Table 2: systematic list of fossil invertebrates. *In* Fossil Treasures of the Anza-Borrego Desert, G.T. Jefferson and L. Lindsay editors, Sunbelt Publications, San Diego, California, pp. 356-260.

Kerr, D.R., and P.L. Abbott, 1996, Miocene sturzstrom deposits, Split Mountain, Anza-Borrego Desert State Park. *In* Sturzstroms and Detachment Faults Anza-Borrego Desert State Park California, P.L. Abbott and D.C. Seymour editors, South Coast Geological Society, Annual Field Trip Guide Book No. 24, pp. 149-163.

Peterson, D.G., Jr., and G.T. Jefferson, 1997, Submarine hydrothermal venting, a reevaluation of the source for the Fish Creek Gypsum, Imperial County, California. *In* San Bernardino County Museum Association, Abstracts from the Proceedings of the 1997 Desert Research Symposium, J. Reynolds, editor, v. 44, no.1, pp. 37-38.

Remeika, P., 1998, Marine invertebrate paleontology and stratigraphy of the Vallecito-Fish Creek Basin: a historic review, synthesis, and revision. *In* Geology and Hydrothermal Resources of the Imperial and Mexicali Valleys, L. Lindsay and W.G. Hample editors, San Diego Association of Geologists Annual Field Trip Guidebook, pp. 59-92.

Remeika, P., 2006, Ancestral woodlands of the Colorado River delta plain. *In* The Fossil Treasures of the Anza-Borrego Desert, G.T. Jefferson and L. Lindsay editors, Sunbelt Publications, San Diego, California, pp. 75-87.

Rogers, J. 1995, A comparative study of manganese nodules off southern Africa. South African Journal of Geology, v. 98, no. 2, pp. 209-215.

Seibold, E., and W.H. Berger, 1996, The sea floor: an introduction to marine geology. Springer Verlag, Berlin, 356 p.

Sharp, R.D., and G.C. Cork, 1995, Geology and mining of the Miocene Fish Creek Gypsum in Imperial County, California. *In* Proceedings of the 29th Forum on the Geology of Industrial Minerals, M. Tabilio and D.L. Dupras editors, California Department of Conservation, Division of Mines and Geology Special Publication, v. 110, pp. 169-180.

Shearer, C.K., J.J. Papike, S.B. Simon, B.L. Davis, and J.C. Laul, 1988, Mineral reactions in altered sediments from California State Well 2-14; variations in the modal mineralogy, mineral chemistry, and bulk composition of the Salton Sea Scientific Drilling Project Core. *In* A Compendium of Reports on the Salton Sea Scientific Drilling Project, W.A. Elders selector and compiler, Center for Geothermal Research, Institute of Geophysics and Planetary Physics, University of California, Riverside, p. 13.

Usui, A., A. Nishimura, and N. Mita editors, 1993, Composition and growth history of surficial and buried manganese nodules in the Penrhyn Basin, Southwestern Pacific. Marine Geology v. 114, pp. 133-153.

Winker, C.D., 1987, Neogene stratigraphy of the Fish Creek-Vallecito section, southern California and Colorado delta. Doctorial Dissertation, Geosciences, University of Arizona, Tucson, 494 p.

Winker, C.D., and S. Kidwell, 1996, Stratigraphy of a marine rift basin: Neogene of the western Salton Trough, California. *In* Field Conference Guide, P.L. Abbott and J.D. Cooper editors, Pacific Section of American Association of Petroleum Geologists, GB 73, Pacific Section Society of Economic Paleontologists and Mineralogists, Book 80, pp. 295-336.

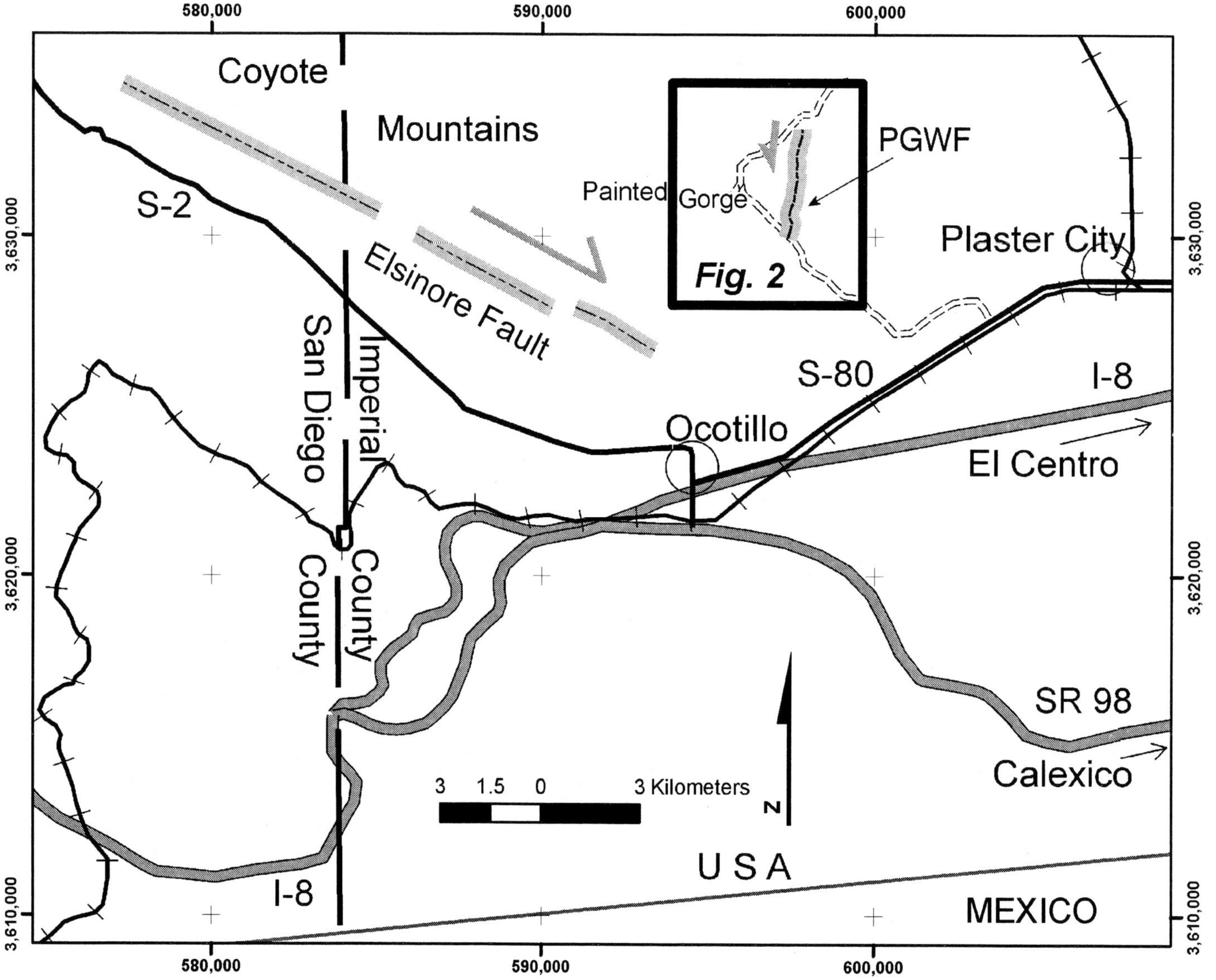

Figure 1. Location Map. PGWF = Painted Gorge Wash Fault. Grid in UTM meters, NAD 83, Zone 11N.

The Painted Gorge Wash Fault: Recent Left-Lateral Faulting, East Side Coyote Mountains, Imperial County, California

George J. Morgan and JR Morgan
4671 Lee Avenue, La Mesa, California 91941-4938
jrmorgan07@worldnet.att.net

ABSTRACT

The Painted Gorge Wash Fault (PGWF) is an active high-angle, left-lateral fault located immediately southeast of the Coyote Mountains, Imperial County, California. The PGWF is marked by prominent discontinuous-scarps, collapsed mole-tracks, offset drainages and other features that can be followed for over 3300 meters. The highest scarp is 3.9 meters and the largest left-lateral offset is 80 meters. The recent movement appears to have taken place on a preexisting right-lateral fault with an apparent displacement of 1700 meters. The preexisting fault is estimated to have been rotated 50 degrees clockwise to its present position.

INTRODUCTION

During mapping of the Coyote Mountains, Imperial County, California, a clearly identifiable linear feature was found on air photographs southeast of the mountains (Figures 1 and 2). This feature was field checked and identified as a fault displaying recent movement. The fault was mapped from a borrow pit north, following discontinuous fault scarps and other features for over 3300 meters. The fault was discovered to cut across the drainage from Painted Gorge and herein we name the fault the Painted Gorge Wash Fault (PGWF).

PAST WORK

Previous workers in the area have also noted the PGWF. Christenson (1957) reported the fault while mapping a portion of the Coyote Mountains. Clark (1982) recognized that the fault cuts Holocene age units and mapped its trace, recorded heights of the fault scarps, and described other features. Our mapping supports the earlier work, especially those features indicating recent movement. Our contribution is documenting recent left-lateral sense of movement on the PGWF and the apparent older right-lateral offset of units across the fault.

REGIONAL FRAMEWORK

The PGWF is located in the regional stress field of the San Andreas Fault system along with the Elsinore, San Jacinto, Superstition Hills Faults, and other faults. These and almost all other active faults in this regional system strike parallel to the San Andreas Fault (~N40°W), and display a prominent right-lateral sense of slip. No historic earthquakes have been associated with the PGWF (Jennings 1992). The left-lateral sense of the PGWF and its geographic relationship with the Elsinore Fault (synthetic member) would make the PGWF the antithetic member of this conjugate-paired system (Figure 1).

FEATURES INDICATING RECENT OFFSET AND SENSE OF MOVEMENT

The PGWF is an active left-lateral, high-angle fault, with a west-side-up component of movement that cuts Holocene and older sediments. The southern part of the fault strikes N17°E and northern part strikes almost north-south. The most noticeable feature of the PGWF is the west-side-up fault scarp. Other features found along the fault are collapsed mole-tracks, grabens, offset drainages, an offset ridge, and a pull-apart basin.

The most prominent features of the PGWF are the fault scarps, which can generally be described as benches, which are generally west-side up, have been slightly eroded back, and are cut by both small and large dry drainages. A few fault scarps are east-side up and tend to be lower than west-side-up scarps. Some east-side-up scarps are associated with grabens. The scarps range from approximately 0.05 to 3.9 meters. Generally, the scarps found on older terraces are higher than scarps found on younger terraces. The scarps are discontinuous and are absent from the youngest alluvium. Double fault scarps are found in two locations along the trace of the PGWF. These double scarps are associated with left stepovers, and could be part of small strike-slip duplex structures.

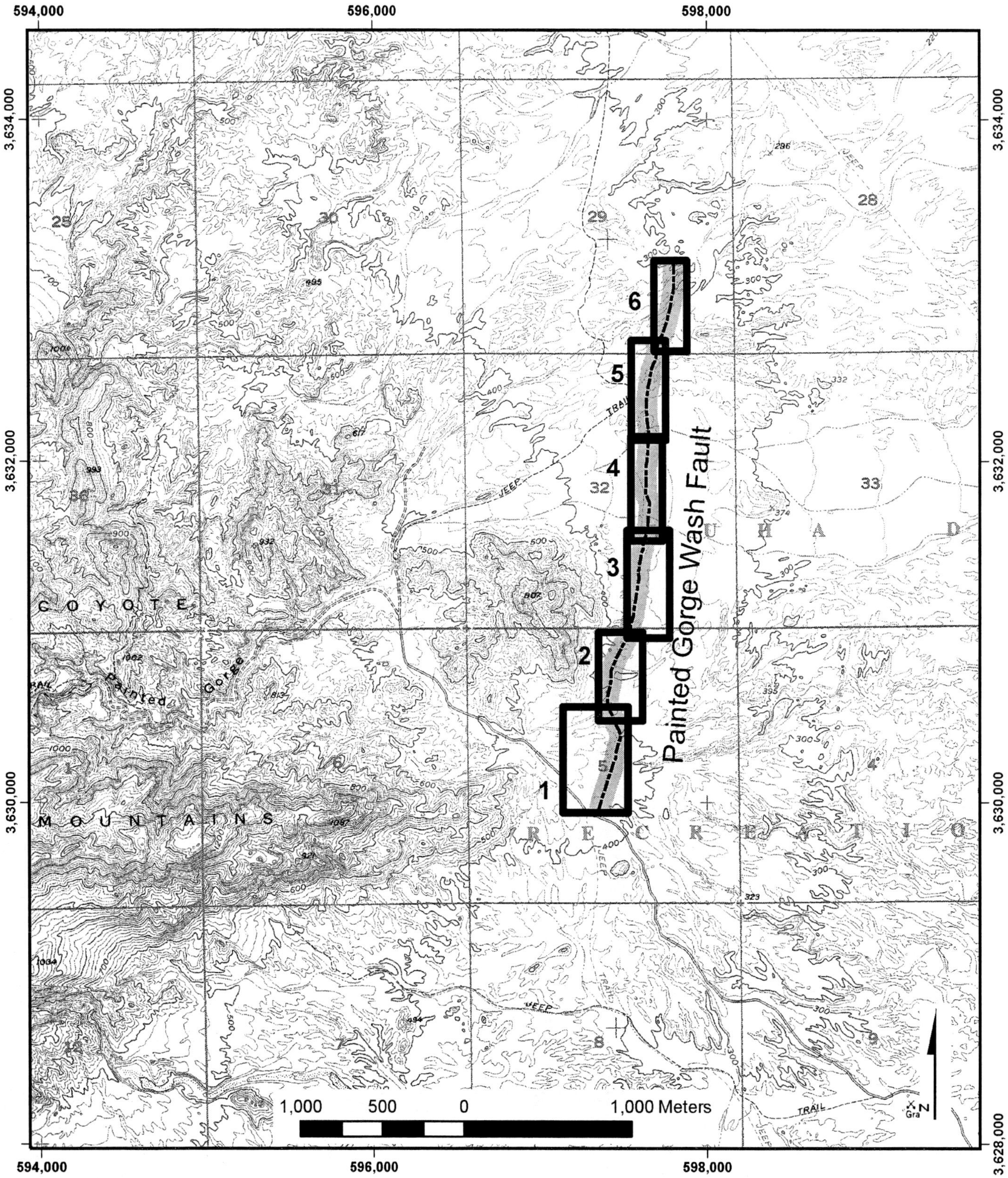

Figure 2. Index to Maps 1 through 6, from south to north along the Painted Gorge Wash Fault.

Collapsed mole-tracks are associated with the PGWF trace and are described as flat, linear, approximately 0.3 meter wide, disturbed ground that represents the erosional remnant of the original mole track. Cobbles found on the collapsed mole-tracks have a noticeably lighter surface color than the desert-varnished cobbles that are outside the collapsed mole track. Collapsed mole-tracks occasionally are found crossing slopes. The collapsed mole tracks are discontinuous and are absent from the youngest alluvium.

A few graben-like structures are found along the trace of the PGWF. These structures have what appear to be west-side-up and east-side-up scarps that oppose each other. The western scarp is almost always higher than the eastern scarp. The width of these graben-like structures ranges up to approximately 4 meters. Some of these structures may have been modified or may have even been created by erosion.

Several dry drainages and one ridge cross the PGWF and are offset (Map 3). All of the mapped offsets indicate that the PGWF is an active left-lateral fault with a west-side-up component of movement. The ridge displays the longest left-lateral offset mapped (80 meters).

A small pull-apart basin is associated with a left step-over of the PGWF and is consistent with the fault having left-lateral motion. A possible pop-up structure is associated with the pull-apart basin (Map 2).

A number of Holocene alluvial terraces are offset by the PGWF. These terraces, with their armored protective surfaces, preserve the fault scarps to a greater extent than the soft and easily erodible, underlying Pliocene Imperial and Palm Spring Formations. These terraces attest to the recent uplift (Dibblee, 2003) and base-level cutting that has and is taking place in the Coyote Mountains and the surrounding area. Currently a drainage north of the Painted Gorge Wash is in the process of capturing the Painted Gorge Wash by stream piracy.

The mapping of the PGWF by the authors is provided as a strip map divided into six parts, referred to as Maps 1 through 6, from south to north. See Figure 2 for an index map. The work of Morton (1977) is the basis for identifying the formations used to define the fault trace. The fault trace, map units, and other features are noted on the maps and discussed in the captions. Figure 3 is the legend that accompanies the six maps. The numbers along the fault trace are the measured height in meters of the fault scarp and are placed on the corresponding up side of the fault.

Local Stratigraphy

The oldest unit mapped is the Pliocene marine Imperial Formation (Pi) (Morton, 1977). The unit is made up of gray-olive claystone that weathers to a yellow, puffy, and easily erodible clay. Interbedded in the claystones are occasional thin, cemented brown sandstones and shell beds. The basal units of this formation are not seen in the area.

Deposited upon the Imperial Formation (Pi) is the Pliocene non-marine Palm Spring Formation (Pcp) (Morton, 1977). Light gray easily eroded fine sandstones and reddish clays make up the unit in the mapped area.

Deposited upon the Palm Spring is the non-marine Canebrake Conglomerate (Pcc) which is made up of coarse sandstones and conglomerate lenses. The Canebrake Conglomerate is considered to be upper Pliocene and with the possibility that upper parts of the unit may be Pleistocene (Morton, 1977).

Several mapped terraces (Qoa, Qoa1, Qoa2, Qoa3, and Qoa4) are deposited uncomfortably upon the Pliocene units. The terraces are more resistant to erosion than the above soft sediments of the underlying formations and thus help to preserve the fault scarps that cut the soft sediments.

Quaternary slope wash (Qt) is recent downward creeping alluvial material that is found covering the above units on the sides of the drainages.

Quaternary alluvium (Qal) fills the bottoms of the drainages and includes the western originating sheet flow deposits.

MAP 1

The PGWF was mapped starting from a barrow pit (located along the road to Painted Gorge) and at this location trends N17°E. The trace is marked by a prominent west-side-up fault scarp, collapsed mole-tracks, left-lateral offset drainages, and a graben. Scarp heights range from 0.12 to 2.4 meters. Below the alluvial terrace 3 (Qoa3), the fault offsets units of the Imperial Formation (Pi). Any recent evidence of faulting that existed in the barrow pit has been destroyed by the quarrying operations.

Southwest of the barrow pit, in an area not completely mapped, we observed a fault that is somewhat on strike with the PGWF; however, features related to recent faulting appear absent from this fault. We noted a shell bed of the Imperial Formation has been wrenched up along or into this other fault zone. Further mapping is needed in the area to the southwest of the barrow pit.

At point A, a 2.3-meter-wide graben-like structure can be found. The western fault scarp is 0.66 meter and the eastern fault scarp is 0.13 meter.

At Point B, the fault trace is marked by fault scarp with collapsed mole-track and a left-lateral offset of a dry drainage (Figure 4).

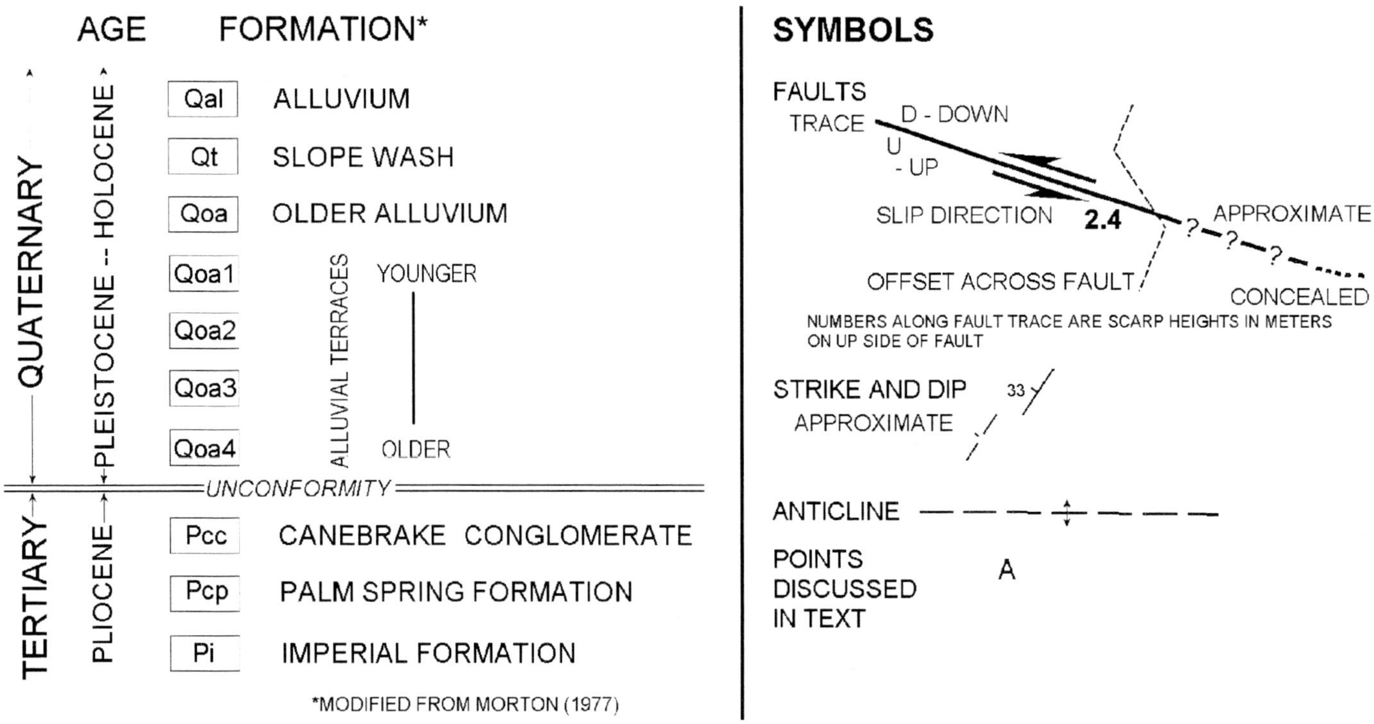

Figure 3. Legend for Maps 1 through 6.

Figure 4. Looking west at the fault scarp in alluvial terrace 3 (Qoa3) with a dry stream that drains off the up side of the terrace and is deflected to the right by the left-lateral slip on the PGWF. South of the offset drainage is a fault scarp with a collapsed mole-track. The "t"s mark the trace of the PGWF.

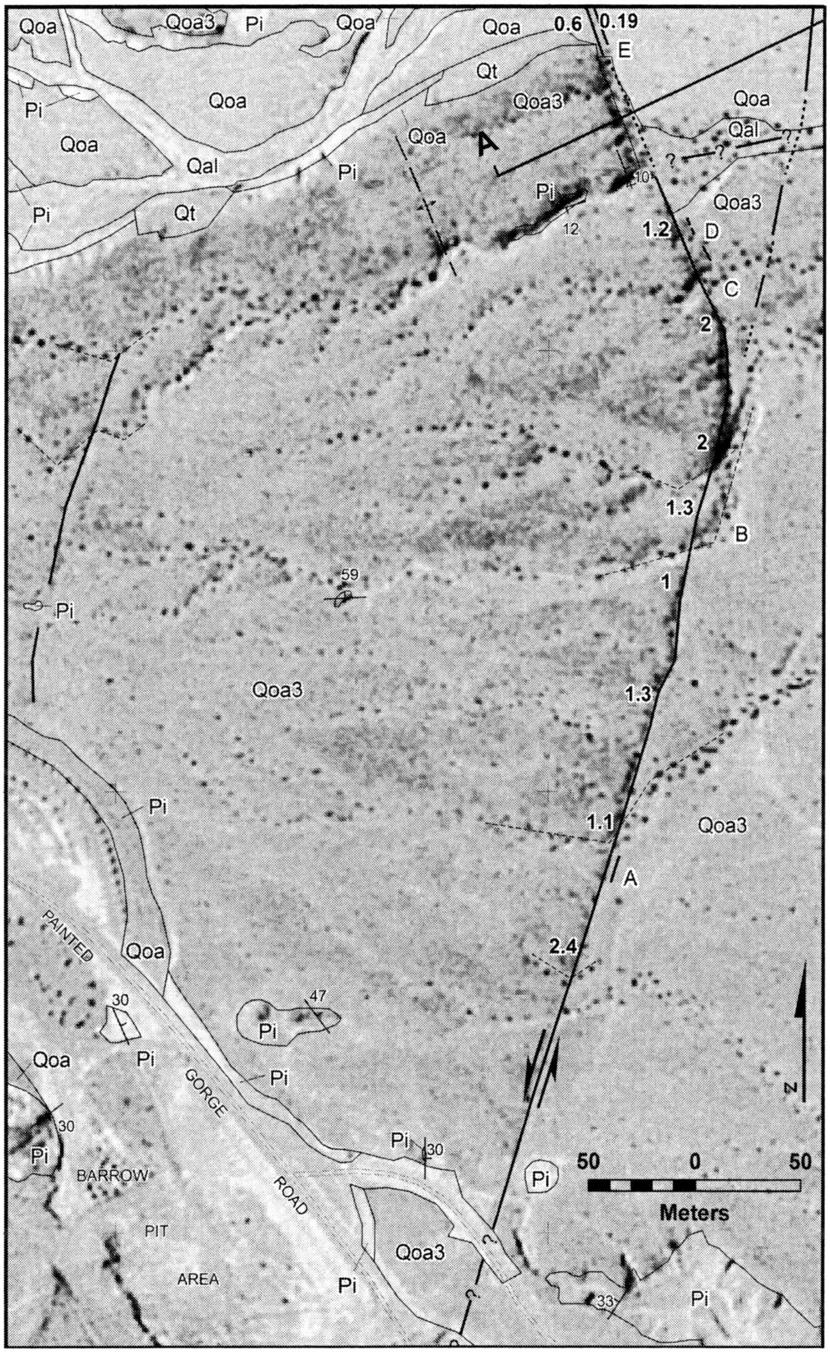
Qoa
Qoa3
Pi
Qoa
Qoa
0.6
0.19
E
Qt
Qoa3
Qoa
Pi
Qoa
Qal
Qoa
A
Qal
Pi
Pi
Qt
Pi
12
Qoa3
1.2
D
C
2
2
1.3
B
1
59
Pi
Qoa3
1.3
Pi
1.1
Qoa3
A
PAINTED
Qoa
2.4
30
47
Pi
Qoa
Pi
GORGE
Pi
30
Pi
Pi
30
BARROW
ROAD
Pi
50
0
50
Meters
PIT
Qoa3
AREA
Pi
33
Pi
N

Map 1.

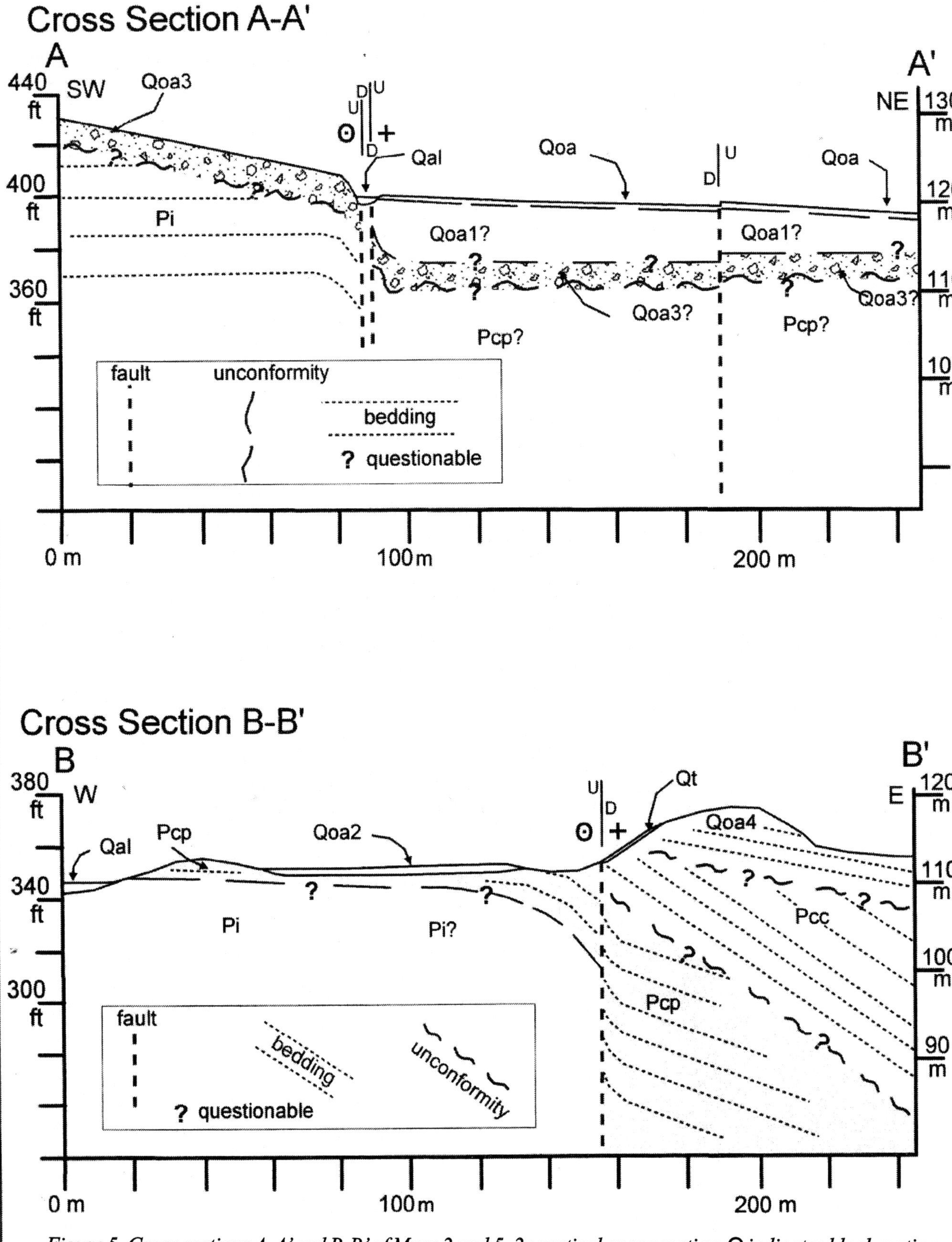

Figure 5. Cross sections A-A' and B-B' of Maps 2 and 5. 2x vertical exaggeration. ⊙ indicates block motion toward the viewer. + indicates block motion away from viewer. D indicates downthrown block. U indicates upthrown block. Pcp = Palm Spring Formation. Pi = Imperial Formation. See text for other unit symbols.

At point C, a recent small cobble fan covers the fault.

At point D, two west-side-up fault scarps are observed. The eastern scarp is 0.56 meter and the western fault scarp is 1.2 meters. The two scarps may be part of a strike-slip duplex. The main fault at this point has already turned to the west and strikes N25°W. This is the beginning of a left stepover of the PGWF and produces what appears to be a small pull-apart basin on the east side of the fault. The formation of the small pull-apart basin is consistent with the PGWF having left-lateral motion. The eastern edge of the pull-apart basin appears to be an eastern branch of the PGWF that strikes N12°E, away from the main branch, east of point C. The pull-apart basin widens to the north (see Map 2 and Cross Section A-A' in Figure 5).

Southeast of point D, the stratigraphic unit under alluvial terrace 3 (Qoa3) changes from the Imperial Formation (Pi) to the younger Palm Spring Formation (Pcp). The contact between the Imperial and the Palm Spring is probably depositional in nature and dips north roughly 20 degrees.

Approximately 280 meters west of the PGWF is another fault that cuts the same alluvial terrace 3 (Qoa3). This other fault trends parallel to the PGWF and has 0.8-meter west-side-up scarps. The scarps associated with this other fault have been eroded farther back than the fault scarps found on the PGWF, making the impression this other fault is older. Dry drainages cross this older-looking fault and indicate left-lateral movement, with a west-side-up component. The older-looking fault was probably produced in the same strain field that produced the PGWF, but we believe at a slightly earlier time. The earlier fault activity is believed to have then moved from the older fault to the PGWF. We consider the older looking fault to be part of the PGWF.

West of point D, the contact between alluvial terrace 3 (Qoa3) and the underlying Imperial Formation (Pi) is warped up into a shallow anticline. This structure was probably formed by drag on the PGWF, or from interaction between the older looking fault and main branch of the PGWF.

MAP 2

Between points E and F, the PGWF is marked by a uniform 4.0-meter-wide graben-like structure and a youthful appearing uniform 0.6-meter west-side-up fault scarp. The eastern scarp is 0.19 meter. The fault crosses a 100-meter-wide wash that drains to the east and cuts all the sediments in this wash except for the youngest alluvium (Qal). The youthful nature and the height of the fault scarps between points E and F imply that the fault trace that cuts alluvial terrace 3 (Qoa3) to the south was possibly made by several fault rupture events. As the fault crosses the wash it changes strike to almost north-south.

East of points E, F and G, the eastern branch of the PGWF strikes generally north, toward point H. This branch of the fault, which may be the eastern edge of a small pull-apart basin, is marked by an east-side-up scarp. Scarp heights range up to 0.20 meter. No left-lateral offsets were observed on the eastern branch. It is noted that the scarps in the drainage decrease in height toward the north.

From point F to point G, the PGWF is marked by erosion scarps and a collapsed mole-track. A ridge that is capped by alluvial terrace 3 (Qoa3) is offset by estimated 80 meters in a left-lateral sense by the PGWF (Figure 6). This offset is the longest offset mapped along the fault. At point G, the offset ridge has an east-side-up scarp of 3.9 meters. An eastward change in strike of the PGWF, north of point G, may be responsible for a part of the upward displacement (pop-up) of the east-side-up scarp and is consistent with the fault having left-lateral displacement.

At Point H, a collapsed mole-track and a 0.41-meter west-side-up scarp is located east of the PGWF and along a ridge capped by alluvial terrace 3 (Qoa3) This collapsed mole-track is similar to the collapsed mole-tracks found along other parts of the PGWF trace; thus it is considered part of the fault. This collapsed mole-track may represent the eastern edge of the pull-apart basin which was first noted at point C (Map 1). If this west-side-up scarp and collapsed mole-track are related to the eastern edge of the pull-apart basin, then the basin changes to a pop-up structure at this location. This structure could explain the reversal of the scarp along the eastern fault branch from east-side-up in the drainage to the south to west-side-up orientation on the ridge at point H to the north. From south to north, decreasing fault-scarp heights of the east branch fault in the drainage are consistent with south to north structural transition from a pull-apart basin to a pop-up structure. The collapsed mole-track could also be part of a strike-slip duplex structure.

North of alluvial terrace 3 (Qoa3) capped ridge the strike of the PGWF appears to change to N23°E and then disappears beneath an east-directed sheet-flow deposit. The Imperial Formation (Pi) and Palm Spring Formation (Pcp) that underlay the thin sheet-flow material probably had fault scarps, but without a protective terrace these easily erodible fault scarps were not preserved.

Possible traces of the PGWF in the area north of the ridge capped by alluvial terrace 3 (Qoa3) are limited fissures related to hydraulic piping at points I and J and a

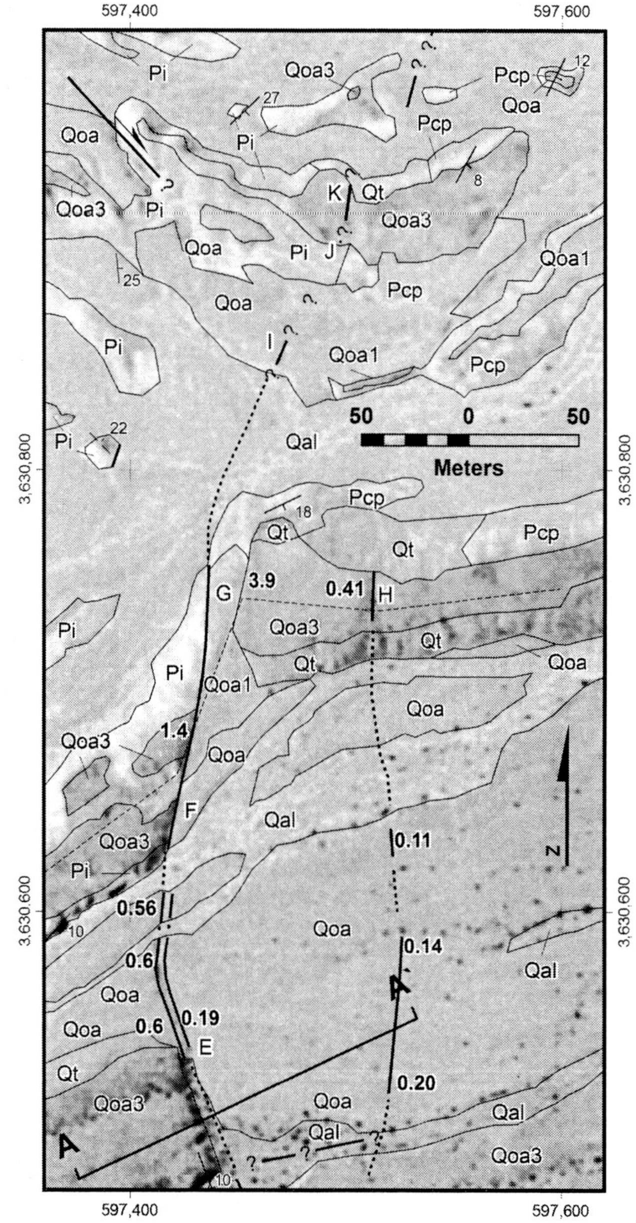
597,400
597,600
3,630,800
3,630,600
50
0
50
Meters
N
Pi
Qoa3
Pcp
Qoa
Qt
Qoa1
Qal
A
A'
E
F
G
H
I
J
K
3.9
0.41
1.4
0.56
0.6
0.19
0.11
0.14
0.20

Map 2.

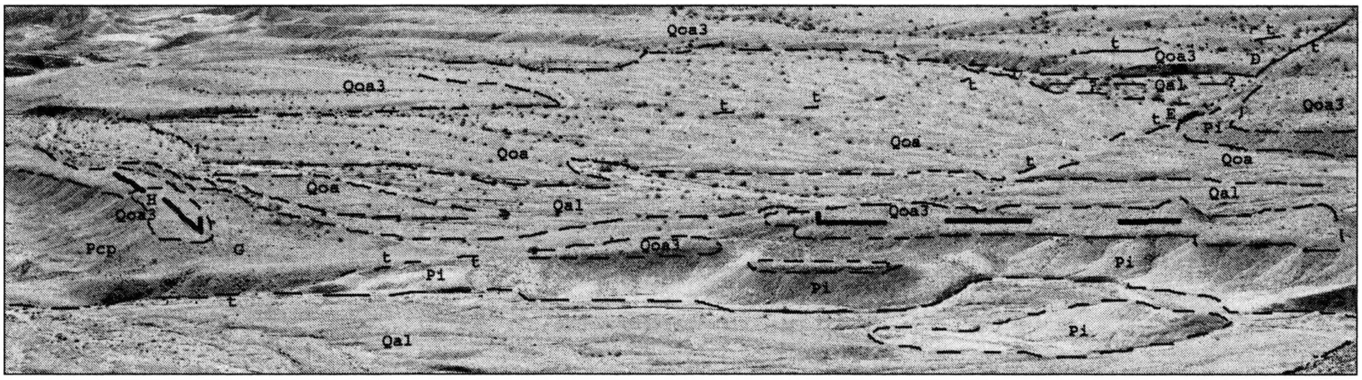

Figure 6. Looking east at the offset-ridge area. The heavy dashed lines are the center lines of the two separated parts of the left-lateral offset ridge that is displaced 80 meters. The "t"s indicates traces of the fault. Letters are referenced in the text and on Map 2. See Figure 3 for unit abbreviations.

possible collapsed mole-track at point K. At points I and J, there are depressions that may be the result of piping along the fault zone. There are no fault scarps associated with the collapsed mole-track or the piping features at these locations. The fault contact between the Imperial and the Palm Spring Formations is poorly exposed in this area. North of point K the fault trace appears again to be buried by sheet flows derived from the west and is obscured until point L (Map 3).

MAP 3

Point L is one of the few places where the PGWF zone can be directly observed. The fault strikes N09°E, dips 66 degrees east and is exposed by fissures that are present along the fault. The fissures are likely the result of hydrologic piping that occurs along the fault. At this point a shell bed of the Imperial Formation has been deformed into or along the fault zone. The Imperial Formation outside the direct influence of the PGWF has a dip of approximately 25 degrees east.

From point L, the PGWF produces scarps and collapsed mole-tracks that can be observed for the next 250 meters to the north (Figure 7). The fault scarps range from 0.23 to 0.76 meter in this interval. The presence of fault scarps indicate that the surfaces of alluvial terrace 1 (Qoa1) stood high enough above the streams to be unaffected by the sheet flows from the west. The range of heights of the fault scarps indicates that alluvial terrace 1 (Qoa1) is younger than alluvial terrace 3 (Qoa3) (Map 1), but is older than the alluvial terrace (Qoa) mapped on Map 2. The heights if the scarps on alluvial terrace 1 (Qoa1) when compared to the heights on older alluvial terrace (Qao) on Map 2 also implies that the scarps at point L and north represent multiple earthquake events.

At point M, there is a 3-meter-wide graben-like structure with a western scarp of 2.3 meters and an eastern scarp of 0.81 meter. A collapsed mole-track without a scarp lies east of the graben-like structure.

At point N, there are two collapsed mole-tracks and a 7.4-meter-wide graben-like structure with an east-side-up fault scarp that is 0.84 meter. A western fault scarp may have existed in the soft Imperial Formation, but with out a protective cap of older alluvium (Qoa), erosion would have easily destroyed the scarp.

Between points N and O, fault scarps appear in the alluvium (Qal). The presence of the fault scarps may be partially due to the increasing influence of Painted Gorge Wash which carries runoff from the west preferentially to sheet-flow from the west.

At point O, there is an east-side-up apparent fault scarp measuring 1.6 meters.

MAP 4

North of point O, the PGWF crosses the 660-meter-wide Painted Gorge Wash and is marked by intermittent fault scarps and collapsed mole-tracks. The fault scarps and collapsed mole-tracks are preserved where flash floods have not destroyed them. Up stream from this area of the Painted Gorge Wash, an unnamed drainage to the north is capturing the wash. This stream piracy may be aiding the preservation of the fault scarps by diverting some of the water from the wash. The fault scarps range from 0.18 to 0.66 meter. The alluvial material (Qal) is thin, and the easily erodible Imperial Formation (Pi to the west) and Palm Spring Formations (Pcp to the east) crop out through the alluvium (Qal), outside of Map 4.

At Point P, a double fault scarp is observed. This double scarp is associated with a left stepover similar to the

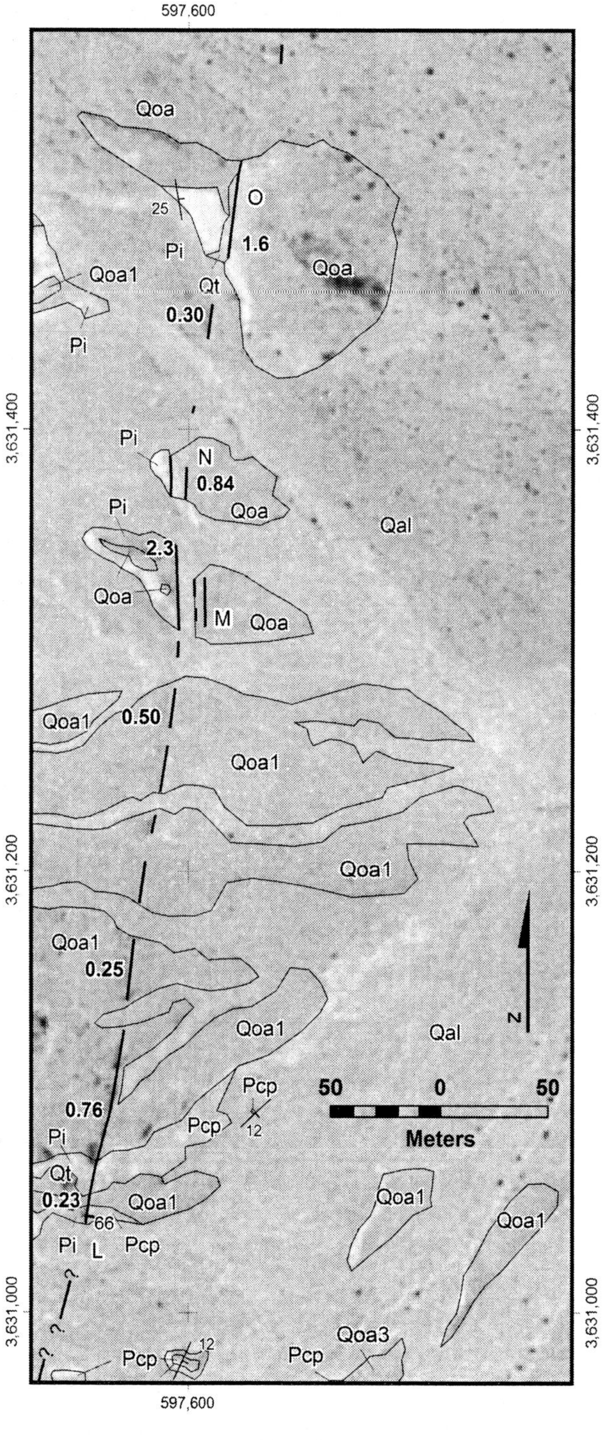
597,600
Qoa
25
O
Pi
1.6
Qoa1
Qt
Qoa
0.30
Pi
3,631,400
Pi
N
0.84
Pi
Qoa
Qal
2.3
Qoa
M
Qoa
Qoa1
0.50
Qoa1
3,631,200
Qoa1
Qoa1
0.25
N
Qoa1
Qal
Pcp
50
0
50
0.76
Pcp
12
Meters
Pi
Qt
0.23
Qoa1
Qoa1
Qoa1
66
Pi
L
Pcp
3,631,000
Qoa3
Pcp
12
Pcp
597,600

Map 3.

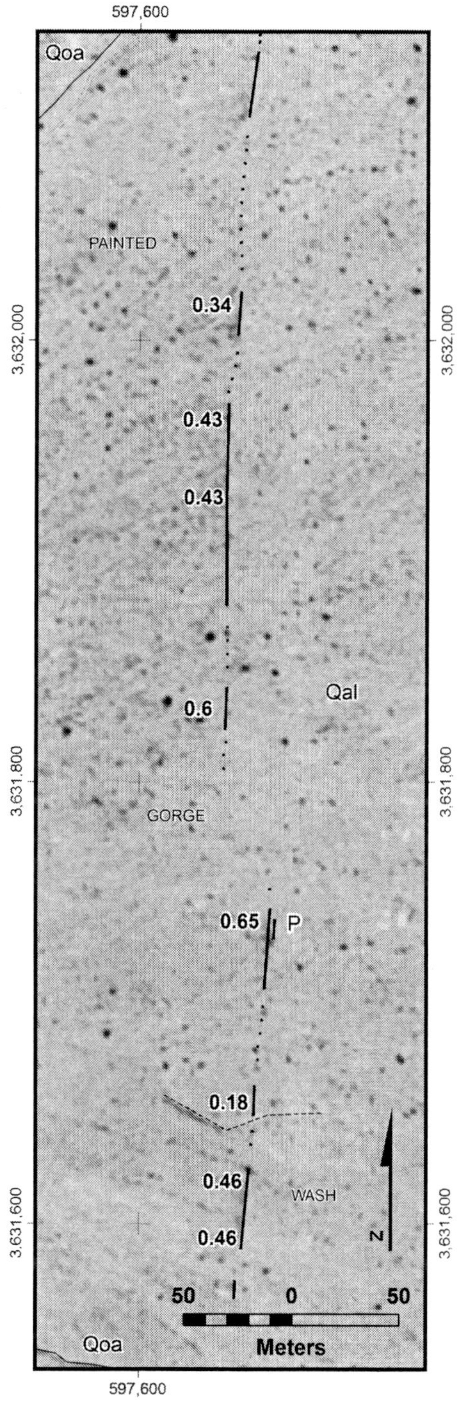
597,600
Qoa
PAINTED
0.34
3,632,000
3,632,000
0.43
0.43
Qal
0.6
3,631,800
3,631,800
GORGE
0.65
P
0.18
0.46
WASH
0.46
3,631,600
3,631,600
N
50
0
50
Meters
Qoa
597,600

Map 4.

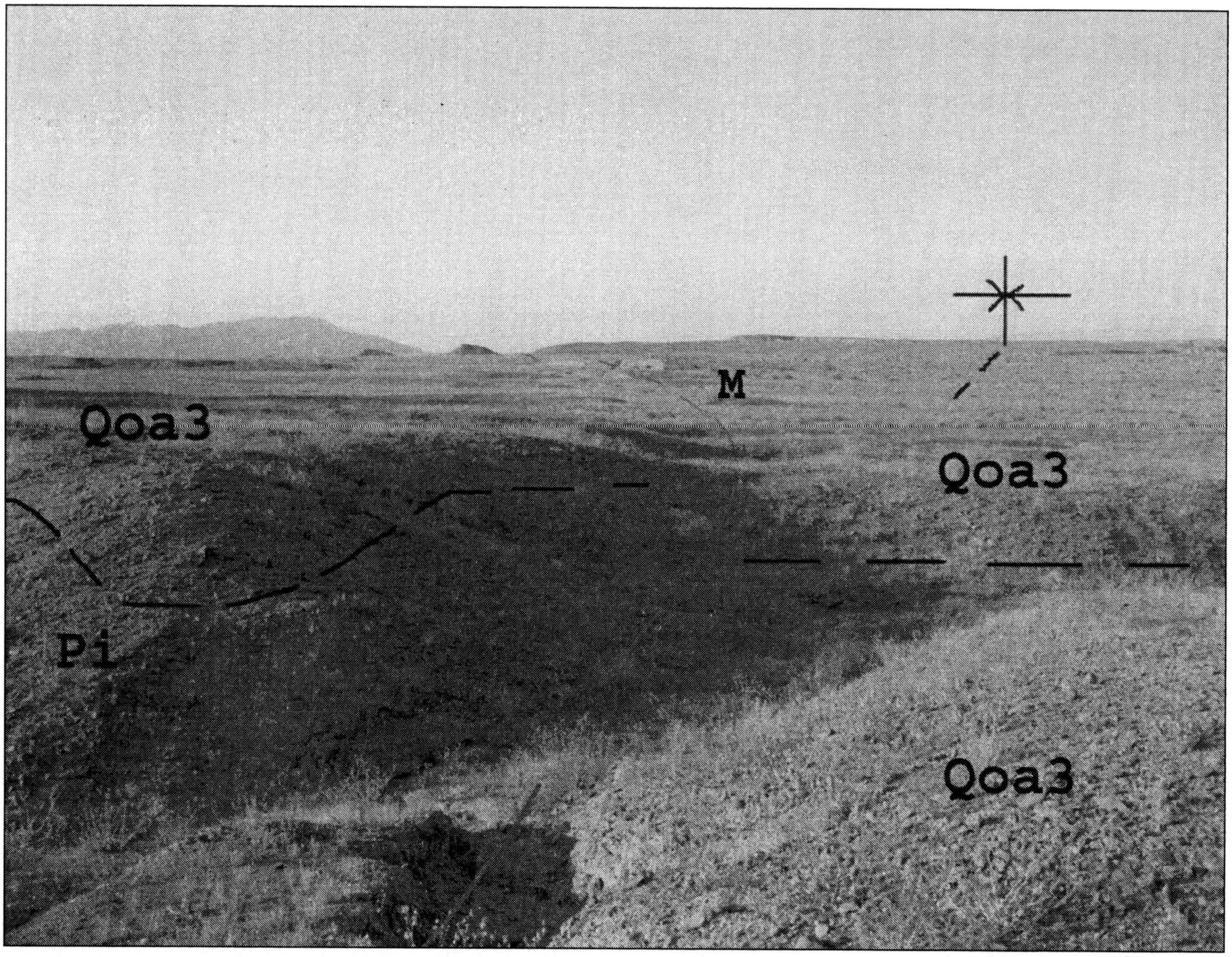

Figure 7. Looking northeast from point L (Map 3) along the scarps of the PGWF toward point M. Fissures that resulted from hydraulic piping along the PGWF form a trench in center foreground of the photo. The syncline on the horizon is the surface of alluvial terrace 4 (Qoa4).

double fault scarp previously observed at point D in Map 1. Both the 0.18-meter eastern scarp and the 0.65-meter western scarp are west-side-up. The left stepover could be a small pull-apart basin that has been filled by the occasional flash flooding of Painted Gorge Wash.

MAP 5

The PGWF trace leaves Painted Gorge Wash and strikes in a northerly direction. The fault is marked by scarps, collapsed mole-tracks, and a graben. The fault scarps range from 0.41 to 1.0 meter. At point Q, the Palm Spring Formation is observed in what appears to be a north dipping, depositional contact with the Imperial Formation (see Figure 8). The Palm Spring is present on both sides of the fault. The displacement of the contact between the Palm Spring and the Imperial Formations on the eastern side of the fault near point D (Map 1), and the contact between these formations on the west side of the fault near point Q, indicates over 1700 meters of apparent right-lateral movement. Ignoring all recent movements on the PGWF, the fault displays apparent right-lateral movement.

At point R, there is a 2.3-meter-wide graben-like structure with 1.0-meter western fault scarp. The eastern fault scarp is 0.41 meter. Erosion may have been partially responsible for the formation of this structure. North of point R, the PGWF trace strikes N07°E and trends up the side of a hill (point S) with west-side-up scarps (see Figures 8 and 9). This hill is capped by alluvial terrace 4 (Qoa4) and may be in part a fault-line scarp. This scarp is on the northern limb of a syncline formed by the top surface of the capping alluvial terrace 4 (Qoa4) (see Figure 7). The fault's

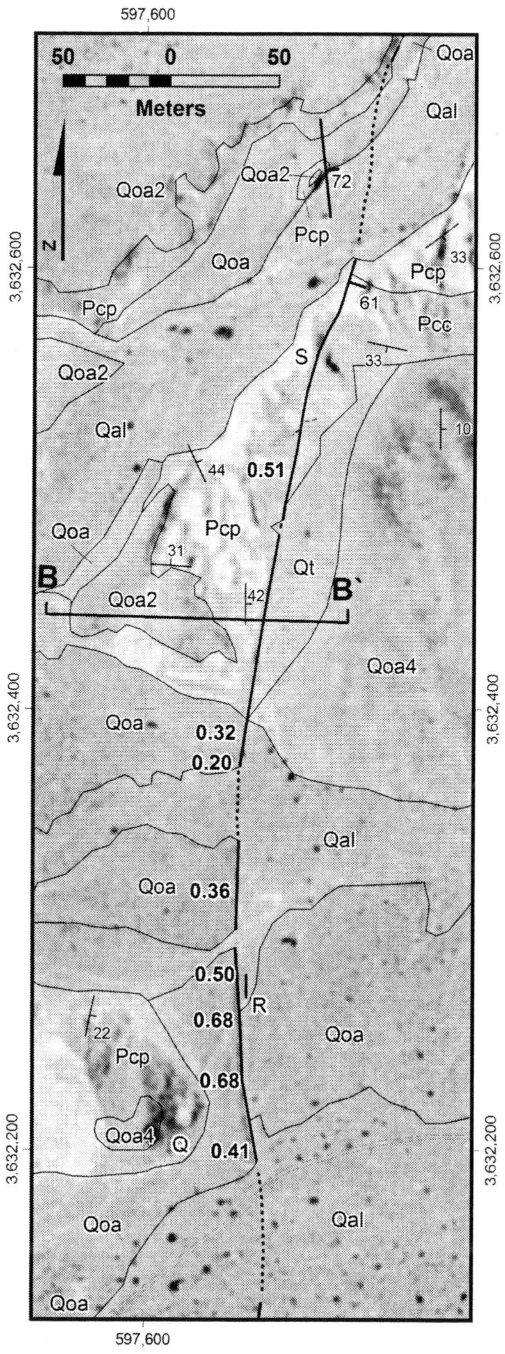
597,600
50 0 50
Meters
N
Qoa
Qal
Qoa2
Qoa2
72
Pcp
Qoa
Pcp
33
61
Pcc
S
33
Qoa2
Qal
10
44
0.51
Pcp
Qoa
31
Qt
B
Qoa2
42
B`
Qoa4
3,632,600
3,632,400
Qoa
0.32
0.20
Qal
Qoa
0.36
0.50
R
0.68
22
Qoa
Pcp
0.68
Qoa4
Q
0.41
3,632,200
Qoa
Qal
Qoa
597,600

Map 5.

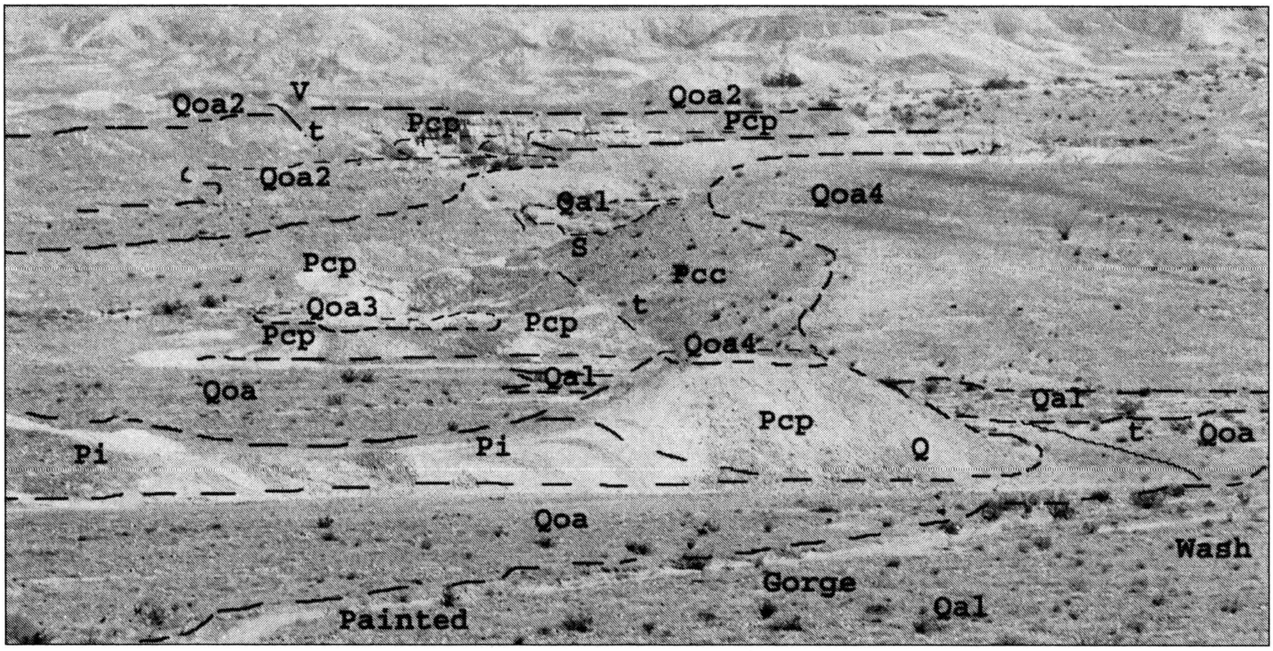

Figure 8. Looking northeast at area covered by Maps 5 and 6, scarps and collapsed mole-tracks cross the northern limb of the syncline in Figure 7. The "t"s mark the fault trace. Letters are indicated on Map 5 and referenced in the captions to that map. See Figure 3 for unit abbreviations.

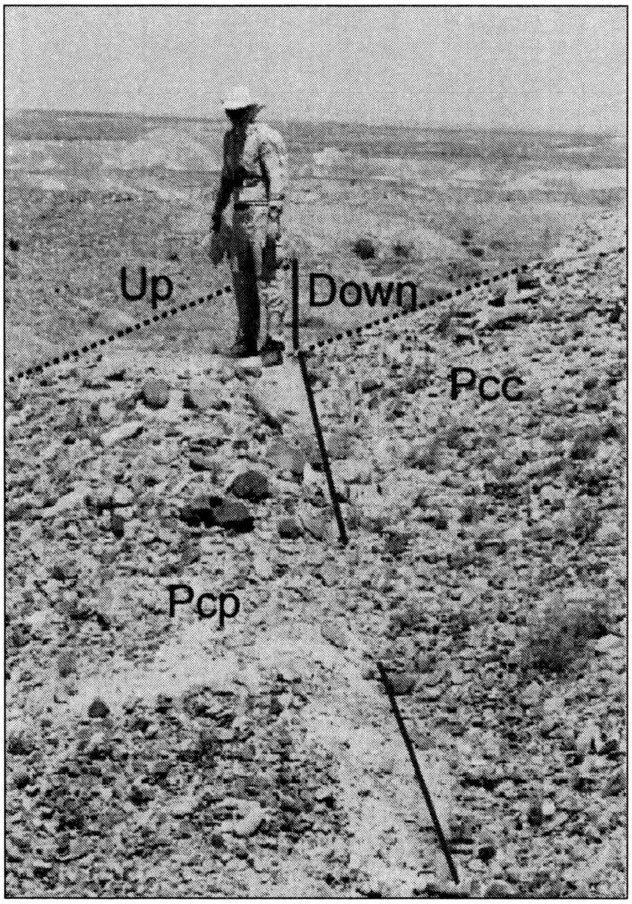

Figure 9. Looking north at point S, on Map 5, the fault separates the Palm Spring (Pcp) and the Canebrake Conglomerate (Pcc) formations with west-side-up scarps and collapsed mole-tracks. Dave Bloom is standing on the fault as it crests a small ridge.

change of strike to the east may provide some upward component of movement to form the hill. The folding of the alluvial terrace 4 (Qoa4) surface may be associated with the same stress field that produced the PGWF. The alluvial terrace 4 (Qoa4) lays unconformity on the Canebrake Conglomerate (Pcc) which is deposited conformably on the Palm Spring Formation (Pcp). See cross section B-B' (Figure 5).

MAP 6

At point T, the fault emerges as a collapsed mole-track along the west side of the drainage and continues to follow the drainage north. Also at point T along the west side of the drainage, the fault cuts alluvial terrace 2 (Qoa2), producing a 0.75-meter east-side-up fault scarp. At point U, the fault strikes N02°E and dips 75 degrees east and older partially calcified fault zone is in close association with PGWF. This zone may represent older movements on the PGWF. Some branches of the PGWF appear not to cut alluvial terrace 2 (Qoa2) in the area of point U.

Between points U and V, traces of the PGWF are obscured. At point V, the last trace of the active fault is a 1.2-meter east-side-up scarp. The fault scarp is eroded back to the east making the impression that the scarp is old. The PGWF strikes north and dips 67 degrees east. North of this point, another fault can be found that is somewhat on strike with PGWF. Evidence of recent faulting has not been found associated with this other fault.

CONCLUSIONS

The PGWF is an active left-lateral, high-angle fault with a west-side-up component of movement. However, the field evidence presented indicates the possibility that the recent movement was preceded by an older and apparently right-lateral sense of movement. The evidence for the older movement consists of: (1) a deformed shell bed (point L in Map 3) that implies more movement on the fault than is indicated by the recent movement, and (2) a calcified fault zone (at point T in Map 6) that indicates the existence of a fault that predates the youthful fault scarps. The faults, north and south, that are on strike of the mapped PGWF appear to be inactive and could be extensions of the preexisting right-lateral fault.

The apparent right-lateral offset of the north-dipping contact between the Imperial and Palm Spring Formations by 1700 meters can indicate that the preexisting fault was a right-lateral fault. A fault with only dip-slip motion would

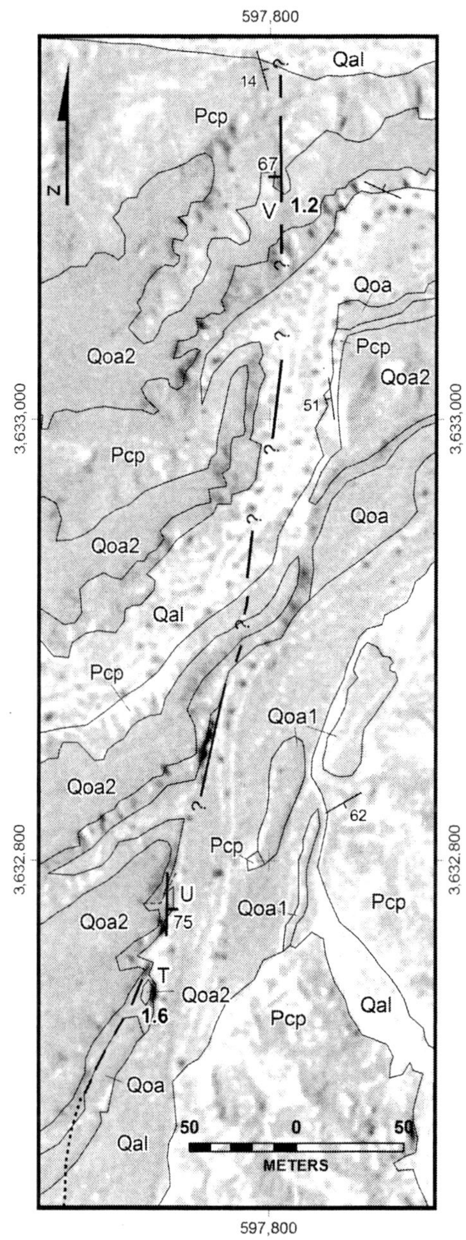

Map 6.

need approximately 618 meters of movement to produce 1700 meters of offset. A dip-slip fault could produce the apparent 1700 meters movement, but in this area of the Salton Trough, dip-slip faults are uncommon and right-lateral faults common. The dip of the bedding that is seen in the Palm Spring Formation (Pcp) west of the fault between points Q and R is consistent with right-lateral movement (drag folding) on the fault. On Map 1, the folding of the Imperial Formation (Pi) along the west side of the PGWF, next to the barrow pit, is also consistent with right-lateral movement on the PGWF.

If a right-lateral fault were rotated clockwise 50 degrees from an original strike of approximate N40°W (parallel to the Elsinore Fault) it would match the strike of the PGWF. Right-lateral movement on the fault would end as the fault was rotated out of the synthetic part of the strain field. Motion on the fault would resume in a left-lateral sense when the fault became oriented in the antithetic part of strain field. This interpretation is consistent with the findings of Mace (1981), who reported that the original magnetic poles of the Miocene Alverson Volcanics found in the southeastern part of the Coyote Mountains have rotated 43 to 75 degrees clockwise.

The apparent right-lateral motion appears to have been active up to the time of the deposition of the Canebrake Conglomerate. The left-lateral motion started sometime after the deposition of alluvial terrace 3 (Qoa3). Further mapping could confirm the apparent right-lateral movement on a preexisting fault and may further constrain the timing of the change from right-lateral to left-lateral motion. Trenching of the PGWF could provide evidence to constrain the timing of the displacement reversal and reveal the recent earthquake history associated with the left-lateral movements of the PGWF.

ACKNOWLEDGMENTS

The authors are grateful to John Prall and Charles Lough for their help with manuscript. Special thanks are also given to David Bloom for his suggestions and editorial help.

REFERENCES CITED

Christensen, A.D., 1957, Part of the geology of the Coyote Mountain area, Imperial County, California, unpublished M. S. Thesis, University of California, Los Angeles, 188 p.

Clark, M.M., 1982, Map showing recently active breaks along the Elsinore and associated faults, California, between Lake Henshaw and Mexico: U.S. Geological Survey Miscellaneous Investigations Series Map 1-1329, 2 sheets, scale 1:24000.

Dibblee, T.W., 2003, Geology of the Coyote Mountains, Imperial and San Diego Counties, California: *in* Murbach, M.L., and Hart, M.W., eds., Geology of the Elsinore fault zone, San Diego region, San Diego Association of Geologists and South Coast Geological Society No. 31, pp. 117-130.

Jennings, C.W., Preliminary fault activity map of California: California Division of Mines and Geology, Open-File Report 92-03, one sheet, scale 1:750,000, 76 p.

Mace, N.W., 1981, A paleomagnetic study of the Miocene Alverson Volcanics of the Coyote Mountains, western Salton Trough: Unpublished M.S. Thesis, San Diego State University, 142 p.

Morton, P.K., 1977, Geology and mineral resources of Imperial County, California: California Division of Mines and Geology, County Report 7, 104 p.

Selected References

Atwater, T.M., 1992, Constraints from plate reconstructions for Cenozoic tectonic regimes of southern and eastern California: *in* Richard, S.M., ed., Deformation associated with the Neogene Eastern California Shear Zone, southwestern Arizona and southeastern California: Redlands, California, San Bernardino County Museum Special Publication, pp. 1-2.

Christensen, A.D., 1957, Part of the geology of the Coyote Mountain area, Imperial County, California: M.S. thesis, University of California, Los Angeles, 188 p.

Dean, M.A., 1988, Genesis, mineralogy and stratigraphy of the Neogene Fish Creek Gypsum, southwestern Salton Trough, California: M.S. thesis, San Diego State University, 150 p.

Dibblee, T.W., Jr., 1954, Geology of the Imperial Valley region, California: *in* Jahns, R.H., ed., Geology of California, Chapter 2: California Division of Mines Bulletin 170, pp. 21-28.

Dockum, M.S., 1982, Greenschist-facies carbonates, eastern Coyote Mountains, western Imperial County, California: M.S. thesis, San Diego State University, 89 p.

Gastil, R.G., and Miller, R.H., 1984, Prebatholithic paleogeography of peninsular California and adjacent Mexico: *in* Frizzell, V.A., Jr., ed., Geology of Baja California Peninsula: Pacific Section, Society of Economic Paleontologists and Mineralogists, v. 39, pp. 9-16.

Girty, G.H., Thomson, C.N., Girty, M.S., Miller, J., and Bracchi, K., 1993, The Cuyamaca-Laguna Mountains shear zone, Late Jurassic plutonic rocks and Early Cretaceous extension, Peninsular Ranges, southern California: *in* Geological investigations in Baja California, Mexico, Abbott, P.L., Sangines, E.M., and Rendina, M.A., eds., South Coast Geological Society Volume, Santa Ana, California, pp. 173-181.

Gjerde, M.W., 1982, Petrography and geochemistry of the Alverson Formation, Imperial County, California: M.S. thesis, San Diego State University, 85 p.

Hart, M.W., 1991, Landslides in the Peninsular Ranges, southern California, *in* Geological excursions in southern California and Mexico, Walawender, M.J., and Hanan, B.B., eds.: Guidebook for the 1991 annual meeting of the Geological Society of America, San Diego, California, pp. 349-365.

Hawkins, J.W., 1970, Petrology and possible tectonic significance of late Cenozoic volcanic rocks, southern California and Baja California: Geological Society of America Bulletin, v. 81, pp. 3323-3338.

Hoggatt, W.C., 1979, Geologic map of the Sweeney Pass quadrangle, San Diego County, California: U.S. Geological Survey Open-File report 79-754, 28 p., scale 1:24,000.

Isaac, S., Rockwell, T.K., and Gastil, G., 1986, Plio-Pleistocene detachment faulting, Yuha Desert region, western Salton Trough, northern Baja California: Geological Society of America Abstracts with Programs, p. 120.

Johnson, N.M., Officer, C.B., Opdyke, N.D., Woodard, G.D., Zeitler, P.K., and Lindsay, E.H., 1983, Rates of late Cenozoic tectonism in the Vallecito-Fish Creek basin, western Imperial Valley, California: Geology, v. 11, pp. 664-667.

Kerr, D.R., 1984, Early Neogene continental sedimentation in the Vallecito and Fish Creek Mountains, western Salton Trough, California: Sedimentary Geology, v. 38, pp. 217-246.

Kerr, D.R., and Kidwell, S.M., 1991, Late Cenozoic sedimentation and tectonics, western Salton Trough, California, *in* Geological excursions in southern California and Mexico, Walawender, M.J., and Hanan, B.B., eds.: Guidebook for the 1991 Annual Meeting of the Geological Society of America, San Diego, California, pp. 397-416.

Kimbrough, D.L., Smith, D.P., Mahoney, B.P., Moore, T.E., Grove, M., Gastil, R.G., Ortega-Rivera, A., and Fanning, C.M., 2001, Forearc-basin sedimentary response to rapid Late Cretaceous batholithic emplacement in the Peninsular Ranges of southern and Baja California: Geology, v. 29, pp. 491-494.

Kimzey, J.A., 1982, Petrography and geochemistry of the La Posta granodiorite: M.S. thesis, San Diego State University, San Diego, California, 64 p.

Miller, R.H., and Dockum, M.S., 1983, Ordovician conodonts from metamorphosed carbonates of the Salton Trough, California: Geology, v. 11, pp. 410-412.

Minch, J.A., and Abbott, P.L., 1973, Post-batholithic geology of the Jacumba area, southeastern San Diego County, California: San Diego Society of Natural History, Transactions, 17(11), pp. 129-136.

Morton, P.K., 1977, Geology and mineral resources of Imperial County, California: California Division of Mines and Geology County Report 7, 104 p.

Ruisaard, C.I., 1979, Stratigraphy of the Miocene Alverson Formation, Imperial County, California: M.S. thesis, San Diego, California, San Diego State University, 125 p.

Shaw, S.E., Todd, V.R., and Grove, M., 2003, Jurassic peraluminous and gneissic granites in the axial zone of the Peninsular Ranges, southern California, *in* Johnson, S.E., Paterson, S.R., Fletcher, J.M., Girty, G.H., Kimbrough, D.L., and Martin-Barajas, A., eds., Tectonic evolution of northwestern Mexico and the southwestern USA: Boulder, Colorado, Geological Society of America Special Paper 374, pp. 157-183.

Strand, R.G., 1962, Geologic map of California, San Diego-El Centro sheet, scale 1:250,000: California Division of Mines and Geology.

Streckeisen, A.L., 1973, Plutonic rocks, classification and nomenclature recommended by the I.U.G.S. Subcommission on the systematics of igneous rocks: Geotimes, v. 18, no. 10, pp. 26-30.

Tarbet, L.A., 1951, Imperial Valley, in possible future petroleum provinces of North America: American Association of Petroleum Geologists Bulletin, v. 35, pp. 260-263.

Todd, V.R., 1977b, Geologic map of the Agua Caliente Springs quadrangle, San Diego County, California: U.S. Geological Survey Open-File Report 77-742, 20 p., scale 1:24,000.

Todd, V.R., Kilburn, J.E., Detra, D.E., Griscom, Andrew, Knepper, Jr., D.H., Kruse, F.A., Cather, Eric, and Lipton, D.A., 1987b, Mineral resources of the Sawtooth Mountains and Carrizo Gorge/eastern McCain Valley Wilderness Study Areas, San Diego County, California: U.S. Geological Survey Bulletin 1711-B, 15 p.

Todd, V.R., Kilburn, J.E., Detra, D.E., Griscom, Andrew, Kruse, F.A., and McHugh, E.L., 1987c, Mineral resources of the Jacumba (In-ko-pah) Wilderness Study Area, Imperial County, California: U.S. Geological Survey Bulletin 1711-D, 18 p.

Todd, V.R., Erskine, B.G., and Morton, D.M., 1988b, Metamorphic and tectonic evolution of the northern Peninsular Ranges batholith, *in* Ernst, W.G., ed., Metamorphism and crustal evolution of the western United States, Rubey Volume VII, Prentice-Hall, Inc., Englewood Cliffs, New Jersey, pp. 894-937.

Todd, V.R., Kimbrough, D.L., and Grove, Marty, 1994b, Volcanic to plutonic transect across the Peninsular Ranges batholith, San Diego County, California, Roadlog for field trip, *in* McGill, S.F., and Ross, T.M., eds., Geological Investigations of an active margin: Geological Society of America Cordilleran Section Guidebook, 27th Annual Meeting, San Bernardino, California, Mar., 1994, pp. 214-226.

Todd, V.R., Shaw, S.E., and Hammarstrom, J.M., 2003, Cretaceous plutons of the Peninsular Ranges batholith, San Diego and westernmost Imperial Counties, California: Intrusion across a Late Jurassic continental margin, *in* Johnson, S.E., Paterson, S.R., Fletcher, J.M., Girty, G.H., Kimbrough, D.L., and Martin-Barajas, A., eds., Tectonic evolution of northwestern Mexico and the southwestern USA: Boulder, Colorado, Geological Society of America Special Paper 374, pp. 185-235.

Walawender, M.J., 1979, Basic plutons of the Peninsular Ranges batholith, southern California, *in* Abbott, P.L., and Todd, V.R., eds., Mesozoic crystalline rocks: San Diego State University, Department of Geological Sciences, Guidebook for the Geological Society of America Annual Meeting, San Diego, California, Nov. 1979, pp. 151-162.

Walawender, M.J., Gastil, R.G., Clinkenbeard, J.P., McCormick, W.V., Eastman, B.G., Wernicke, R.S., Wardlaw, M.S., and Gunn, S.H., 1990, Origin and evolution of the zoned La Posta-type plutons, eastern Peninsular Ranges batholith, southern and Baja California, *in* Anderson, J.L., ed., The nature and origin of Cordilleran magmatism: Geological Society of America Memoir 174, pp. 1-18.

Weber, F.H., Jr., 1963, Geology and mineral resources of San Diego County, California: California Division of Mines and Geology County Report 3, 309 p.

Winker, C.D., 1987, Neogene stratigraphy of the Fish Creek-Vallecito section, southern California: Implications for early history of the northern Gulf of California and Colorado Delta: Ph.D. dissertation, Tucson, Arizona, University of Arizona, 494 p., including "Geologic Map of Arroyo Tapiado, Carrizo Mountain NE, and parts of Harper Canyon, Borrego Mountain SE, and Sweeney Pass, 7.5 Quadrangles," scale 1:24,000.

Woodard, G.D., 1963, The Cenozoic succession of the west Colorado Desert, San Diego and Imperial Counties, southern California: Ph.D. dissertation, Berkeley, California, University of California, Berkeley, 173 p.

Woodard, G.D., 1974, Redefinition of Cenozoic stratigraphic column in Split Mountain Gorge, Imperial Valley, California: American Association of Petroleum Geologists Bulletin, v. 58, pp. 521-526.

Woodring, W.P., 1932, Distribution and age of the Tertiary deposits of the Colorado Desert: Carnegie Institution of Washington Publication 148, pp. 1-25.

Stratigraphic Column

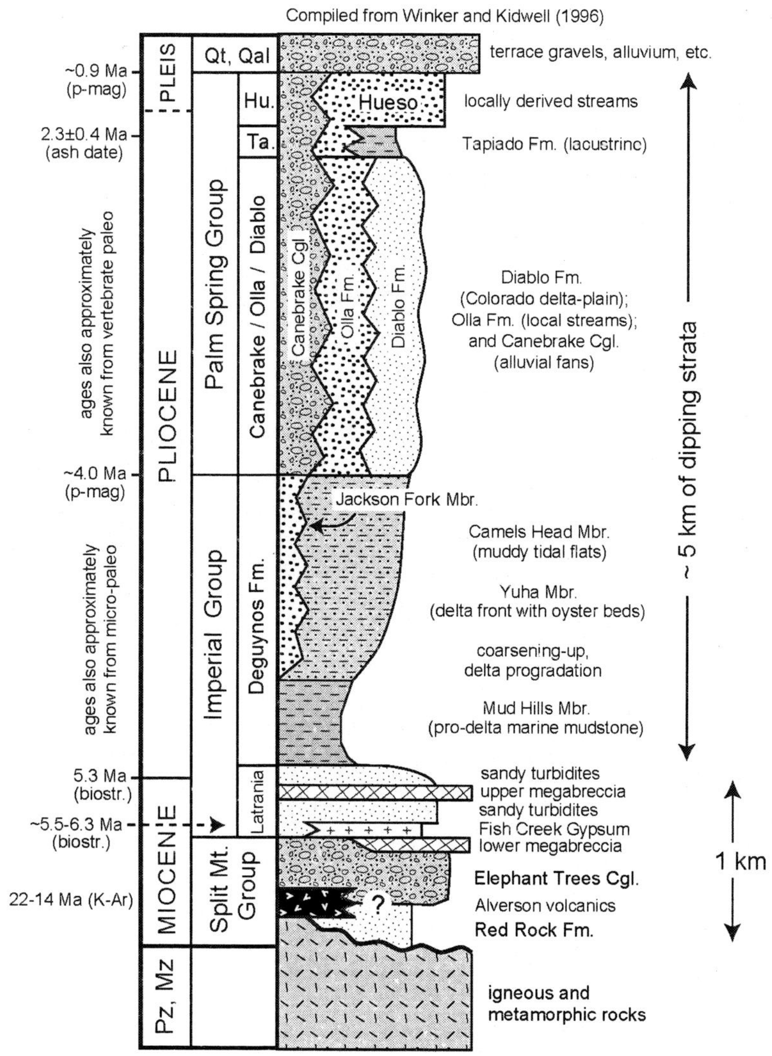

Geologic Explanation for Winker (1987), facing page

Qal Recent Alluvium

Qt Terrace deposits
Older alluvial deposits capping older units

—Palm Spring Group

PQc Canebrake Conglomerate
Conglomerate, locally derived

PQh Hueso Formation
Fluvial sandstone with conglomerate and siltstone, locally derived

Po Olla Formation
Fluvial sandstone, siltstone, claystone, and conglomerate; interbedded local and Colorado River sources

Pd Arroyo Diablo Formation
Nonmarine deltaic sandstone, siltstone, and claystone of Colorado River

—Imperial Group

——Deguynos Formation (Upper Imperial)

Pch Camels Head Member
Marginal marine sandstone, siltstone, and claystone; tidal flats of Colorado River Delta

MPla Latrania Formation (Lower Imperial)
Marine clastics, mainly of local origin

—Split Mountain Group

Ma Alverson Formation (or Alverson Canyon)
Alkaline to tholeiitic basalt with volcaniclastics

Mr Red Rock Formation
Alluvial sandstone with conglomerate

b basement